Advances in bioprotection of plants against diseases

It is widely recognised that agriculture is a significant contributor to global warming and climate change. Agriculture needs to reduce its environmental impact and adapt to current climate change whilst still feeding a growing population, i.e. become more 'climate-smart'. Burleigh Dodds Science Publishing is playing its part in achieving this by bringing together key research on making the production of the world's most important crops and livestock products more sustainable.

Based on extensive research, our publications specifically target the challenge of climate-smart agriculture. In this way we are using 'smart publishing' to help achieve climate-smart agriculture.

Burleigh Dodds Science Publishing is an independent and innovative publisher delivering high quality customer-focused agricultural science content in both print and online formats for the academic and research communities. Our aim is to build a foundation of knowledge on which researchers can build to meet the challenge of climate-smart agriculture.

For more information about Burleigh Dodds Science Publishing simply call us on +44 (0) 1223 839365, email info@bdspublishing.com or alternatively please visit our website at www.bdspublishing.com.

Related titles:

Advances in biocontrol of crop insect pests
Print (ISBN 978-1-80146-839-8); Online (ISBN 978-1-80146-840-4, 978-1-80146-841-1)

Biopesticides for sustainable agriculture
Print (ISBN 978-1-78676-356-3); Online (ISBN 978-1-78676-358-7, 978-1-78676-359-4)

Microbial bioprotectants for plant disease management
Print (ISBN 978-1-78676-813-1); Online (ISBN 978-1-78676-815-5, 978-1-78676-816-2)

Understanding and minimising fungicide resistance
Print (ISBN 978-1-80146-198-6); Online (ISBN 978-1-80146-200-6, 978-1-80146-201-3)

Chapters are available individually from our online bookshop:
https://shop.bdspublishing.com

climate•SMART•publishing
IN AGRICULTURAL SCIENCE

BURLEIGH DODDS SERIES IN AGRICULTURAL SCIENCE
NUMBER 158

Advances in bioprotection of plants against diseases

Edited by Professor Shashi Sharma, Murdoch University, Australia and Dr Minshad Ansari, Bionema, UK

burleigh dodds
SCIENCE PUBLISHING

Published by Burleigh Dodds Science Publishing Limited
82 High Street, Sawston, Cambridge CB22 3HJ, UK
www.bdspublishing.com

Burleigh Dodds Science Publishing, 1518 Walnut Street, Suite 900, Philadelphia, PA 19102-3406, USA

First published 2025 by Burleigh Dodds Science Publishing Limited

Library of Congress Control Number: 2025939903

British Library Cataloguing in Publication Data
A catalogue record for this book is available from the British Library

ISBN 978-1-80146-728-5 (print)
ISBN 978-1-80146-729-2 (ePub)
ISBN 978-1-80146-730-8 (PDF)
ISSN 2059-6936 (print)
ISSN 2059-6944 (online)

DOI: 10.19103/AS.2024.0142

Typeset by Deanta Global Publishing Services, Dublin, Ireland

Contents

Series list

Title	Series number
Achieving sustainable cultivation of maize - Vol 1 From improved varieties to local applications Edited by: Dr Dave Watson, CGIAR Maize Research Program Manager, CIMMYT, Mexico	001
Achieving sustainable cultivation of maize - Vol 2 Cultivation techniques, pest and disease control Edited by: Dr Dave Watson, CGIAR Maize Research Program Manager, CIMMYT, Mexico	002
Achieving sustainable cultivation of rice - Vol 1 Breeding for higher yield and quality Edited by: Prof. Takuji Sasaki, Tokyo University of Agriculture, Japan	003
Achieving sustainable cultivation of rice - Vol 2 Cultivation, pest and disease management Edited by: Prof. Takuji Sasaki, Tokyo University of Agriculture, Japan	004
Achieving sustainable cultivation of wheat - Vol 1 Breeding, quality traits, pests and diseases Edited by: Prof. Peter Langridge, The University of Adelaide, Australia	005
Achieving sustainable cultivation of wheat - Vol 2 Cultivation techniques Edited by: Prof. Peter Langridge, The University of Adelaide, Australia	006
Achieving sustainable cultivation of tomatoes Edited by: Dr Autar Mattoo, USDA-ARS, USA and Prof. Avtar Handa, Purdue University, USA	007
Achieving sustainable production of milk - Vol 1 Milk composition, genetics and breeding Edited by: Dr Nico van Belzen, International Dairy Federation (IDF), Belgium	008
Achieving sustainable production of milk - Vol 2 Safety, quality and sustainability Edited by: Dr Nico van Belzen, International Dairy Federation (IDF), Belgium	009
Achieving sustainable production of milk - Vol 3 Dairy herd management and welfare Edited by: Prof. John Webster, University of Bristol, UK	010
Ensuring safety and quality in the production of beef - Vol 1 Safety Edited by: Prof. Gary Acuff, Texas A&M University, USA and Prof. James Dickson, Iowa State University, USA	011
Ensuring safety and quality in the production of beef - Vol 2 Quality Edited by: Prof. Michael Dikeman, Kansas State University, USA	012
Achieving sustainable production of poultry meat - Vol 1 Safety, quality and sustainability Edited by: Prof. Steven C. Ricke, University of Arkansas, USA	013
Achieving sustainable production of poultry meat - Vol 2 Breeding and nutrition Edited by: Prof. Todd Applegate, University of Georgia, USA	014

Cover image information

- Top left: Bunch of black grapes ready for harvest
- Top right: Actinomycetes colonies
- Bottom left: Growth of Actinobacteria colonies on agar plates, depicting different colony textures and colours
- Bottom right: Ripe maize crop ready for harvest

Introduction

With increasing regulation of synthetic fungicides to minimise their negative environmental impacts, fewer new products coming onto the market, and a growing level of resistance to conventional crop protection products, there is an heightened focus on alternative bioprotectant products as an essential tool in crop disease control. Building on and expanding the 2021 Burleigh Dodds Science title, *Microbial bioprotectants for plant disease management*, this new collection reviews the latest trends in this evolving sector, exploring advances in techniques, emerging bioprotectants or biocontrol agents (BCA) and the growing range of applications.

Part 1 Advances in techniques for identifying and developing bioprotectants

Chapter 1 sets the stage for the book by providing an overview of methods for understanding and modifying phytobiome microbial activity to enhance plant disease control. It reviews both traditional and advanced techniques for analysing soil microbial communities, including molecular, culture-based and microscopic approaches. The chapter then explores strategies to harness soil microbial activity for suppressing plant disease and enhancing plant disease resistance, including mechanisms such as competition for resources, antibiosis, hyperparasitism and induced plant resistance. Additionally, it discusses the emerging fields of phytobiome engineering, which involves manipulating microbial activity to control plant disease either by designing synthetic microbial communities or by breeding plants optimized for microbial interactions. The chapter highlights both opportunities and challenges in this field and identifies future directions for translating phytobiome research into practical field application.

Chapter 2 examines the critical issue of ensuring that bioprotection strategies do not have unintended effects on non-target organisms, particularly bee species that have been significantly impacted by pesticide use. Understanding the potential impacts of bioprotectants and other plant protection products (PPPs) on bumblebees and their populations is key to ensuring their safety and conservation while maintaining their essential roles in pollinating agricultural crops and wild plant communities.

Biosafety assessment methods for bumblebees continue to evolve, with an increasing emphasis on semi-field- and field-based risk assessments. This chapter discusses current best practices, including a three-tier biosafety testing approach focussed on foraging activity, which directly correlates with pollination services. Key areas for improvement include extending the duration of exposure in risk assessments, semi-field- and field-based risk evaluations and assessing the cumulative impact of multiple PPPs.

Building on the previous chapter, Chapter 3 reviews the fragmented development of pesticide regulations across different regions and countries and the challenges of establishing a more harmonised and global approach to approving new crop protection products. An effective regulatory framework must balance multiple criteria – including efficacy, safety, environmental impact, cost, and approval speed – while also addressing the distinctive characteristics of biological products. The chapter examines the work of the Organisation for Economic Cooperation and Development (OECD) in creating a harmonised regulatory framework to meet these objectives, highlighting the role of global joint reviews in fostering an integrated approach for regulatory approvals.

Part 2 Advances in the use of bioprotection agents

Chapters in Part 2 begin by exploring advances in developing specific bioprotectants. Chapter 4 reviews the use of *Actinobacteria* as microbial bioprotectants (also referred to as microbial biocontrol agents - MBCAs) for managing crop diseases. The development of metagenomic and meta-transcriptomic tools over the past decade, has significantly advanced the characterisation of *Actinobacteria* including taxonomic and functional insights into unculturable strains. Recent bioprospecting efforts have revealed new *Actinobacteria* with bioactive properties applicable in antibiotics, fungicides, insecticides, and herbicides. This renewed interest has driven the development of *Actinobacteria*-based inoculants, composed of microbial cells and cell-free microbial cultured extracts. The chapter reviews their ecological roles, modes of action and case studies highlighting their practical applications. It also discusses challenges in large-scale production and improving the field durability of *Actinobacteria*-based MBCAs, alongside emerging trends and future opportunities.

Chapter 5 focuses on *Clonostachys rosea*, a versatile fungal bioprotectant or fungal biocontrol agent with applications in mitigating plant diseases such as damping-off, root rot, botrytis blight, and fusarium head blight. Given increasing concerns about the environmental and health risks of chemical pesticides, *C. rosea* presents an eco-friendly alternative, reducing the need for chemical interventions in agriculture. However, its effectiveness varies depending on environmental conditions and the genetic interactions between the pathogen, host plant, and the biocontrol agent. Understanding the diverse modes of action employed by *C. rosea* – including mycoparasitism, antibiosis and plant defence induction – can help optimize its application with plant diseases, as well as pests like nematodes and insects. This chapter explores *C. rosea's*, biocontrol mechanisms, application strategies, and factors influencing its efficacy.

Chapter 6 examines *Pythium oligandrum* and its contributions to plant disease control, particularly through immunity-inducing proteins that help manage diseases caused by oomycetes such as *Phytophthora*. By leveraging its natural ability to engage in competitive interactions with pathogens, *P. oligandrum* enhances plant resilience while reducing reliance on chemical pesticides. As the agricultural industry increasingly seeks eco-friendly alternatives to chemical controls, *Pythium oligandrum* represents a promising tool for sustainable disease management. The chapter explores the biocontrol mechanisms of *P. oligandrum*, its role in disease management, and the potential of its immunity-inducing proteins in the control of diseases like *Phytophthora*.

The remaining chapters in Part 2 focus on bioprotectant applications in specific crops.

Chapter 7 addresses grapevines diseases, which can severely affect both yield and quality. Traditionally, these diseases have been managed with chemical fungicides but concerns over their environmental and health impacts have spurred interest in alternative plant protection products. Despite extensive research on bioprotectants or biocontrol agents, practical guidance for their application remains limited due to inconsistent efficacy across different environments and pathogen strains. This chapter synthesizes existing literature on biocontrol efficacy in grapevine disease management, particularly for powdery mildew, trunk diseases, and ochratoxin A-producing species of *Aspergillus* species. Through a meta-analysis of available data, it evaluates the performance of biocontrol agents and explores strategies to enhance their consistency and effectiveness considering pathogen, plant, and environmental interactions. The chapter includes case studies on *Plasmopora viticola*, *Erysiphe necator* and *Botrytis cinerea*, with a focus on selecting microbial agents capable of thriving in the same environment as the target pathogen.

Chapter 8 explores the role of bioprotectants in disease management in cereals. While chemical interventions have traditionally dominated, there is increasing recognition of how plant-associated microorganisms mitigate the effects of plant pathogens. Plants interact with diverse microbial communities residing in the endosphere, phyllosphere, rhizosphere, and bulk soil - that influence disease resistance through priming plant defences, outcompeting pathogens, or directly attacking harmful microorganisms. Some beneficial microbes also produce antibiotics, modify plant-produced secondary metabolites, or transfer pathogen-specific viruses to induce hypovirulence. Even bacteriophages in the soil can target plant pathogenic bacteria. This chapter delves into these complex interactions between beneficial microorganisms and cereals, highlighting the mechanisms by which these organisms enhance plant resilience to disease. By deepening our understanding of these biological entities and how plants manipulate them, we can pave the way for the development of disease-resistant cultivars, supporting a more sustainable approach to cereal crop protection.

Chapter 9 examines biological control strategies for managing potato diseases. While synthetic fungicides have been widely used, they pose environmental risks, contribute to pathogen resistance, and raise public health concerns. As a result, there is a growing interest in bioprotectants or biological control agents (BCAs) including bacteria, fungi, viruses, nematodes and other organisms, as sustainable alternatives. These BCAs leverage natural antagonistic interactions and enhance plant defence mechanisms. In addition to microorganisms, substances derived from biological materials also fall under the category of BCAs, expanding their potential applications. This chapter reviews the mechanisms by which BCAs control potato diseases such as competition, antibiosis, and the induction of systemic resistance. It also examines recent advances in BCA research, as well as field applications, highlighting the growing role of biological control in achieving sustainable agricultural practices. By focusing on eco-friendly, efficient, and natural solutions, biological control represents a crucial step toward meeting the demands for safer and more sustainable farming practices.

The final chapter (Chapter 10) addresses the critical role of weed control in crop production. Overuse of synthetic herbicides has led to the development of herbicide-resistant weed species and negative environmental impacts. To reduce dependence on chemical herbicides, sustainable weed management strategies are needed. Bioherbicides, derived from natural organisms such as bacteria, fungi, and plant-based substances, present a promising alternative. These bio-based agents offer the potential to reduce chemical inputs while maintaining effective weed control. However, further research is required to optimize their efficacy, commercial viability and application consistency. This chapter evaluates the potential of bioherbicides in sustainable weed control, highlighting their benefits, challenges and future role in integrated weed management.

Part 1

Advances in techniques for identifying and developing bioprotectants

Chapter 1

Techniques for understanding and modifying microbial activity in the phytobiome to suppress disease/promote plant disease resistance

Angela Records, Foundation for Food & Agriculture Research, USA

1 Introduction

Phytobiomes consist of plants, their environment, and the intricate network of micro- and macroorganisms that reside within and around them (Leach et al., 2016). These complex communities have a profound influence on plant health, growth and resilience, and their impact extends far beyond an individual plant to influence ecosystems, agriculture, and even human health.

Microbes – viruses, bacteria, fungi, oomycetes, and algae – play a significant and multifaceted role in the phytobiome. Key aspects of phytobiome microbial activity in the context of plant health include the following:

- *Biodegradation:* Certain microbes can break down and detoxify harmful compounds in the soil, reducing the persistence of pollutants and promoting a healthier environment for plants.
- *Biological pest control:* Some beneficial microbes, like entomopathogenic fungi in the *Metarhizium*, *Beauveria*, and *Isaria* genera and the bacterium *Bacillus thuringiensis*, play a significant role in controlling insect pests of

http://dx.doi.org/10.19103/AS.2024.0142.01

plants through colonizing insect bodies, the production of toxins, or by interfering with the pests' life cycles.

- *Crop yield:* The collective impact of beneficial microbial activity on plant health can lead to crop yield increases that translate to economic growth and food security.
- *Disease induction:* When pathogenic microbes are in intimate association with a susceptible host plant under 'favorable' environmental conditions, disease may occur.
- *Enhanced stress tolerance:* Certain microbial communities, such as endophytes, can help plants withstand various environmental stresses, including drought, salinity, and extreme temperatures.
- *Improved soil structure:* Microbial activity, particularly by mycorrhizal fungi, enhances soil structure by producing glomalin, a glycoprotein that binds soil particles into stable aggregates. This improves soil aeration, water retention, and root penetration.
- *Nutrient cycling:* Microbes are instrumental in nutrient cycling and mineralization. They break down organic matter, making essential nutrients available to plants. This microbial activity helps maintain soil fertility and ensures plants have access to the nutrients they need to thrive.
- *Pathogen suppression:* Certain microbes can act as natural antagonists to plant pathogens. They can compete with pathogens for resources, produce antimicrobial compounds, and/or induce systemic resistance in plants.
- *Plant growth promotion:* Plant growth-promoting rhizobacteria stimulate plant growth by producing plant growth hormones, facilitating nutrient uptake, and enhancing root development.
- *Symbiotic relationships:* Mycorrhizal fungi and nitrogen-fixing bacteria form symbiotic relationships with plants. Mycorrhizae extend the plant's root system and improve nutrient uptake, while nitrogen-fixing bacteria convert atmospheric nitrogen into a form usable by plants. These interactions enhance plant health and reduce the need for synthetic fertilizers.

Compared to their solitary counterparts, members of microbial communities, or consortia, are more robust to environmental challenges; exhibit reduced metabolic burden due to a division of labor and exchange of resources; possess expanded metabolic capabilities; and can communicate with one another (Bassler and Losick, 2006; Stenuit and Agathos, 2015; Tsoi et al., 2018). Manipulation of microbial consortia within the phytobiome offers the potential to exploit these community-based capacities for the improvement of plant health and increase in crop yields, with limited impact on the environment. Understanding microbial communities is essential to harnessing the phytobiome to manage plant disease.

2 Techniques for analyzing microbial communities

Understanding microbial communities requires determining not only 'who' is present in the community (through taxonomic profiling) but also 'what' they could be doing (through functional profiling). Various techniques have been used to this end, ranging from classical isolation and cultivation in culture media to amplicon-based methods to microscopy. These approaches have revealed numerous fundamental details about microbes, and when used in concert, these tools can provide a holistic view of microbial consortia (Orozco-Mosqueda et al., 2022).

2.1 Molecular approaches

The era of high-throughput functional genomics has ushered in new approaches for microbial community research. Advances in DNA sequencing now enable researchers to probe microbial community composition and function in a high-resolution and culture-independent manner (Franzosa et al., 2015). Omics-based exploration of microbial communities through marker gene amplicon sequencing, whole-metagenome shotgun sequencing, metatranscriptomics, and metaproteomics has generated a wealth of data, expanding our knowledge of plant microbial community structure and the interactions between phytobiome members under various biotic and abiotic contexts (Sohrabi et al., 2023).

2.1.1 Marker gene amplicon sequencing

Marker gene amplicon sequencing, which is used to amplify a specific genomic locus common to a target group of organisms – usually the 16S rRNA gene for bacteria and the internal transcribed spacer region for fungi – is a commonly used tool for elucidating the taxonomic composition of a microbiome (Regalado et al., 2020). This approach bypasses the need for culturing microorganisms, which are instead studied directly from phytobiome samples. DNA is extracted from all cells in the microbial community, and the reads that result from sequencing the sample are either directly matched to reference taxa or first grouped into operational taxonomic units (OTUs) that share a fixed level of sequence identity. Individual reads or OTUs are then 'binned' to specific taxa based on sequence homology to a reference genomic sequence (Franzosa et al., 2015).

Amplicon sequencing is usually very informative, but because it relies on oligonucleotide primers that can deemphasize or exclude some organisms, some taxa are systematically ignored. Moreover, 'universal' primers amplify genes from different taxonomic lineages with varying efficiency, 16S genes with long introns might be missed by typical PCR design due to their length, and differing numbers of rRNA gene clusters per genome have direct impact in estimating the relative abundance of individual bacterial taxa (Lucaciu et al., 2019). Finally, amplicon sequencing is limited to the analysis of taxa for which taxonomically informative genomic markers are known and can be amplified. Novel or highly diverged microbes, especially viruses, are difficult to study using this approach (Sharpton, 2014).

Whole-metagenome shotgun sequencing of DNA extracts has become a popular alternative to amplicon sequencing for dissecting complex microbial consortia. This method supplies information on microbes' total DNA content as opposed to a specific locus, which, in principle, enables the functional characterization of reads, *de novo* assembly of genes and can lead to the recovery of entire metagenome assembled genomes (Regalado et al., 2020).

2.1.2 Whole-metagenome shotgun sequencing

Whole-metagenome shotgun sequencing makes the elucidation of all microbial DNA within a phytobiome sample possible, yielding more detailed genomic information and taxonomic resolution than marker gene amplicon sequencing alone (Knight et al., 2018). As in the marker gene amplicon sequencing approach, DNA is extracted from all cells in a microbial community as a first step for whole-metagenome shotgun sequencing. The two approaches diverge, however, at the second step. Rather than the targeting of a specific genomic locus for amplification that occurs in amplicon sequencing, in whole-metagenome shotgun sequencing, all DNA is sheared into tiny fragments that are independently sequenced (Sharpton, 2014). This results in reads that align to various locations on the myriad genomes present in the sample, including non-microbes. Given adequate sequencing depth (the number of reads per sample), whole-metagenome shotgun sequencing can reveal taxonomic resolution to species or strain level, and the assembly of whole microbial genomes from short DNA sequence reads is possible (Knight et al., 2018). Some of the reads will align with taxonomically informative genomic loci, like the 16S rRNA gene in bacteria, and others will align with coding sequences, providing insight into the biological functions encoded in the genomes present in the phytobiome sample (Sharpton, 2014). Moving well beyond the limits of marker gene amplicon sequencing, whole-metagenomic shotgun sequencing profiles the functional capacity of an entire community at the gene level.

An important limitation of whole-metagenome shotgun sequencing is that it does not distinguish between active and inactive members of a microbial community and thus cannot reveal which members are contributing to observed community behavior and which are merely present, presumably awaiting more favorable conditions (Shakya et al., 2019). Moreover, most common models for microbial community study involve cross-sectional but not longitudinal sampling, and thus fail to capture the communities' dynamic behavior (Franzosa et al., 2015).

2.1.3 Metatranscriptomics

Sequencing methods reveal the genomic content of a community and its functional potential, but the extent to which functional *potential* dictates functional *activity* is not well understood (Franzosa et al., 2015). A full understanding of the determinants of function requires multi-omic analysis, including metatranscriptomics, metaproteomics, and metabolomics.

In a metatranscriptomic analysis, the total RNA of a microbial community is sequenced. This profiles transcription levels, providing information on gene expression and the active functional output of the microbiome (Knight et al., 2018). From this functional data, active metabolic pathways in the microbial community can be identified and associated to particular environmental conditions (Bashiardes et al., 2016). Indeed, metatranscriptomics can reveal microbial community responses to perturbations like drought (Ossowicki et al., 2021; Chowdhury et al., 2019; Veach et al., 2020).

The process involves first enriching microbial mRNAs by depleting rRNAs, which vastly outnumber mRNAs in the total microbial RNA pool (Franzosa et al., 2015). Microbial mRNA is then converted into cDNA and sequenced. Functional profiling begins by associating metagenomic and metatranscriptomic (collectively, 'meta-omic') sequence data with known gene families by mapping the DNA or RNA reads to databases of gene sequences that have been clustered at the family level or by assembling the reads into contigs to determine putative protein-coding sequences (CDSs), which are then assigned to gene families (Franzosa et al., 2015). Both strategies profile the presence or absence of a gene family and its relative abundance within a meta-omic sample.

As with most transcriptomic methods, experimental design is critical for metatranscriptomics work. Collection of sufficient material for sequencing (or coupled experiments) is required, and RNA must be carefully preserved to avoid degradation (Shakya et al., 2019; Knight et al., 2018). Certain sample types may warrant specialized protocols for RNA purification, including soil samples, which require the removal of enzyme-inhibiting humic substances (Wang et al., 2012; Knight et al., 2018).

Due to the complexity (high diversity and relative ratios of members) that characterizes many microbial communities, the short half-life of RNA and several other limitations to the technology, metatranscriptomics is not always able to capture the entire metatranscriptome (Shakya et al., 2019). Likewise, the sequencing of mRNA is biased toward organisms with higher rates of transcription (Knight et al., 2018). Nonetheless, metatranscriptomics is a powerful tool for phytobiome analysis.

2.1.4 Metaproteomics

Metagenomics and metatrascriptomics provide useful information about 'who' is present in a microbial community (measured through DNA) and 'what' they *could be* doing (measured through RNA), but these approaches do not fully capture community functioning. Proteins ultimately carry out cellular processes, and while mRNA presence can be a predictor of protein presence, the abundance of mRNA and that of protein are often weakly correlated. Several factors contribute to this discrepancy, including the regulatory control of translation in alternative splicing of mRNA, codon bias, mRNA degradation, post-translational modification of proteins, and protein turnover rate (Abiraami et al., 2020).

Metaproteomics is a powerful tool for profiling proteins expressed by all the organisms present in diverse biological and ecological samples (Miura and Okuda, 2023). Metaproteomic profiles include information about the proteins, post-translational modifications, protein–protein interactions, and protein turnover in a microbial community, giving clues to community structure, dynamics, and metabolic activities (Lucaciu et al., 2019; Hettich et al., 2013). This information, in turn, leads to a better understanding of various aspects of community life, including microbial recruitment, member cooperation or competition for nutrient resources, and how members distribute metabolic activities, including defense, across the community (Hettich et al., 2013).

Proteomic methods rely on mass spectrometry (MS)-based shotgun quantification of peptide mass and abundance. Extracted proteins are digested to peptides followed by tandem mass spectrometry to obtain mass fragmentation spectra data (Abiraami et al., 2020). The fragmentation pattern of a peptide reveals its amino acid sequence and any post-translational modifications, such as phosphorylation. Peptides identified by this approach are then associated with full-length proteins by sequence homology-based searches against reference databases (Franzosa et al., 2015).

Metaproteomics is still in its infancy and can be limited by low protein quality and concentration; low sensitivity due to host proteins and microbial complexity; *de novo* protein identification if a (meta)genome reference sequence is lacking; difficulty in constructing amino acid sequence databases for MS due to the diversity of microbiota in a sample; false identification in database searches; and issues with spectrum acquisition (Levy et al., 2018; Miura and Okuda, 2023). The current application of proteomics to describe phytobiome microbial communities is limited due to relatively low microbial protein expression levels in complex plant-associated samples and consequent detection limits, and the need for a comprehensive peptide reference database (Levy et al., 2018).

2.1.5 Metabolomics

Interactions among microorganisms and between microorganisms and their hosts are often mediated through shared metabolite pools or metabolite exchanges (Franzosa et al., 2015). Thus, metabolomics – the detection, identification, and quantification of all (or a subset of) the metabolites and other small molecules in a biological specimen – is an important tool for understanding phytobiome functional activities (Parida et al., 2018).

Metabolomics relies on chromatography techniques to separate compounds, which are then identified and quantified using MS (Franzosa et al., 2015). MS combined with gas chromatography (GC) or liquid chromatography (LC) (GC–MS and LC–MS), direct injection mass spectrometry, and nuclear magnetic resonance spectroscopy are frequently used for the study of plant metabolomics (Parida et al., 2018). Like other meta-omic methods, metabolomics analyzes a broad array of potential features that occur with high dynamic range. Metabolomics studies are further complicated, however, by the non-uniformity of the features (molecules) that are profiled. Whereas all transcripts identified during genomic analyses belong to the same class of biomolecules (RNAs), metabolites span a broad range that includes amino acids, peptides, lipids, nucleic acids, carbohydrates, organic acids, ketones, aldehydes, amines, steroids, vitamins, hormones, antibiotics, signaling molecules, and specialized metabolites like polyphenols and flavonoids (Parida et al., 2018).

In general, the metabolome consists of primary and specialized metabolites. Phytobiomes, as compared to other biological systems, feature especially complex metabolomes, largely because plants evolved a high diversity of specialized metabolites as part of their defense system against pests and pathogenic microbes (Wink, 2003) and, conversely, to serve as chemo-attractants for beneficial microbes (Reinhold-Hurek et al., 2015). Indeed, an estimated 200 000 to 1 million metabolites are associated with the plant kingdom (Wang et al., 2019), but these large, chemically diverse metabolomes are often under-represented in public databases (Freund and Hegeman, 2017).

2.1.6 Holo-omics: integrating multi-omic data

Several studies have emerged in recent years wherein multiple omics techniques are used in parallel (Dugourd et al., 2021). Taken together, DNA,

RNA, protein, and metabolite data provide valuable clues to the functional activity of a microbial community. However, simply collecting numerous types of data stops short of realizing the full potential multi-omics analysis. Multi-omic data is most powerful when it is integrated (Franzosa et al., 2015).

Data integration is one of several requirements for high-quality multi-omics studies, which include proper experimental design; thoughtful selection, preparation, and storage of samples; careful collection of quantitative multi-omics data and associated meta-data; better tools for integration and interpretation of the data; minimum standards for multi-omics methods and meta-data; and new resources for the deposition of intact multi-omics data sets (Pinu et al., 2019). Data scientists are attempting to integrate multi-omic data sets to create new and meaningful biological knowledge, but the analytical methodology required for such data integration remains nascent (Ferrocino et al., 2023).

Higher order integration of multi-omic data has recently been described as 'holo-omics,' which (Nyholm et al., 2020) defined as 'the incorporation of data from multiple omic levels from both host and microbiota domains.' Holo-omics incorporates multi-omic data from both the host and its associated microbiomes to allow untangling of the interactions between the two. Xu et al. (2021) recently reviewed this concept in context of the phytobiome and emphasized that such holistic approaches have the power to resolve the functionality of the phytobiome ecosystem by presenting a complete picture of what is expressed, translated and produced during plant-microbiome interactions (Orozco-Mosqueda et al., 2022).

2.2 Culture-based techniques

The marked expansion of omics-derived data in recent years has led to a plethora of well-supported inferences about microbial consortia, including conclusions based on the reconstructed genomes of uncultured organisms. Thorough understanding of these communities and their members is limited, however, to information derived from a minority of well-studied, cultured lineages (Lewis et al., 2021). Members of microbial consortia must be isolated and cultured for genome-based predictions about their cell biology, physiology, and ecological roles to be tested. Indeed, there are several examples of novel enzymatic reactions and pathways that were discovered through experimental testing of microbial enrichments or cultures while being undetectable by genomic methods alone (Nunoura et al., 2018; Mulder et al., 1995; Daims et al., 2015; Lewis et al., 2021).

Unfortunately, only a small subset of the microorganisms present in an environmental sample can be grown and maintained axenically. It is estimated that only 0.1–1.0% of soil bacteria are cultivable using standard cultivation methods (Daniel, 2005). To overcome challenges associated with standard techniques, several innovative cultivation methods have been developed in recent years, including microfluidics (Ma et al., 2014; Boitard et al., 2015), cultivation chips (Ingham et al., 2007; Hesselman et al., 2012; Gao et al., 2013), manipulation of single cells (Ben-Dov et al., 2009; Park et al., 2011), and high-throughput 'culturomics' (Lagier et al., 2012). These methods expanded the number of novel species brought into culture immensely, but the rate of discovery of microbes that are as yet uncultivable outpaces the rate of isolating novel species (Gutleben et al., 2018).

2.3 Microscopy

Microscopy has long been used in the study of plant–microbe interactions, allowing visualization of microorganisms as they associate with plant tissues (Kandel et al., 2017; Steinberg et al., 2021; Bandara et al., 2021; Berg et al., 2014; Liu et al., 2021). Several microscopic techniques are employed in phytobiome-related analysis, including light microscopy and electron microscopy, and newly emerging imaging technologies demonstrate significant and unrealized potential to reveal fundamental information about the phytobiome (Romano et al., 2020; Czymmek et al., 2023).

2.3.1 Light microscopy

Light microscopy is the most common microscopic technique for assessing microbial communities due to the relatively low cost of purchasing, maintaining and servicing the necessary equipment (Romano et al., 2020). Light microscopy techniques have progressed substantially since the first observations of microscopic organisms in the late 1600s (Leewenhoeck, 1673). Variations on light microscopy methods range from elementary staining techniques that allow differentiation between microbes and plant tissues (Hulse, 2018; Vierheilig et al., 1998; White et al., 2014) to advanced approaches like those that involve 'optical sectioning' to extract in-focus information from the objective focal plane and enable high-contrast, high-resolution three-dimensional (3D) perspectives (Czymmek et al., 2023). Such advanced platforms include confocal (Cardinale and Berg, 2015; Hepler and Gunning, 1998; Czymmek et al., 1994) and spinning disk microscopy (Henty-Ridilla et al., 2013; Oreopoulos et al., 2014), which reject out-of-focus light; multi-photon excitation (Mizuta, 2021; Bourett et al., 2002; Czymmek et al., 2007) and light sheet fluorescence microscopy (Ovečka et al., 2022; Keller and Dodt, 2012), which restrict illumination to a thin plane; and approaches that avail computational extraction (Sibarita, 2005).

Genetically encoded fluorescent proteins are often used with light microscopy to allow visualization of specific components in living cells of a plant, an associated microbe or both (Bloemberg et al., 2000; Rufián et al., 2018), while providing opportunities to trace microbe progression in plant tissue and study plant–microbe interactions (Berg and Beachy, 2008; Ckurshumova et al., 2011; Wu et al., 2013). Fluorescence microscopy has been used in the plant–pathogen context (Bloemberg et al., 2000; Kankanala et al., 2007; Ovečka et al., 2022; Czymmek et al., 2007; Kim et al., 2022), including to determine the dynamic temporal and spatial distribution of pathogen ramification, virulence factors, elicitors, and effector proteins (Cui et al., 2021; Khang et al., 2010; Czymmek et al., 2023). These approaches have also been used extensively to study beneficial plant–microbe interactions, such as *Rhizobium*-legume symbioses (Libault et al., 2011; Mendoza-Suárez et al., 2020), arbuscular mycorrhizal fungi root colonization (Hardham, 2012; Ivanov et al., 2019; Martino et al., 2007), and microorganisms that provide protective and growth-stimulating effects (Bloemberg et al., 2000; Jabusch et al., 2021).

A recent development in fluorescence imaging technology, termed super-resolution microscopy (SRM), allows up to ten-fold better resolution with fluorescent probes than traditional widefield and laser-based imaging (Schubert, 2017; Sydor et al., 2015; Komis et al., 2015). SRM platforms include structured illumination microscopy – which acquires multiple-patterned images to access new frequencies that double optical resolution (Gustafsson, 2000; Wu and Shroff, 2018), as well as single-molecule localization microscopy – which captures a series of thousands of over-sampled images of sparsely spaced single molecules, finds the centroid of each molecule and recreates a new image with a ten-fold resolution improvement (Czymmek et al., 2023; Huang et al., 2010).

2.3.2 Electron microscopy

Since its use in the visualization of the first known virus – *Tobacco mosaic virus*, isolated from plant sap (Kausche et al., 1939) – electron microscopy has been an important tool for the study of plant-associated microbes. Electron microscopy offers advantages over other visualization approaches, including better resolution of fine detail than light microscopy and freedom from the fluorescent reporters required for confocal scanning and other optical sectioning methods. Various forms of electron microscopy have emerged over the years, and all have been employed to some extent in phytobiome analysis (Czymmek et al., 2023; Mendgen and Lesemann, 1991).

Transmission electron microscopy (TEM) is a high-resolution imaging technique that passes a beam of electrons through a thin sample to produce a 2D image. TEM is capable of unparalleled resolution, routinely capturing images at the sub-nanometer scale and providing invaluable information on the internal structure of a sample (Romano et al., 2020). For example, TEM has been used to establish endophytic interaction in roots (Singh and Sharma,

2016). While the resolution afforded by TEM is superior to other approaches, a drawback to this technology is its requirement for laborious sample preparation.

Like TEM, scanning electron microscopy (SEM) employs an electron beam to provide an image and has become an important tool for the visualization of plant–microbe interactions. The two approaches differ in that while TEM passes electrons *through* a sample, SEM electrons move *across* the sample and are reflected back to detectors that collect and process the electrons into a 3D image of the sample surface (Caldwell and Iyer-Pascuzzi, 2019; Pathan et al., 2008).

Multiple types of SEM are available, with key differences in their requirements for sample preparation. In ambient temperature SEM, samples are dehydrated by critical point drying (CPD), freeze-drying, or chemical methods (Caldwell and Iyer-Pascuzzi, 2019). CPD works well for plant surface features that are resilient to dehydration (Caldwell and Iyer-Pascuzzi, 2019) and is commonly used to observe phytobiome phenomena (Fritschi et al., 2008; Koutsoudis et al., 2006; Mbofung et al., 2016; Sun et al., 2013). Ambient temperature SEM has led to many important findings in plant–microbe interactions, including subcellular chemical immunolocalization (Benhamou and Bélanger, 1995) and elemental composition (Eder and Lütz-Meindl, 2008).

Low-temperature SEM, also known as cryo-scanning electron microscopy (CSEM), is useful for specimens that would be damaged if dehydrated (Read and Jeffree, 1990). For CSEM, samples are placed into subcooled liquid nitrogen for cryofixation, then fractured, coated and viewed on a cold stage (Caldwell and Iyer-Pascuzzi, 2019). CSEM can be ideal for surface and cytoplasmic ultrastructural studies (Howard, 1981; Wightman, 2022; Gilkey and Staehelin, 1986; Jeffree and Read, 1991), especially for the study of labile or dynamic cellular phenomena or when the plant cuticle or air spaces impede penetration of chemical fixatives (Czymmek et al., 2023).

Environmental scanning electron microscopy (ESEM) enables the investigation of non-conductive and hydrated samples, bypassing the preparation steps required for conventional SEM (Danilatos, 1981; Danilatos, 1993). During ESEM analysis, gas (usually water vapor) is present in the microscope chamber creating humid conditions wherein gas atoms collide with electrons, amplifying the signal (Stabentheiner et al., 2010). This approach was recently used to evaluate the colonization of wheat by a bacterial consortium in a commercial biofertilizer (Dal Cortivo et al., 2017). ESEM imaging revealed external and internal colonization of leaf and root tissues.

A newly described approach, paraffin scanning electron microscopy, enables light microscopy and SEM from the same sample (Caldwell and Iyer-Pascuzzi, 2019). Paraffin-embedded samples are sectioned and viewed relatively quickly under a light microscope to identify a region of interest.

Subsequently the sample is recut, deparaffinized and scanned via SEM to generate high-resolution images of pre-identified target regions. This method was used to visualize fungal colonization of tomato and was effective in other plant pathosystems, as well (Caldwell and Iyer-Pascuzzi, 2019).

Scanning transmission electron microscopy (STEM) combines the principles of TEM and SEM, leveraging the powerful resolution of the former with the broad, 3D surface visualization made possible by the latter. When a standard transmission electron microscope operates in 'STEM mode,' the electron beam is finely focused and scans the sample area (as happens with SEM), while the image is generated by the transmitted electrons (as is the case with TEM). This approach was used to show that plant growth-promoting rhizobacteria inoculants successfully colonized the surface and within root tissues of axenic mandarin orange seedlings (Thokchom et al., 2017).

Lastly, volume electron microscopy (vEM) is a set of high-resolution imaging techniques used to reveal the 3D structure of cells at nanometer resolution. vEM approaches leverage the unique ability of electron microscopes to reveal the structure of cells and tissues with the contrast and resolution to distinguish cell membranes and membrane bilayers (Peddie et al., 2023). vEM was recognized by Nature as one of 'seven technologies to watch in 2023' (Einstein, 2023), and while phytobiome-related vEM studies have not yet been reported, the potential for these approaches to be widely used in the generation of full 3D volumes of plant–microbe interaction sites cannot be overstated (Czymmek et al., 2023).

3 Biological control: exploiting microbial activity to suppress plant disease/promote disease resistance

Every year, up to 40% of crops are lost due to plant pests and diseases costing the global economy $290 billion (FAO, 2022). Addressing this challenge effectively and responsibly will require a range of management strategies beyond traditional approaches like the use of resistant plant varieties and chemical control (Oerke and Dehne, 2004; Rahman et al., 2018). Methods that harness the capacities of microbes are especially attractive for sustainable plant disease management. 'Biological control', or 'biocontrol', is defined as 'the exploitation of living agents (including viruses) to combat pestilential organisms (including pathogens, pests, and weeds) for diverse purposes to provide human benefits' (Stenberg et al., 2021). Four primary modes of action underly biological control of plant diseases: (1) competition for resources; (2) antibiosis – pathogen inhibition via specialized metabolites from the microbial biocontrol agent (MBA); (3) hyperparasitism – MBA predation on the pathogen; and (4) MBA-induced plant resistance to the pathogen (Collinge et al., 2022).

3.1 Competition for resources

Substrate colonization by an MBA is the first step in a cascade of events that may lead to the deployment of biological control mechanisms like antibiosis, hyperparasitism or the induction of host plant disease resistance (Jaroszuk-Ściseł et al., 2019; Lasinio et al., 2021). Successful colonization requires access to vital resources, and within the phytobiome context, may lead to interspecific competition when there is simultaneous demand for a resource that is in limited supply. This 'exploitation competition' happens when one organism, by exploiting the resource, reduces its availability to another organism (Holomuzki et al., 2010). Such competition among microbes can occur across space and time, when spatial niche differentiation and/or temporal separation fail (Lasinio et al., 2021; Fitt et al., 2006).

Fast-colonizing yeasts provide an example of this efficient mode of action, as they are known to limit wound invasion by fruit pathogens (Spadaro and Droby, 2016). Yeasts consume a broad range of carbohydrates, as well as various nitrogen sources, and as single-celled organisms, they can multiply rapidly under favorable conditions – like those in nutrient-rich wounds of fruits – utilizing the available nutrients. Antagonistic interactions between the yeast *Pichia guilliemodii* and pathogens *Penicillium digitatum*, *Penicillium expansum*, *Botrytis cinerea*, or *Colletotrichum* spp. in wounds of various fruits and between the yeast *Aureobasidium pullulans* and *P. expansum* in apple wounds demonstrate the importance of competition in biological control of plant pathogens (Spadaro and Droby, 2016).

3.2 Antibiosis

Microbial specialized metabolites – antibiotics, pigments, toxins, enzyme inhibitors, immunomodulators, effectors of ecological competition or symbiosis, and more – are products of secondary metabolism that takes place out of step with the main microbial growth phase (Fouillaud and Dufossé, 2022). Many MBAs employ specialized metabolites in their interactions with pathogens. For example, *Bacillus subtilis*, a soil bacterium used in several biocontrol formulations (Wang et al., 2015), is known to produce at least 76 specialized metabolites (Kaspar et al., 2019). These include surfactin, which disrupts the cell membranes of organisms by integrating into their lipid layers (Ongena and Jacques, 2007). *B. subtilis* 9407 exerts strong surfactin-mediated anti-bacterial activity against *Acidovorax citrulli* (Fan et al., 2017). *Bacillus* spp. also produce exoenzymes, such as proteases and chitinases that can decompose fungal cell walls (Blake et al., 2021), as well as volatiles that can inhibit the spore germination and hyphal growth of phytopathogens (Grahovac et al., 2023; Lee et al., 2023; Raza et al., 2016; Yu et al., 2012).

3.3 Hyperparasitism

Hyperparasitism, the phenomenon in which parasites are themselves infected with parasites, is common in nature and extends to plant-associated microbes (Parratt and Laine, 2016). Infection by a hyperparasite can attenuate a pathogen's virulence against plants by directly debilitating its growth or impeding its ability to produce and deploy virulence factors. It is the ability of micro-hyperparasites to reduce realized pathogen virulence in this way that has led to their use as MBAs.

Necrotrophic parasitic fungi are especially well-suited for roles in biological control because they compete well in the environment (e.g. in the rhizosphere or in root colonization) and survive without a living host–pathogen (Collinge et al., 2022). Examples are cosmopolitan species of *Trichoderma* and *Clonostachys* that can survive both as mycoparasites and in other ways (Benítez et al., 2004).

Biotrophic hyperparasites require living hosts for activity and growth and a pre-existing host population to parasitize. This makes them effective in the short term only if applied inundatively (Collinge et al., 2022). Nonetheless, there are examples of biotrophic hyperparasites used in commercial biocontrol products, including *Ampelomyces quisqualis* against powdery mildew (Karlsson et al., 2017) and *Coniothyrium minitans*, which parasitizes several sclerotia-forming plant pathogens (Whipps et al., 2008).

All viruses are parasites, and several are hyperparasites that may be useful in the biological control of plant disease. Examples include Cryphonectria hypovirus-1 (CHV1), which reduces *Cryphonectria parasitica* growth rate and virulence in chestnut trees (Kazmierczak et al., 2012); RsPod1EGY, a phage that shows promise as a biocontrol agent against *Ralstonia solanacearum* on tomato (Elhalag et al., 2018); and virus cocktails – like a combination of P4L, P43M, P23M1 phage against *Xanthomonas oryzae* pv. *oryzae* in rice (Chae et al., 2014).

3.4 Induced plant resistance

Plants have several lines of defense against would-be pathogens, including structural barriers (e.g. cell walls and waxy epidermal cuticles), constitutive secondary metabolites (e.g. phytoanticipins) and inducible defense mechanisms that are triggered when plant receptors recognize microbe-associated molecular patterns (MAMPs) (Freeman and Beattie, 2008). When inducible defenses are triggered, plant cells become fortified against attack, and plants become 'primed' for future responses to low levels of a stimulus in a faster and stronger manner than unprimed plants (Conrath et al., 2015). Such priming can be maintained throughout a plant's life cycle and can even be transmitted to subsequent generations (Mauch-Mani et al., 2017).

Non-pathogens, as well as pathogens, are capable of triggering this induced systemic resistance due to the widespread presence of the triggering molecular components common across microbial species (Freeman and Beattie, 2008). The most studied MAMPs are the bacterial proteins flagellin and elongation factor EF-Tu (Köhl et al., 2019). Other known MAMPs include glucan, chitin and xylan produced by fungi; proteinaceous MAMPs from bacteria (e.g. superoxide dismutase and 23 amino-acid peptide) or from oomycetes (e.g. pep13 trans-glutaminase, sterol-binding elicitins, and cellulose-binding lectins); lipophilic MAMPs (e.g. ergosterol and arachidonic acid); oligosaccharide MAMPs (e.g. N-glycosylated yeast peptides, cell wall components); and lipooligosaccharides produced by Gram-negative bacteria (Boller and Felix, 2009). In addition, siderophores, antibiotics, biosurfactants, and volatiles can act as elicitors inducing systemic resistance (Pieterse et al., 2014). Mild viruses can also function as elicitors.

In the context of biocontrol, MAMPs have several positive characteristics: their effects are long-lasting, they can simultaneously address a broad range of targets and they do not cause the manifestation of resistant pathogen isolates (Romanazzi et al., 2016). On the other hand, screening for MBAs with high potential for induction of resistance is complicated because this mode of action depends on a complex sequence of events. First, the MBA must become established on the host plant, which requires that the microbe be ecologically competent for the context (Köhl et al., 2019). Next, the plant must harbor the appropriate genetic and physiological status to recognize inducers from the MBA and trigger pathways that result in the onset of defense reactions or priming. The selection of MBAs that use induced plant resistance as a mode of action requires complex bioassays that are not well-standardized (Romanazzi et al., 2016).

4 Phytobiome engineering: manipulating microbial activity to suppress plant disease/promote disease resistance

Plant-associated microbes have been the subject of intense study for many years – from their structure to their metabolic activities to their interactions with host plants. This research has led to various approaches for controlling plant disease, including the employment of microbes as MBAs, which has seen limited success (Ke et al., 2021). It has become increasingly apparent that the benefits of culturing and introducing a single microbial species into a plant's environment, as is done in standard biocontrol approaches, could be compounded dramatically by phytobiome engineering – designing and deploying synthetic microbial communities to enhance plant resilience to pathogens (Vorholt et al., 2017).

4.1 Designing synthetic microbial communities

Conventional inoculants for plant disease control are typically limited to a single microbial strain isolated by *in vitro* screening assays or inoculation experiments under controlled conditions. These strategies have seen some success, but they fail to capture important aspects of the interactions that take place in the phytobiome. Efforts to improve plant health and control disease using synthetic communities (SynComs) have gained momentum in recent years (Moreira et al., 2023; de Souza et al., 2020).

In the plant context, SynComs are small consortia of microorganisms designed to mimic aspects of the observed function and structure of the phytobiome. Importantly, SynComs are intended to recapitulate key attributes of plant–microbe and microbe–microbe interactions that take place in the phytobiome, providing functions and synergies that would not be achievable by a single MBA (de Souza et al., 2020). SynCom design efforts benefit from an understanding of a plant's 'core microbiome' – microbes that are systematically and consistently associated with a particular plant species under various environmental conditions (Lemanceau et al., 2017). The core microbiomes of various crops, including maize, rice and sugarcane, have been characterized – each with up to hundreds of microbial taxa as members (Peiffer et al., 2013; Edwards et al., 2015; Hamonts et al., 2018). Studies using both abundant and prevalent core microbes in SynComs have demonstrated that core microbial groups are extremely efficient at colonizing plants (Niu et al., 2017; Armanhi et al., 2018), demonstrating that dominant groups are likely involved in critical plant growth and survival functions (de Souza et al., 2020).

'Hub' members of a plant's core microbiome are those that can form strong facilitative and mutualistic interactions (Toju et al., 2018) and are central to microbial community assembly (Bulgarelli et al., 2013; Shade and Handelsman, 2012; Qiu et al., 2019). Marshalling the capabilities of the formidable hub contingent of the phytobiome to recruit beneficial microbes could allow the selection of novel and improved microbial functions. The relationships and identities of hub microbes can be decoded via network analysis (van der Heijden et al., 2016), providing critical information for *in situ* phytobiome engineering. Ultimately, it may be possible for SynComs to be vertically transmitted across generations. Qiu et al. (2019) described an approach that starts with profiling the phytobiome to identify core and hub microbes using next-generation sequencing and statistical and network analysis. Then, strongly linked microbes harboring critical functions (e.g. defense against pathogens) are identified using metagenomic sequencing. The highly connected hub microbes are isolated from the plant and sprayed, as a cocktail, on parental flowers, resulting in seeds enriched with a microbiome that has increased capability to recruit beneficial microbes).

SynCom design for plant disease control is promising but challenging. Natural systems are complex, making it difficult to predict which microbial combinations will be effective in the field (Moreira et al., 2023). Indeed, a complete understanding of a phytobiome microbial community would require not only an understanding of members' behaviors as individuals, but also knowledge of their pairwise and higher order interactions and their consequences. Pairwise interactions do not necessarily predict high-order outcomes, demonstrating that the whole of the phytobiome is different from the sum of its parts (Vogel et al., 2021; Sanchez-Gorostiaga et al., 2019). Furthermore, priority effects (the order in which microbes are introduced) impact both the final microbial composition and, ultimately, disease outcomes (Leopold and Busby, 2020; Wei et al., 2019).

4.2 Breeding microbe-optimized plants

Plants play a role in selecting their phytobiome. For example, plants engineer their rhizosphere microbial communities by secreting exudates into the soil that improve nutrient availability and attract specific beneficial microbes. This suggests that a plant's ability to support a beneficial microbiome is a plant trait under selection (Wallenstein, 2017). Such plant traits influence microbial community composition, abundance or function (Turner et al., 2013; Horton et al., 2014), and many plant genes are correlated with variation in the phytobiome across environmental conditions (Brachi et al., 2017). Thus, it may be possible to breed 'microbe-optimized plants' with bespoke capacity to attract and maintain beneficial microbes (Abhilash et al., 2012).

5 Conclusion and future trends

The phytobiome is a complex, dynamic network of micro- and macro-organisms engaged in collective activities that determine plant health and resilience. Since the 1600s, an array of increasingly sophisticated tools – omics technologies, culture-based techniques, various types of microscopy – has offered glimpses into the microbial world. Many questions remain, however, about phytobiome behavior in the field. For example, how do natural phytobiome communities respond to environmental changes that occur in the field? Do plants with a particular genotype consistently recruit a particular core microbiota irrespective of environmental conditions?

Future research directions should address these questions and further include efforts to understand the intersections between microbial interactions, plant genotype variability and wider ecosystem functionality. Advancements in phytobiome engineering will leverage this understanding, along with computational methods, including machine learning and AI, to make SynCom design predictable.

As knowledge about phytobiomes mounts and tools continue to evolve, opportunities to harness these communities for plant disease management will become increasingly evident. It is not difficult to imagine a future in which farmers have phytobiome analysis tools in hand to make quick assessments of their fields and tailor solutions to their needs. One can envision a tool that detects pathogens, microbes with pathogenic potential and beneficial microbes in a field, allowing a farmer – based on phytobiome profiling – to calculate the field's pathogen/antagonist ratio. Armed with this information, the farmer could choose to rely on native microbiota to manage pathogen populations or opt to intervene by stimulating pathogen-suppressive taxa or introducing MBAs or synthetic microbial communities (Bell et al., 2019). A 'designer phytobiome' future is fast approaching.

6 References

Abhilash PC, Powell JR, Singh HB and Singh BK (2012) Plant–microbe interactions: novel applications for exploitation in multipurpose remediation technologies. *Trends Biotechnol.* 30(8): 416–420.

Abiraami TV, Singh S and Nain L (2020) Soil metaproteomics as a tool for monitoring functional microbial communities: promises and challenges. *Rev. Environ. Sci. Biotechnol.* 19: 73–102.

Armanhi JSL, de Souza RSC, Damasceno NB, et al. (2018) A community-based culture collection for targeting novel plant growth-promoting bacteria from the sugarcane microbiome. *8*. 2191.

Bandara CD, Schmidt M, Davoudpour Y, et al. (2021) Microbial identification, high-resolution microscopy and spectrometry of the rhizosphere in its native spatial context. *Front. Plant Sci.* 12: 668929.

Bashiardes S, Zilberman-Schapira G and Elinav E (2016) Use of metatranscriptomics in microbiome research. *Bioinform. Biol. Insights* 10: 19–25.

Bassler BL and Losick R (2006) Bacterially speaking. *Cell* 125: 237–246.

Bell TH, Hockett KL, Alcalá-Briseño RI, et al. (2019) Manipulating wild and tamed phytobiomes: challenges and opportunities. *Phytobiomes J.* 3(1): 3–21.

Ben-Dov E, Kramarsky-Winter E and Kushmaro A (2009) An *in situ* method for cultivating microorganisms using a double encapculation technique. *FEMS Microbiol. Ecol.* 68: 363–371.

Benhamou N and Bélanger R (1995) Immunoelectron microscopy. In: Singh US and Singh RP (eds) *Molecular Methods in Plant Pathology*. Boca Raton: CRC Press.

Benítez T, Rincón AM, Limón MC and Codón AC (2004) Biocontrol mechanisms of *Trichoderma* strains. *Int. Microbiol.* 7: 249–260.

Berg G, Grube M, Schloter M and Smalla K (2014) Unraveling the plant microbiome: looking back and future perspectives. *Front. Microbiol.* 5: 148.

Berg RH and Beachy RN (2008) Fluorescent protein applications in plants. *Methods Cell Biol.* 85: 153–177.

Blake C, Christensen MN and Kovács ÁT (2021) Molecular aspects of plant growth promotion and protection by *Bacillus subtilis*. *Mol. Plant-Microbe Interact.* 34(1): 15–25.

Bloemberg GV, Wijfjes AHM, Lamers GEM, et al. (2000) Simultaneous imaging of *Pseudomonas fluorescens* WCS365 populations expressing three different autofluorescent proteins in the rhizosphere: new perspectives for studying microbial communities. *Mol. Plant-Microbe Interact.* 13(11): 1170–1176.

Boitard L, Cottinet D, Bremond N, et al. (2015) Growing microbes in millifluidic droplets. *Eng. Life Sci.* 15: 318–326.

Boller T and Felix G (2009) A renaissance of elicitors: perception of microbe-associated molecular patterns and danger signals by pattern-recognition receptors. *Annu. Rev. Plant Biol.* 60: 379–406.

Bourett TM, Sweigard JA, Czymmek KJ, et al. (2002) Reef coral fluorescent proteins for visualizing fungal pathogens. *Fungal Genet. Biol.* 37: 211–220.

Brachi B, Filiault D, Darme P, et al. (2017) Plant genes influence microbial hubs that shape beneficial leaf communities. *bioRxiv*. DOI: 10.1101/181198.

Bulgarelli D, Schlaeppi K, Spaepen S, et al. (2013) Structure and functions of the bacterial microbiota of plants. *Annu. Rev. Plant Biol.* 64: 807–838.

Caldwell D and Iyer-Pascuzzi AS (2019) A scanning electron microscopy technique for viewing plant-microbe interactions at tissue and cell-type resolution. *Phytopathology* 109: 1302–1311.

Cardinale M and Berg G (2015) Visualization of plant-microbe interactions. In: Lugtenberg B (ed) *Principles of Plant-Microbe Interactions*. Cham, Switzerland: Springer International Publishing, pp.299–306.

Chae J-C, Hung NB, Yu S-M, et al. (2014) Diversity of bacteriophages infecting *Xanthomonas oryzae* pv. *oryzae* in paddy fields and its potential to control bacterial leaf blight of rice *J. Microbiol. Biotechnol.* 24(6): 740–747.

Chowdhury TR, Lee J-Y, Bottos EM, et al. (2019) Metaphenomic responses of a native prairie soil microbiome to moisture perturbations. *mSystems* 4(4): e00061-00019.

Ckurshumova W, Caragea AE, Goldstein RS and Berleth T (2011) Glow in the dark: fluorescent proteins as cell and tissue-specific markers in plants. *Mol. Plant* 4(5): 794–804.

Collinge DB, Jensen DF, Rabiey M, et al. (2022) Biological control of plant diseases – what has been achieved and what is the direction? *Plant Pathol.* 71: 1024–1047.

Conrath U, Beckers GJM, Langenbach CJG and Jaskiewicz MR (2015) Priming for enhanced defense. *Annu. Rev. Phytopathol.* 53: 97–119.

Cui Y, Zhao Y, Lu Y, et al. (2021) In vivo single-particle tracking of the aquaporin AtPIP2;1 in stomata reveals cell type-specific dynamics. *Plant Physiol.* 185: 1666–1681.

Czymmek KJ, Duncan KE and Berg H (2023) Realizing the full potential of advanced microscopy approaches for interrogating plant-microbe interactions. *Mol. Plant-Microbe Interact.* 36(4): 245–255.

Czymmek KJ, Fogg M, Powell DH, et al. (2007) *In vivo* time-lapse documentation using confocal and multi-photon microscopy reveals the mechanisms of invasion into the *Arabidopsis* root vascular system by *Fusarium oxysporum*. *Fungal Genet. Biol.* 44: 1011–1023.

Czymmek KJ, Whallon JH and Klomparens KL (1994) Confocal micrsocopy in mycological research. *Exp. Biol.* 18: 275–293.

Daims H, Lebedeva EV, Pjevac P, et al. (2015) Complete nitrification by *Nitrospira* bacteria. *Nature* 528: 504–509.

Dal Cortivo C, Barion G, Visioli G, et al. (2017) Increased root growth and nitrogen accumulation in common wheat following PGPR inoculation: assessment of plant-microbe interactions by ESEM. *Agric., Ecosyst. Environ.* 247: 396–408.

Daniel R (2005) The metagenomics of soil. *Nat. Rev. Microbiol.* 3: 470–478.

Danilatos GD (1981) The examination of fresh or living plant material in an environmental scanning electron microscope. *J. Microsc.* 121(2): 235–238.

Danilatos GD (1993) Introduction to the ESEM instrument. *Microsc. Res. Tech.* 25: 354–361.

de Souza RSC, Armanhi JSL and Arruda P (2020) From microbiome to traits: designing synthetic microbial communities for improved crop resiliency. *Front. Plant Sci.* 11: 1179.

Dugourd A, Kuppe C, Sciacovelli M, et al. (2021) Causal integration of multi-omics data with prior knowledge to generate mechanistic hypotheses. *Mol. Syst. Biol.* 17: e9730.

Eder M and Lütz-Meindl U (2008) Pectin-like carbohydrates in the green alga *Micrasterias* characterized by cytochemical analysis and energy filtering TEM. *J. Microsc.* 231(2): 201–214.

Edwards J, Johnson C, Santos-Medellín C, et al. (2015) Structure, variation, and assembly of the root-associated microbiomes of rice. *Proc. Natl. Acad. Sci. U.S.A.*: E911–E920.

Einstein M (2023) Seven technologies to watch in 2023. *Nature* 613: 794–797.

Elhalag K, Nasr-Eldin M, Hussein A and Ahmad A (2018) Potential use of soilborne lytic Podoviridae phage as a biocontrol agent against *Ralstonia solanacearum*. *J. Basic Microbiol.* 58: 658–699.

Fan H, Zhang Z, Li Y, et al. (2017) Biocontrol of bacterial fruit blotch by *Bacillus subtilis* 9407 via surfactin-mediated antibacterial activity and colonization. *Front. Microbiol.* 8: 1973.

FAO (2022) *FAO's Plant Production and Protection Division.* Rome: FAO.

Ferrocino I, Rantsiou K, McClure R, et al. (2023) The need for an integrated multi-OMICs approach in microbiome science in the food system. *Compr. Rev. Food Sci. Food Saf.* 22: 1082–1103.

Fitt BDL, Huang Y-J, van den Bosch F and West JS (2006) Coexistence of related pathogen species on arable crops in space and time. *Annu. Rev. Phytopathol.* 44: 163–182.

Fouillaud M and Dufossé L (2022) Microbial secondary metabolism and biotechnology. *Microorganisms* 10: 123.

Franzosa EA, Hsu T, Sirota-Madi A, et al. (2015) Sequencing and beyond: integrating molecular 'omics' for microbial community profiling. *Nat. Rev. Microbiol.* 13: 360–372.

Freeman BC and Beattie GA (2008) An overview of plant defenses against pathogens and herbivores. *Plant Health Insp.* https://doi.org/10.1094/PHI-I-2008-0226-01.

Freund DM and Hegeman AD (2017) Recent advances in stable isotope-enabled mass spectrometry-based plant metabolomics. *Curr. Opin. Biotechnol.* 43: 41–48.

Fritschi FB, Lin H and Walker MA (2008) Scanning electron microscopy reveals different response pattern of four *Vitis* genotypes to *Xylella fastidiosa* infection. *Plant Dis.* 92: 276–286.

Gao W, Navarroli D, Naimark J, et al. (2013) Microbe observation and cultivation array (MOCA) for cultivating and analyzing environmental microbiota. *Microbiome* 1: 4.

Gilkey JC and Staehelin LA (1986) Advances in ultrarapid freezing for the preservation of cellular ultrastructure. *J. Electron Microsc. Tech.* 3: 177–210.

Grahovac J, Pajčin I and Vlajkov V (2023) *Bacillus* VOCs in the context of biological control. *Antibiotics* 12: 581.
Gustafsson MGL (2000) Surpassing the lateral resolution limit by a factor using structured illumination microscopy. *J. Microsc.* 198(2): 82–87.
Gutleben J, De Mares MC, van Elsas JD, et al. (2018) The multi-omics promise in context: from sequence to microbial isolate. *Crit. Rev. Microbiol.* 44: 212–229.
Hamonts K, Trivedi P, Garg A, et al. (2018) Field study reveals core plant microbiota and relative importance of their drivers. *Environ. Microbiol.* 20(1): 124–140.
Hardham AR (2012) Confocal microscopy in plant-pathogen interactions. *Methods Mol. Biol.* 835: 295–309.
Henty-Ridilla JL, Shimono M, Li J, et al. (2013) The plant actin cytoskeleton responds to signals from microbe-associated molecular patterns. *PLoS Pathogens* 9(4): e1003290.
Hepler PK and Gunning BES (1998) Confocal fluorescence microscopy of plant cells. *Protoplasma* 201: 121–157.
Hesselman MC, Odoni DI, Ryback BM, et al. (2012) A multi-platform flow device for microbial (co-) cultivation and microscopic analysis. *PLoS One* 7(5): e36982.
Hettich RL, Pan C, Chourey K and Giannone RJ (2013) Metaproteomics: harnessing the power of high performance mass spectrometry to identify the suite of proteins that control metabolic activities in microbial communities. *Anal. Chem.* 85: 4203–4214.
Holomuzki JR, Feminella JW and Power ME (2010) Biotic interactions in freshwater benthic habitats. *J. N. Am. Benthol. Soc.* 29(1): 220–244.
Horton MW, Bodenhausen N, Beilsmith K, et al. (2014) Genome-wide association study of *Arabidopsis thaliana*'s leaf microbial community. *Nat. Commun.* 5: 5320.
Howard RJ (1981) Ultrastructural analysis of hyphal tip cell growth in fungi: Spitzenkörper, cytoskeleton and endomembranes after freeze-substitution. *J. Cell Sci.* 48: 89–103.
Huang B, Babcock H and Zhuang X (2010) Breaking the diffraction barrier: super-resolution imaging of cells. *Cell* 143: 1047–1058.
Hulse JD (2018) Review of comprehensive staining techniques used to differentiate arbuscular mycorrhizal fungi from plant root tissues. *Acta Sci. Agric.* 2(7): 39–44.
Ingham CJ, Sprenkels A, Bomer J, et al. (2007) The micro-Petri dish, a million-well growth chip for the culture and high-throughput screening of microorganisms. *Proc. Natl. Acad. Sci. U.S.A.* 104(46): 18217–18222.
Ivanov S, Austin J, Berg RH and Harrison MJ (2019) Extensive membrane systems at the host-arbuscular mycorrhizal fungus interface. *Nat. Plants* 5: 194–203.
Jabusch LK, Kim PW, Chiniquy D, et al. (2021) Microfabrication of a chamber for high-resolution, in situ imaging of the whole root for plant-microbe interactions. *Int. J. Mol. Sci.* 22: 7880.
Jaroszuk-Ściseł J, Tyśkiewicz R, Nowak A, et al. (2019) Phytohormones (auxin, gibberellin) and ACC deaminase in vitro synthesized by the mycoparasitic *Trichoderma* DEMTkZ3A0 strain and changes in the level of auxin and plant resistance markers in wheat seedlings inoculated with this strain conidia. *Int. J. Mol. Sci.* 20: 4923.
Jeffree CE and Read NDR (1991) Ambient- and low-temperature scanning electron microscopy. In: Hall C and Hawes JL (eds) *Electron Microscopy of Plant Cells*. London: Academic Press.
Kandel SL, Joubert PM and Doty SL (2017) Bacterial endophyte colonization and distribution within plants. *Microorganisms* 5: 77.

Kankanala P, Czymmek K and Valent B (2007) Roles for rice membrane dynamics and plasmodesmata during biotrophic invasion by the blast fungus. *Plant Cell* 19: 706–724.

Karlsson M, Atanasova L, Jensen DF and Zeilinger S (2017) Necrotrophic mycoparasites and their genomes. *Microbiol. Spectrum* 5(2): FUNK-0016-2016.

Kaspar F, Neubauer P and Gimpel M (2019) Bioactive secondary metabolites from *Bacillus subtilis*: a comprehensive review. *J. Nat. Prod.* 82: 2038–2053.

Kausche GA, Pfankuch E and Ruska H (1939) Die sichtbarmachung von pflanzlichem virus im übermikroskop (In German). *Naturwissenschaften* 27: 292–299.

Kazmierczak P, McCabe P, Turina M, et al. (2012) The mycovirus CHV1 disrupts secretion of a developmentally regulated protein in *Cryphonectria parasitica*. *J. Virol.* 86(11): 6067–6074.

Ke J, Wang B and Yoshikuni Y (2021) Microbiome engineering: synthetic biology of plant-associated microbiomes in sustainable agriculture. *Trends Biotechnol.* 39(3): 244–261.

Keller PJ and Dodt H-U (2012) Light sheet microscopy of living or cleared specimens. *Curr. Opin. Neurobiol.* 22: 138–143.

Khang CH, Berruyer R, Giraldo MC, et al. (2010) Translocation of *Magnaporthe oryzae* effectors into rice cells and their subsequent cell-to-cell movement. *Plant Cell* 22: 1388–1403.

Kim H-S, Park S-Y, Kang S and Czymmek KJ (2022) Time-lapse imaging of root pathogensis and fungal proliferation without physically disrupting roots. In: Coleman J (ed) *Fusarium Wilt: Methods and Protocols*. New York: Humana, pp.153–170.

Knight R, Vrbanac A, Taylor BC, et al. (2018) Best practices for analysing microbiomes. *Nat. Rev. Microbiol.* 16: 410–422.

Köhl J, Kolnaar R and Ravensberg WJ (2019) Mode of action of microbial biological control agents against plant diseases: relevance beyond efficacy. *Front. Plant Sci.* 10: 845.

Komis G, Šamajová O, Ovečka M and Šamaj J (2015) Super-resolution microscopy in plant cell imaging. *Trends Plant Sci.* 20(12): 834–843.

Koutsoudis MD, Tsaltas D, Minogue TD and von Bodman SB (2006) Quorum-sensing regulation governs bacterial adhesion, biofilm development, and host colonization in *Pantoea stewartii* subspecies *stewartii*. *Proc. Natl. Acad. Sci. U.S.A.* 103(15): 5983–5988.

Lagier J-C, Armougom F, Million M, et al. (2012) Microbial culturomics: paradigm shift in the human gut microbiome study. *Clin. Microbiol. Infect.* 18(12): 1185–1193.

Lasinio GJ, Pollice A, Pappalettere L, et al. (2021) A statistical protocol to describe differences among nutrient utilization patterns of *Fusarium* spp. and *Trichoderma gamsii*. *Plant Pathol.* 70: 1146–1157.

Leach JE, Beattie GA and Eversole K (2016) *Phytobiomes: A Roadmap for Research and Translation.* St. Paul, MN: American Phytopathological Society.

Lee J, Kim S, Jung H, et al. (2023) Exploiting bacterial genera as biocontrol agents: mechanisms, interactions and applications in sustainable agriculture. *J. Plant Biol.* https://doi.org/10.1007/s12374-023-09404-6.

Leewenhoeck M (1673) A specimen of some observations made by a microscope, contrived by M. Leewenhoeck in Holland, lately communicated by Dr. Regnerus de Graaf. *Phil. Trans.* 1673: 6037–6038.

Lemanceau P, Blouin M, Muller D and Moënne-Loccoz Y (2017) Let the core microbiota be functional. *Trends Plant Sci.* 22(7): 583–595.

Leopold DR and Busby PE (2020) Host genotype and colonist arrival order jointly govern plant microbiome composition and function. *Curr. Biol.* 30: 3260–3266.

Levy A, Conway JM, Dangl JL and Woyke T (2018) Elucidating bacterial gene functions in the plant microbiome. *Cell Host Microbe* 24: 475–485.

Lewis WH, Tahon G, Geesink P, et al. (2021) Innovations to culturing the uncultured microbial majority. *Nat. Rev. Microbiol.* 19: 225–240.

Libault M, Govindarajulu M, Berg RH, et al. (2011) A dual-targeted soybean protein is involved in *Bradyrhizobium japonicum* infection of soybean root hair and cortical cells. *Mol. Plant-Microbe Interact.* 24(9): 1051–1060.

Liu Y, Patko D, Engelhardt I, et al. (2021) Plant-environment microscopy tracks interactions of *Bacillus subtilis* with plant roots across the entire rhizosphere. *Proc. Natl. Acad. Sci. U.S.A.* 118(48): e2109176118.

Lucaciu R, Pelikan C, Gerner SM, et al. (2019) A bioinformatics guide to plant microbiome analysis. *Front. Plant Sci.* 10: 1313.

Ma L, Kim J, Hatzenpichler R, et al. (2014) Gene-targeted microfulidic cultivation validated by isolation of a gut bacterium listed in Human Microbiome Project's Most Wanted taxa. *Proc. Natl. Acad. Sci. U.S.A.* 111(27): 9768–9773.

Martino E, Murat C, Vallino M, et al. (2007) Imaging mycorrhizal fungal transformants that express EGFP during ericoid endosymbiosis. *Curr. Genet.* 52: 65–75.

Mauch-Mani B, Baccelli I, Luna E and Flors V (2017) Defense priming: an adaptive part of induced resistance. *Annu. Rev. Plant Biol.* 68: 485–512.

Mbofung GCY, Sernett J, Horner HT and Robertson AE (2016) Comparison of susceptible and resistant maize hybrids to colonization by *Clavibacter michiganensis* subsp. *nebraskensis*. *Plant Dis.* 100(4): 711–717.

Mendgen K and Lesemann D-E (1991) *Electron Microscopy of Plant Pathogens*. Heidelberg: Springer Berlin.

Mendoza-Suárez MA, Geddes BA, Sánchez-Cañizares C, et al. (2020) Optimizing *Rhizobium*-legume symbioses by simultaneous measurement of rhizobial competitiveness and N_2 fixation in nodules. *Proc. Natl. Acad. Sci. U.S.A.* 117(18): 9822–9831.

Miura N and Okuda S (2023) Current progress and critical challenges to overcome in the bioinformatics of mass spectrometry-based metaproteomics. *Comput. Struct. Biotechnol.* 21: 1140–1150.

Mizuta Y (2021) Advances in two-photon imaging in plants. *Plant Cell Physiol.* 62(8): 1224–1230.

Moreira ZPM, Chen MY, Ortuno DLY and Haney CH (2023) Engineering plant microbiomes by integrating eco- evolutionary principles into current strategies. *Curr. Opin. Plant Biol.* 71: 102316.

Mulder A, van de Graaf AA, Robertson LA and Kuenen JG (1995) Anaerobic ammonium oxidation discovered in a denitrifying fluidized bed reactor. *FEMS Microbiol. Ecol.* 16: 177–184.

Niu B, Paulson JN, Zheng X and Kolter R (2017) Simplified and representative bacterial community of maize roots. *Proc. Natl. Acad. Sci. U.S.A.*: E2450-E2459.

Nunoura T, Chikaraishi Y, Izaki R, et al. (2018) A primordial and reversible TCA cycle in a facultatively chemolithoautotrophic thermophile. *Science* 359: 559–563.

Nyholm L, Koziol A, Marcos S, et al. (2020) Holo-omics: integrated host-microbiota multi-omics for basic and applied biological research. *iScience* 23: 101414.

Oerke E-C and Dehne E-W (2004) Safeguarding production–losses in major crops and role of crop protection. *Crop Prot.* 23: 275–285.

Ongena M and Jacques P (2007) *Bacillus* lipopeptides: versatile weapons for plant disease biocontrol. *Trends Microbiol.* 16(3): 115–125.

Oreopoulos J, Berman R and Browne M (2014) Spinning-disk confocal microscopy: present technology and future trends. *Methods Cell Biol.* 123: 153–175.

Orozco-Mosqueda MC, Fadiji AE, Babalola OO, et al. (2022) Rhizobiome engineering: unveiling complex rhizosphere interactions to enhance plant growth and health. *Microbiol. Res.* 263: 127137.

Ossowicki A, Raaijmakers JM and Garbeva P (2021) Disentangling soil microbiome functions by perturbation. *Environ. Microbiol. Rep.* 13(5): 582–590.

Ovečka M, Sojka J, Tichá M, et al. (2022) Imaging plant cells and organs with light-sheet and super-resolution microscopy. *Plant Physiol.* 188: 683–702.

Parida AK, Panda A and Rangani J (2018) Metabolomics-guided elucidation of abiotic stress tolerance mechanisms in plants. In: Ahmad P, Ahanger MA, Singh VP, et al. (eds) *Plant Metabolites and Regulation Under Environmental Stress*. Academic Press, pp.89–131.

Park J, Kerner A, Burns MA and Lin XN (2011) Microdroplet-enabled highly parallel co-cultivation of microbial communities. *PLoS One* 6(2): e17019.

Parratt SR and Laine A-L (2016) The role of hyperparasitism in microbial pathogen ecology and evolution. *ISME J.* 10: 1815–1822.

Pathan AK, Bond J and Gaskin RE (2008) Sample preparation for scanning electron microscopy of plant surfaces—horses for courses. *Micron* 39: 1049–1061.

Peddie CJ, Genoud C, Kreshuk A, et al. (2023) Volume electron microscopy. *Nat. Rev. Methods Primers*. https://doi.org/10.1038/s43586-43022-00131-43589.

Peiffer JA, Spor A, Koren O, et al. (2013) Diversity and heritability of the maize rhizosphere microbiome under field conditions. *Proc. Natl. Acad. Sci. U.S.A.* 110(16): 6548–6553.

Pieterse CMJ, Zamioudis C, Berendsen RL, et al. (2014) Induced systemic resistance by beneficial microbes. *Annu. Rev. Phytopathol.* 52: 347–375.

Pinu FR, Beale DJ, Paten AM, et al. (2019) Systems biology and multi-omics integration: viewpoints from the metabolomics research community. *Metabolites* 9: 76.

Qiu Z, Egidi E, Liu H, et al. (2019) New frontiers in agriculture productivity: optimised microbial inoculants and in situ microbiome engineering. *Biotechnol. Adv.* 37: 107371.

Rahman SFSA, Singh E, Pieterse CMJ and Schenk PM (2018) Emerging microbial biocontrol strategies for plant pathogens. *Plant Sci.* 267: 102–111.

Raza W, Wei Z, Ling N, et al. (2016) Effect of organic fertilizers prepared from organic waste materials on the production of antibacterial volatile organic compounds by two biocontrol *Bacillus amyloliquefaciens* strains. *J. Biotechnol.* 227: 43–53.

Read ND and Jeffree CE (1990) Low-temperature scanning electron microscopy in biology. *J. Microsc.* 161(1): 59–72.

Regalado J, Lundberg DS, Deusch O, et al. (2020) Combining whole-genome shotgun sequencing and rRNA gene amplicon analyses to improve detection of microbe–microbe interaction networks in plant leaves. *ISME J.* 14: 2116–2130.

Reinhold-Hurek B, Bünger W, Burbano CS, et al. (2015) Roots shaping their microbiome: global hotspots for microbial activity. *Annu. Rev. Phytopathol.* 53: 403–424.

Romanazzi G, Sanzani SM, Bi Y, et al. (2016) Induced resistance to control postharvet decay of fruit and vegetables. *Postharvest Biol. Technol.* 122: 82–94.

Romano I, Ventorino V and Pepe O (2020) Effectiveness of plant beneficial microbes: overview of the methodological approaches for assessment of root colonization and persistence. *Front. Plant Sci.* 11: 6.

Rufián JS, Macho AP, Corry DS, et al. (2018) Confocal microscopy reveals *in planta* dynamic interactions between pathogenic, avirulent and non-pathogenic *Pseudomonas syringae* strains. *Mol. Plant Pathol.* 19(3): 537–551.

Sanchez-Gorostiaga A, Bajić D, Osborne ML, et al. (2019) High-order interactions distort the functional landscape of microbial consortia. *PLoS Biol.* 17(12): e3000550.

Schubert V (2017) Super-resolution microscopy – applications in plant cell research. *Front. Plant Sci.* 8: 531.

Shade A and Handelsman J (2012) Beyond the Venn diagram: the hunt for a core microbiome. *Environ. Microbiol.* 14(1): 4–12.

Shakya M, Lo C-C and Chain PSG (2019) Advances and challenges in metatranscriptomic analysis. *Front. Genet.* 10: 904.

Sharpton TJ (2014) An introduction to the analysis of shotgun metagenomic data. *Front. Plant Sci.* 5: 209.

Sibarita J-B (2005) Deconvolution microscopy. *Adv. Biochem. Engin. Biotechnol.* 95: 201–243.

Singh B and Sharma RA (2016) Yield enhancement of phytochemicals by *Azotobacter chroococcum* biotization in hairy roots of *Arnebia hispidissima*. *Ind. Crops Prod.* 81: 169–175.

Sohrabi R, Paasch BC, Liber JA and He SY (2023) Phyllosphere microbiome. *Annu. Rev. Plant Biol.* 74: 539–568.

Spadaro D and Droby S (2016) Development of biocontrol products for postharvest diseases of fruit: The importance of elucidating the mechanisms of action of yeast antagonists. *Trends Food Sci. Tech.* 47: 39–49.

Stabentheiner E, Zankel A and Pölt P (2010) Environmental scanning electron microscopy (ESEM)–a versatile tool in studying plants. *Protoplasma* 246: 89–99.

Steinberg S, Grinberg M, Beitelman M, et al. (2021) Two-way microscale interactions between immigrant bacteria and plant leaf microbiota as revealed by live imaging. *ISME J.* 15: 409–420.

Stenberg JA, Sundh I, Becher PG, et al. (2021) When is it biological control? A framework of definitions, mechanisms, and classifications. *J. Pest Sci.* 94: 665–676.

Stenuit B and Agathos SN (2015) Deciphering microbial community robustness through synthetic ecology and molecular systems synecology. *Curr. Opin. Biotechnol.* 33: 305–317.

Sun Q, Sun Y, Walker MA and Labavitch JM (2013) Vascular occlusions in grapevines with Pierce's disease make disease symptom development worse. *Plant Physiol.* 161: 1529–1541.

Sydor AM, Czymmek KJ, Puchner EM and Mennella V (2015) Super-resolution microscopy: from single molecules to supramolecular assemblies. *Trends Cell Biol.* 25(12): 730–748.

Thokchom E, Thakuria D, Kalita MC, et al. (2017) Root colonization by host-specific rhizobacteria alters indigenous root endophyte and rhizosphere soil bacterial communities and promotes the growth of mandarin orange. *Eur. J. Soil Biol.* 79: 48–56.

Toju H, Peay KG, Yamamichi M, et al. (2018) Core microbiomes for sustainable agroecosystems. *Nat. Plants* 4: 247–257.

Tsoi R, Wu F, Zhang C, et al. (2018) Metabolic division of labor in microbial systems. *Proc. Natl. Acad. Sci. U.S.A.* 115: 2526–2531.

Turner TR, James EK and Poole PS (2013) The plant microbiome. *Genome Biol.* 14: 209.

van der Heijden MGA, de Bruin S, Luckerhoff L, et al. (2016) A widespread plant-fungal-bacterial symbiosis promotes plant biodiversity, plant nutrition and seedling recruitment. *ISME J.* 10: 389–399.

Veach AM, Chen H, Yang ZK, et al. (2020) Plant hosts modify belowground microbial community response to extreme drought. *mSystems* 5(3): e00092-00020.

Vierheilig H, Coughlan AP, Wyss U and Piché Y (1998) Ink and vinegar, a simple staining technique for arbuscular-mycorrhizal fungi. *Appl. Environ. Microbiol.* 64(12): 5004–5007.

Vogel CM, Potthoff DB, Schäfer M, et al. (2021) Protective role of the *Arabidopsis* leaf microbiota against a bacterial pathogen. *Nat. Microbiol.* 6.

Vorholt JA, Vogel C, Carlström CI and Müller DB (2017) Establishing causality: opportunities of synthetic communities for plant microbiome research. *Cell Host Microbe* 22: 142–155.

Wallenstein MD (2017) Managing and manipulating the rhizosphere microbiome for plant health: a systems approach. *Rhizosphere* 3: 230–232.

Wang S, Alseekh S, Fernie AR and Luo J (2019) The structure and function of major plant metabolite modifications. *Mol. Plant* 12: 899–919.

Wang T, Liang Y, Wu M, et al. (2015) Natural products from *Bacillus subtilis* with antimicrobial properties. *Chin. J. Chem. Eng.* 23: 744–754.

Wang Y, Hayatsu M and Fujii T (2012) Extraction of bacterial RNA from soil: challenges and solutions. *Microbes Environ.* 27(2): 111–121.

Wei Z, Gu Y, Friman V-P, et al. (2019) Initial soil microbiome composition and functioning predetermine future plant health. *Sci. Adv.* 5: eaaw0759.

Whipps JM, Sreenivasaprasad S, Muthumeenakshi S, et al. (2008) Use of *Coniothyrium minitans* as a biocontrol agent and some molecular aspects of sclerotial mycoparasitism. *Eur. J. Plant Pathol.* 121: 323–330.

White JF, Torres MS, Somu MP, et al. (2014) Hydrogen peroxide staining to visualize intracellular bacterial infections of seedling root cells. *Microsc. Res. Tech.* 77: 566–573.

Wightman R (2022) An overview of cryo-scanning electron microscopy techniques for plant imaging. *Plants* 11: 1113.

Wink M (2003) Evolution of secondary metabolites from an ecological and molecular phylogenetic perspective. *Phytochemistry* 64(1): 3–19.

Wu Q, Luo A, Zadrozny T, et al. (2013) Fluorescent protein marker lines in maize: generation and applications. *Int. J. Dev. Biol.* 57: 535–543.

Wu Y and Shroff H (2018) Faster, sharper, and deeper: structured illumination microscopy for biological imaging. *Nat. Methods* 15: 1011–1019.

Xu L, Pierroz G, Wipf HM-L, et al. (2021) Holo-omics for deciphering plant-microbiome interactions. *Microbiome* 9: 69.

Yu S-M, Oh B-T and Lee YH (2012) Biocontrol of green and blue molds in postharvest satsuma mandarin using *Bacillus amyloliquefaciens* JBC36. *Biocontrol Sci. Technol.* 22(10): 1181–1197.

Chapter 2

Improving biosafety testing of plant protection products (PPPs) and bioprotectants

Jonathan Willow, Pheasants Forever and Quail Forever, USA; and Guy Smagghe, Vrije Universiteit Brussel (VUB)/Ghent University, Belgium/Southwest University/Guizhou University, China

1 Introduction

Wild bee populations are globally important for the provision of pollination services. The Intergovernmental Platform on Biodiversity and Ecosystem Services (IPBES) reported that almost 90% of wild plant species and over 75% of crops that people use for food depend, in part, on pollination by bees, butterflies and other animals. The report estimates the annual economic value of pollinators at *$235 to $577 billion* (IPBES, 2016).

Within agricultural landscapes, one of the primary threats to populations of pollinators (and other non-target organisms (NTOs) such as natural enemies useful for biological control), is the application of plant protection products (PPPs) that are meant to protect crops from insect/arthropod pests as well as pathogens and weed encroachment. PPPs can be used in combination with beneficial insects, such as pollinators and natural enemies, in biological control and integrated pest management (IPM) programs. Regulations exist to minimize

http://dx.doi.org/10.19103/AS.2024.0142.07

the impacts that applications of PPPs of both chemical and biological origin can have on both managed- and wild bee populations that are useful for pollination in agricultural crops. Despite this, pollinators remain negatively affected by individual PPP residues on crops and drift to non-target areas, as well as the effects of the application of tank-mixed PPPs. To preserve bee health and help wild bee populations within their native landscapes, we need to understand the effects that various types of PPPs have on bees and alter our application practices accordingly.

Today, there is awareness to reduce the pressure on the environment through classical and modern PPPs, both of chemical and biological origin, and support the preservation of biodiversity and sustainability. This is in accordance with policies worldwide in many countries and continents. A good example is the "Green Deal" in Europe (EU 2024a, 2024b; https://commission.europa.eu/strategy-and-policy/priorities-2019-2024/european-green-deal_en) where the main focus is to increase sustainability and biodiversity. As part of the "Farm to Fork" strategy, one of the ambitions of the European Commission is that (by 2030) the use and risk of chemical and more hazardous PPPs in the EU should be reduced by 50%. At the EU level, close to €2.1 billion from the European Agricultural Fund for Rural Development (EAFRD) is allocated to facilitate knowledge, exchange, and information.

The EU "Nature Restoration Law" also combines an overarching restoration objective for the long-term recovery of nature in the EU's land and sea areas with binding restoration targets for specific habitats and species, and these measures should cover at least 20% of the EU's land and sea areas by 2030, and ultimately all ecosystems in need of restoration by 2050 (EU 2024a, 2024b). In addition, the Kunming-Montreal Global Biodiversity Framework (GBF; https://www.cbd.int/doc/decisions/cop-15/cop-15-dec-04-en.pdf) is aimed at enhancing sustainability and biodiversity, and is supported by 196 nations; and the initiative is committed to restoring 30% of terrestrial and marine ecosystems, with the primary goal of halting and reversing biodiversity loss by the year 2030 (Hughes and Grumbine 2023).

In this chapter, we will discuss the current standard testing with the three-tier approach and future directions of bioassays with classic and modern PPPs, both of chemical and biological origin. There is a specific emphasis on examining the risk of negative side-effects of PPPs on foraging in bumblebees. The chapter discusses here bioprotectants that are used in biological control programs to provide relevant, effective, risk-based assessment. As bumblebees are important pollinators, special attention will also go to the applications of entomovectoring for enhanced pollination and biological control where bumblebees are used as "flying doctors", and the need for adequate technique for biosafety testing of bioprotectants. Furthermore, we discuss key areas of interest for forwarding relevant and effective risk-based assessment in bumblebees and factors that are at play herein. Foraging behavior/success represents a vital endpoint for biosafety testing given its direct relationship with the provision of pollination services. All in all, we aim to present an overview to evaluate, map, and summarize the accumulated evidence from existing and novel techniques in the risk assessment of PPPs, both of chemical and biological origin, with bumblebees of *B. terrestris*; to identify research gaps, common challenges, and factors that are at play in the environment of bumblebees; and to provide recommendations for future studies and policies on this topic. We hope the chapter will be a valuable guide for researchers, practitioners, and policymakers dealing with the complex dynamics of environmental risk assessment of pollinators particularly and NTOs in general.

2 Studying the range of wild bee species

Methods for understanding field-realistic toxicity of PPPs in bees have come a long way in recent years. For example, it is no longer thought acceptable to extrapolate honeybee (*Apis mellifera*) toxicology data across other bee species, even those within the same family (Apidae), such as bumblebees (*Bombus* spp.) (Mommaerts and Smagghe, 2011b; Gradish et al., 2019; Kueh Tai et al., 2022). The potential effects of exposure to PPPs in wild bees are not only limited to acute mortality. There are numerous measurable endpoints that are highly relevant to their ability to provide pollination services in agroecosystems. The most used wild bee species used in toxicity studies with PPPs are bumblebees, particularly:

- *Bombus terrestris* (the buff-tailed bumblebee or large earth bumblebee) in Europe (Figure 1a)
- *Bombus impatiens* (the common eastern bumblebee) in North America (Figure 1b)

These are discussed below.

B. terrestris is the most intensely studied non-*Apis* bee. It is one of the most abundant and widespread bumblebee species in the western Palearctic (the biogeographic region covering Europe, Central Asia, and North Africa). It is a geographically variable species with a wide distribution in Europe, present in all the countries around the Mediterranean Sea (except Egypt), the Mediterranean Islands, Canary Islands and Madeira, and east to Afghanistan. In the north, it is found and extends up to the latitude of Helsinki (in Finland) and east to Altai (in Siberia) (Rasmont, 1983, 1984). *B. terrestris* workers are highly valued pollinators worldwide and substantially add to the value of crop production (Goulson 2003a; Velthuis and van Doorn 2006). *B. terrestris* is a highly polylectic bumblebee, collecting pollen from hundreds of different

Figure 1 Two major bumblebee species: (a) the buff-tailed earth bumblebee (*Bombus terrestris*) that is the most common bumblebee in Europe (from https://www.freenatureimages.eu/) and (b) the common eastern bumblebee (*Bombus impatiens*), the most common on North America (from https://www.inaturalist.org/). (c) Life cycle of bumblebee of *B. terrestris*: hibernating queens start foraging in spring and initiate a new bumblebee colony; the foundress queen starts building a brood nest and starts laying eggs; the foundress queen is still foraging and takes care of the brood; as soon as the first workers emerge, they take over the brood care and start foraging, the eusocial phase in the life cycle starts, and the foundress queen stays in the brood nest; the colony grows and produces workers; in the reproduction phase, males and gynes are produced; the foundress queen dies and the young queens and males start foraging; gynes are mating; mated queens search for a place to overwinter and hibernate. From Klein et al. (2022)

flowering plant species. Outside of Europe, *B. terrestris* has been introduced into New Zealand, Tasmania, Brazil, Chile, Mexico, and Japan. It appears to have been introduced to mainland Australia (New South Wales) without persisting, and it has spread from Chile to Argentina (CABI 2024).

B. impatiens, the common eastern bumblebee, is the most commonly encountered bumblebee across much of eastern North America. They can be found in the temperate forest region of the eastern United States and southern Canada, and in the eastern Great Plains (Williams et al., 2014). Because of their great adaptability, they can live in rural areas, suburbs, and even urban cities. This adaptability makes them a great pollinator species, leading to an increase in their commercial use by the greenhouse industry. This increase consequently led to their further spread outside their previous distribution. They are considered one of the most important bee species in North America.

As described in detail in Goulson (2003a), the world's bumblebee fauna consists of approximately 250 known species. Most of them are found in temperate parts of the northern hemisphere. Bumblebees are social insects (eusocial), and the vast majority have an annual life cycle. In their life cycle, mated queens emerge from hibernation in the spring and attempt to find a nest in which to rear daughter workers. In optimal conditions, the number of workers may reach 300 to 500 in some species by mid-summer, when new queens and males are reared. These leave the nest and mate, and the new queens enter hibernation while the rest of the population dies off. *B. terrestris* form annual colonies, and new colonies are initiated each spring by mated solitary queens (Figure 1c) (Goulson 2003a, Klein et al 2022).

Bumblebees are adapted to a diversity of climates and habitats and are active even when light intensity is low. Due to their relatively large body sizes and dense pile, they can continue foraging even at temperatures as low as 10°C and as high as 32°C, with observations of *Bombus terrestris dalmatinus* at temperatures as low as 2°C (CABI, 2010). Their increased motility allows them to continue flower visits for most of the year, unlike honeybees which are mostly inactive at temperatures below 16°C (Heinrich, 1979). Bumblebees can forage during adverse climatic conditions, even flying during light rain, visiting from 20 to 50 flowers per minute with high pollination efficiency. Consequently, many early spring- and late autumn-flowering plants benefit from pollination services provided by members of this hardy genus.

Typically, bumblebees exhibit tremendous variation in body size and proboscis length, and forage on a variety of floral resources. Most bumblebees have a longer tongue than honeybees, allowing them to reach nectar even in deep, tubular flowers. They also exhibit a distinctive behavior of sonication, or "buzz pollination", which vibrates pollen from the poricidal anthers of plants such as tomatoes (Buchmann, 1983). This is why they have been chosen for the greenhouse pollination of Solanaceae crops (Dag and Kammer, 2001).

Bumblebees rapidly contract their indirect flight muscles while curled around a tomato flower androecium and transmit vibrations into a flower's anthers, resulting in rapid pollen ejection from its apical pores. Bumblebees can harvest pollen from "buzzed" tomato flowers 400 times faster than honeybees can. Whereas managed honeybees are also generalists that can pollinate a wide variety of native plants and managed crops, they are less efficient and more temperature-restricted than bumblebees for many crops (Alford 1975; Goulson 2003a, 2003b; Velthuis and van Doorn, 2006).

3 The three-tiered approach to assessing impacts of plant protection products (PPPs) on bumblebees

3.1 Tier-1 risk assessment

There are different levels to examining the effects of classic and modern PPPs, both of chemical and biological origin, on bumblebees. The United States-Environmental Protection Agency (US-EPA), the Organisation for Economic Co-operation and Development (OECD), and the European Food Safety Authority (EFSA) all adopt a three-tiered approach to determining the potential impact that a PPP may have on bees. Tier-1 represents the most basic toxicological study, often determining LD_{50} or LC_{50} values (the dose or concentration that is killing 50% of the bumblebees tested compared to the control), or determining the onset of certain sublethal endpoints. Techniques for Tier-1 studies vary between agencies, laboratories, and industries. These tests are intended as screening tools, employing assumptions that are likely to overestimate exposure, and using the most sensitive toxicity estimates to calculate risk. Tier-1 risk assessments generally rely upon laboratory-based toxicity data from studies using individual bees. If potential risks of concern persist after a Tier-1 risk assessment, Tier-2 (e.g. semi-field, greenhouse) and Tier-3 (e.g. outdoor, field) risk assessments may be conducted.

In detail, the Tier-1 tests of OECD, as developed by the "Non-Apis working group" of the International Commission for Plant-Pollinator Relationships (ICPPR), have been ring-validated (OECD, 2017a, 2017b) (Figure 2a). These two test guidelines are laboratory test methods designed to assess the acute contact- and acute oral toxicity of PPPs and other chemicals, as bioprotectants of chemical and biological origin, to adult worker bumblebees. The tests are based on OECD 213-Honeybees, Acute Oral Toxicity Test (OECD 1998a), OECD 214-Honeybees, Acute Contact Toxicity Test (OECD, 1998b), van der Steen et al. (1996), and Hanewald et al. (2013). The test methods were ring-tested first by an international ICPPR ring-test group in 2014 and second by an international OECD ring-test group in 2015. In the ring-tests, the following species were successfully used: *B. terrestris* and *B. impatiens*. The tests are also applicable to other bumblebee species, but this has not been documented yet.

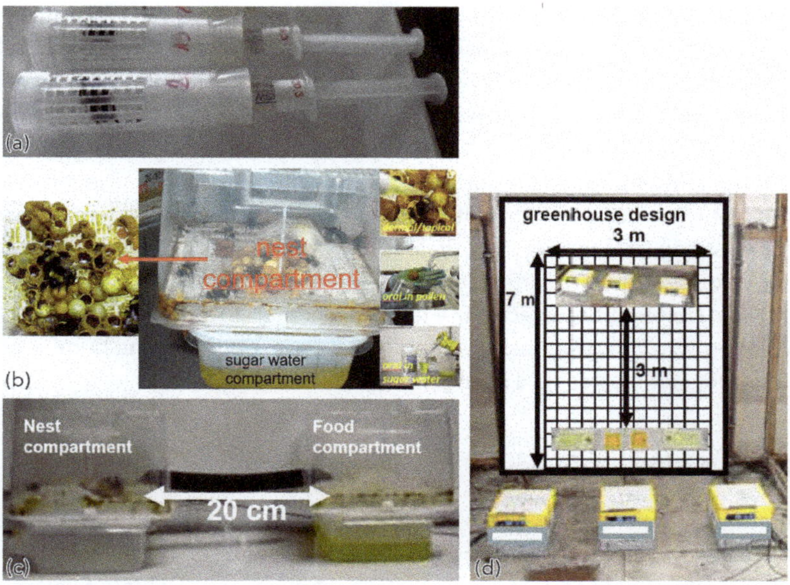

Figure 2 Schematic overview of the different "tier" levels with bumblebees *B. terrestris*. (a) "Single housing" cages to test lethal effects of PPPs and other chemicals, as bioprotectants of chemical and biological origin, in individual bumblebees at Tier-1 level, as developed by the "Non-Apis" working group of ICPPR, and validated as OECD 246 and 247 guideline for testing of chemicals. From OECD (2017a, 2017b). (b) Setup with "microcolony" to assess for side-effects at Tier-2 level, for lethal and sublethal effects of PPPs and other chemicals, as bioprotectants of chemical and biological origin. The microcolonies were started with 5 new workers and placed in transparent, plastic nest boxes (15 cm x 15 cm x 10 cm). At the center of the nest box, there was a drinking place and brood area. Under the nest box, a container with 500 ml of sugar water was provided. After 1 week, the dominant worker started to produce eggs that developed into males. The microcolonies were kept under standardized laboratory conditions in the dark, and commercial pollen and sugar water were provided as food. The compounds were applied to the bumblebees via direct dermal/topical contact, or orally via drinking sugar water or eating pollen, which were supplemented with the compound. With his method, the risks for lethal and sublethal effects in bumblebee survival and reproduction have been evaluated with nearly all groups of classic and modern PPPs with insecticide, acaricide, and fungicide activities, and of both chemical and microbial origin and many used in biological control and IPM programs. See Mommaerts et al. (2010d). (c) "The foraging behavior test" is making use of micro-colonies of *B. terrestris* to assess, at the Tier-2 level, for lethal and sublethal effects, including foraging behavior, by different classic and modern PPPs of both chemical and microbial origin. A box containing a micro-colony was connected by a tube of 20 cm in length with an empty nest containing the food (pollen and sugar water). This experimental setup allows the evaluation of interferences with the orientation capacity of the adult bumblebee workers. Also, other endpoints important for the foraging process can become affected after exposure. From Mommaerts et al. (2010c). (d) Bioassay with full queen-right colonies of *B. terrestris* in small greenhouse compartments with free-flying bumblebees that can be used to assess for lethal and sublethal effects, including foraging behavior, by different classic and modern PPPs of both chemical and microbial origin. The inset gives the greenhouse design (3 x 7 m) with the bumblebee colonies placed at 3 m from the food. From Mommaerts and Smagghe (2011b).

In detail, as shown in Figure 2a, bumblebees are kept individually in cages, described as "single housing". Single housing prevents hierarchy fights (among the queen-less bumblebee workers), potentially introducing mortality, and it allows a precise assessment of affected or non-affected bumblebees. Furthermore, bumblebees do not share food via trophallaxis and need to be fed individually. Mortality is recorded within 4–5 h after the start of administration of the test chemical, as well as after 24 h and 48 h; and when different doses/concentrations are tested, LD_{50} values can be calculated. If a prolonged observation is required, further assessments should be made after 72 h and 96 h. Also, bumblebees used in the test are weighed individually which allows calculation of toxicity per fresh weight. During the experiment, the bumblebees are offered sugar water via a syringe.

3.2 Tier-2 risk assessment

Mommaerts et al. (2006a, 2006b) developed a standard method with the use of a "microcolony" of *B. terrestris* to evaluate the risks of different classes of PPPs and other chemicals, as bioprotectants of chemical and biological origin. This was to answer criticism on the aforementioned laboratory risk assessment tests "Tier-1" with individual bees over 72 h, since the potential side-effects of PPPs might take a longer time (>72 h) before becoming visible under practical conditions and, secondly, that bumblebee workers show a social organization with the building of a nest (brood) and with foraging behavior to transport food from outside to inside the nest. It is therefore recommended to conduct an extended laboratory test as a second step of the risk assessment ("Tier-2") with a miniature colony of bumblebees. To cover all potential side-effects, bumblebees should then be exposed, as in the insect laboratory test with individual workers ("Tier-1"), to compounds at concentrations recovered in the field to concentrations as recommended for use, or to the maximum field recommended concentration (MFRC).

In detail, the microcolonies were started with 5 new workers and placed in transparent plastic nest boxes (15 cm x 15 cm x 10 cm) at standard laboratory conditions in full darkness. At the center of the nest box, there was a drinking place and brood area. Under the nest box, a container with 500 ml of sugar water was provided. After 1 week, the dominant worker started to produce eggs that developed into males. The microcolonies were kept under standardized laboratory conditions in the dark, and commercial pollen and sugar water were provided as food, as shown in Figure 2b. The bumblebees were exposed to the compound to be tested via the different possible routes of exposure as in the field (topically via direct contact after spraying, or orally via the consumption of contaminated food, i.e. nectar and pollen). In the microcolonies, the bumblebees were treated with the compound via direct dermal/topical contact, or orally via drinking sugar water or eating pollen, which were supplemented with the compound; and the microcolonies were followed for 1´ weeks. The compound was treated at its MFRC and a series of dilutions. Besides worker mortality (i.e. lethal side-effects), this setup with "microcolonies" allowed to test for potential sublethal side-effects on bumblebee reproduction and life cycle development, particularly egg production and larval, pupal, and adult development. These parameters are of crucial importance to assess risks in the life cycle and so to guarantee pollination.

With this bioassay setup, the risks of different classic and modern PPPs with insecticide, acaricide, and fungicide activities and of both chemical and microbial origin–and many used in biological control programs–for lethal and sublethal effects in bumblebee survival and reproduction have been evaluated (Mommaerts et al., 2006a, 2006b, 2008, 2009, 2010a, 2011b; Besard et al., 2010). Importantly, these data help to provide guidance to practitioners, farmers, and advisers on the combined use of these products with bumblebees. Figure 3 presents an overview of the different classes such as carbamates, pyrethroids, organophosphates, neonicotinoids, insect growth regulators (IGRs), biological insecticides, acaricides, chemical fungicides, and biological fungicides. For each group, the bars represent the percentage of compounds which are non-toxic (green), weakly/moderately toxic (yellow-orange), and toxic (red). The division in toxicity levels is based on the recommendations made by the side-effect list when available, or on the obtained toxicity with microcolonies (Tier-2). More details with responses on survival and reproduction to the different compounds and the concentrations and formulation type tested are published in Mommaerts and Smagghe (2011b).

A note with the risk assessment of PPPs in bumblebees is that, so far, the side-effects classification for arthropods and beneficial organisms by the "International Organization for Biological Control of Noxious Animals and Plants" (IOBC) has been used: "class 1": <25% effect, non-toxic; "class 2": 25–50% effect, weakly toxic; "class 3": 50–75% effect, moderately toxic; and "class 4": >75% effect, highly toxic. There is still no validation of this classification at present. For example, when a product causes a loss of <25%, it is considered as not toxic. However, Goulson (2010) argued that the effect of a loss on the colony depends on the colony size. It can therefore be suggested that in the future, these classifications should be defined in relation to the range of the colony size. In addition, it should be noted here that for microbial products as biofungicides, the exposure can change after treatment and during the exposure period with the growth of the biofungicide or bioinsecticide microbial organisms and potential presence of secondary metabolites. Microbial products can be diverse and based on an organism of bacterial, fungal, or viral origin. The development of relevant and effective risk-based assessment protocols and standards is today under discussion in the "Microbials" working group of

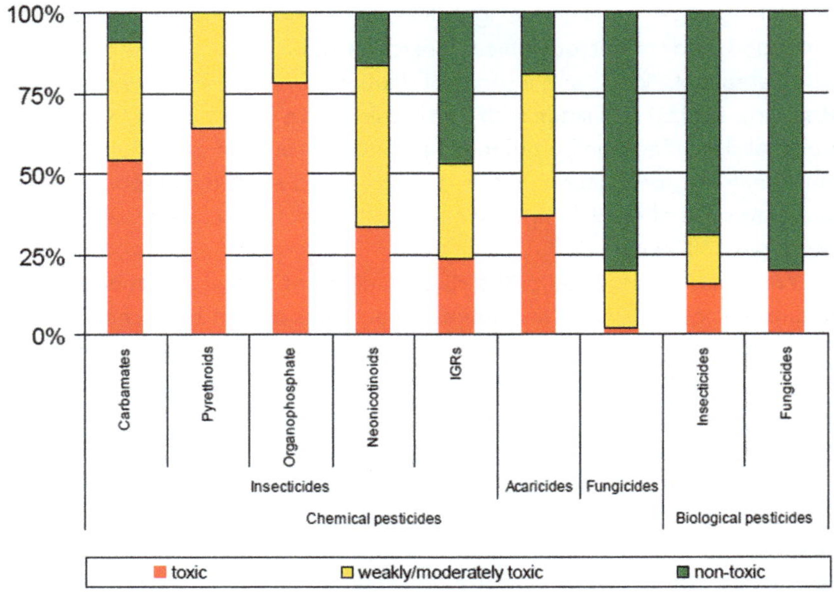

Figure 3 Overview of the toxicity of different classic and modern products with insecticide, acaricide, and fungicide activities and of both chemical (classic and modern PPPs) and microbial origin (i.e. bioprotectants as biological insecticides and fungicides) toward bumblebees (*Bombus terrestris*). For each PPP group, the bars represent the percentage of compounds which are non-toxic (green), weakly/moderately toxic (yellow-orange), and toxic (red). The division in toxicity levels is based on the recommendations made by the side-effect list when available, or on the obtained toxicity with micro-colonies ("tTer-2"). The number of compounds considered per group are *n* = 11 for carbamates, *n* = 14 for pyrethroids, *n* = 32 for organophosphates, *n* = 6 for neonicotinoids, *n* = 17 for insect growth regulators (IGRs), *n* = 13 for bioprotectants as biological insecticides (for instance, *Bacillus thuringiensis* preparations), *n* = 27 for acaricides, *n* = 66 for chemical fungicides, and *n* = 5 for biological fungicides (for instance, microbial biofungicide preparations). For more details, see Mommaerts et al. (2010b).

the ICPPR. For instance, representative standard microorganism species for testing can include one species of bacteria and one species of fungus (e.g. *Bacillus thuringiensis*, *Beauveria bassiana,* or *Metarhizium anisopliae*).

With the development and use of neonicotinoid insecticides, accumulation in the environment and reports on failures of foraging (Blacquière et al., 2012), Mommaerts et al. (2010c) developed an advanced semi-field (Tier-2 risk assessment) technique, which they called "the foraging behavior test", for examining the effect of PPPs on both foraging behavior and colony performance in the bumblebee *B. terrestris*, as shown in Figure 2c. This setup used a microcolony in a plastic box, but now two plastic boxes were used and connected by a tube of 20 cm in length. In one box, the worker bees constructed their nest. After the appearance of third- and fourth-instar larvae, sugar water was relocated to the second box, and the bees were allowed two days of foraging on untreated food. After this, bumblebee colonies foraged on neonicotinoid-treated sugar water, and worker survival and drone production were observed weekly for 11 weeks. With this "laboratory foraging test" method, including foraging behavior, the 50% effect value for mortality (LC_{50} value) was 20 ppb (=1/9850 of the MFRC), while this was 59 ppb (=1/3390 of the MFRC) with the classic microcolony test. For the 50% effect on nest reproduction (EC_{50}), the respective values were 3.7 ppb (=1/54100) and 37 ppb (1/5410). The no-observable effect concentration (NOEC) for imidacloprid was 10 ppb for lethal effects and <2.5 ppb for sublethal effects (nest reproduction) when using the laboratory foraging test.

Interestingly, although this "laboratory foraging behavior test" has been developed in light of the risks for chemical neonicotinoid insecticides, it is also useful to assess the risks for microbial products, for instance, products of bacterial, fungal, or viral origin. Indeed today, there is much attention to synergistic activities between chemical and microbial stressors. The exposure of microbial products can change with, for instance, the growth of the microbial organism and the potential presence of secondary metabolites and fermentation contaminants, and in turn, these microorganism and products can affect the fitness of pollinators and so their pollination service. How to set up techniques and develop guidelines and standards for relevant and effective risk-based assessment with pollinators are today topics of the ICPPR "Microbials" working group. In this line, for instance, Mommaerts et al. (2010a) already reported on the impact of *Bacillus thuringiensis* strains on survival, reproduction and foraging behavior in bumblebees (*B. terrestris*) using the "laboratory foraging behavior test". These results with *Bt kurstaki* and *Bt aizawai* demonstrated that, in general, the *Bt* strains are safe for *B. terrestris* bumblebees, although in some cases, there were detrimental effects that depended on strain and route of exposure. In addition, these authors stated that to draw firm conclusions regarding the hazards of *Bt* to bumblebees, this would require more information on relevant concentrations of *Bt* products in the environment. Hence, routine testing for lethal and sublethal effects is recommended to ascertain the combined use of *Bt* products and other bioprotectants and pollinators in modern agriculture.

As stated by Mommaerts et al. (2010c), this newly developed behavior test has a great practical advantage in that a flight cage is not necessary, because the 20 cm long plastic tube suffices to reveal impairments of foraging behavior, allowing a relatively easy first screening assay to be performed in the laboratory. These results can then be used in a tiered approach. Although a long-term toxicity test should be performed under more natural conditions as in the field, bumblebees need to collect food over a longer distance than 20 cm; according to Wolf and Moritz (2008), *B. terrestris* forages a maximum of 800 m from their nest.

In continuation of their efforts, Mommaerts et al. (2010d) performed a similar study in the greenhouse with bumblebees exposed to imidacloprid, but now they used free-flying bumblebees of the species *B. terrestris*, as shown in Figure 2d. This study with queen-right colonies examined the effects of different imidacloprid concentrations (in sugar water) on foraging behavior, colony growth, and brood growth in a greenhouse setting, where separate 3 x 7 x 2 m areas, each containing a *B. terrestris* hive, represented experimental mesocosms. In the latter experiment, 2 L containers of treatments were placed 3 m from the hive. Imidacloprid at 20 and 10 ppb provoked 62 ± 1% (class M) and 92 ± 5% (class T) worker mortality. However, with 20 ppb, nearly all dead bumblebee workers were found around the sugar water and the pollen; whereas with 10 ppb, all dead workers were found inside the hives, indicating that with 10 ppb of imidacloprid, the worker bees were able to fly back to their hive. In contrast, at a lower concentration of 2 ppb imidacloprid, the percentage of worker mortality (class N) was equal to that in the controls (6 ± 1%) after 2 weeks. There were also obvious sublethal effects from 20 and 10 ppb treatments, with a total loss of reproduction. It was clear that only with 2 ppb, there were no harmful effects on the colony performance over the 2 weeks of the experiment. While imidacloprid at 20 and 10 ppb provoked 62 ± 1% (class M) and 92 ± 5% (class T) worker mortality. Also, with 20 ppb, nearly all dead bumblebee workers were found around the sugar water and the pollen, whereas with 10 ppb, all dead workers were found inside the hives, indicating that with 10 ppb of imidacloprid, the worker bees were able to fly back to their hive. In contrast, at a lower concentration of 2 ppb imidacloprid, the percentage of worker mortality (class N) was equal to that in the controls (6 ± 1%) after 2 weeks. There were also obvious sublethal effects by 20 and 10 ppb, with a total loss of reproduction. Only with 2 ppb, there were no harmful sublethal effects on the colony performance over the 2 weeks of the experiment. In addition, Mommaerts et al. (2010d) scored the foraging behavior and potential sublethal effects of imidacloprid as the numbers of workers in/out the hive during 30-min intervals. At the start, the foraging behavior was equal in all the hives used for 20, 10, and 2 ppb, as well as controls; whereas at the end, only for 2 ppb imidacloprid, the foraging behavior did not differ from the controls.

In the context of entomovectoring where bumblebees are used as "flying doctors, as a strategy for enhanced pollination together with protection strategy, this also demanded the design of an appropriate bioassay to assess the risks of potential side-effects of these powder formulations, especially bioprotectants of microbial origin, in the dispenser toward the vectoring insect. Mommaerts et al. (2012b) developed a laboratory "miniature-dispenser-based bioassay" (Figure 4). They exposed microcolonies of *B. terrestris* workers to five model products, consisting of chemical and biological fungicides and also carrier powders: Prestop-Mix, Signum, kaolin, wheat flour, and cellulose.

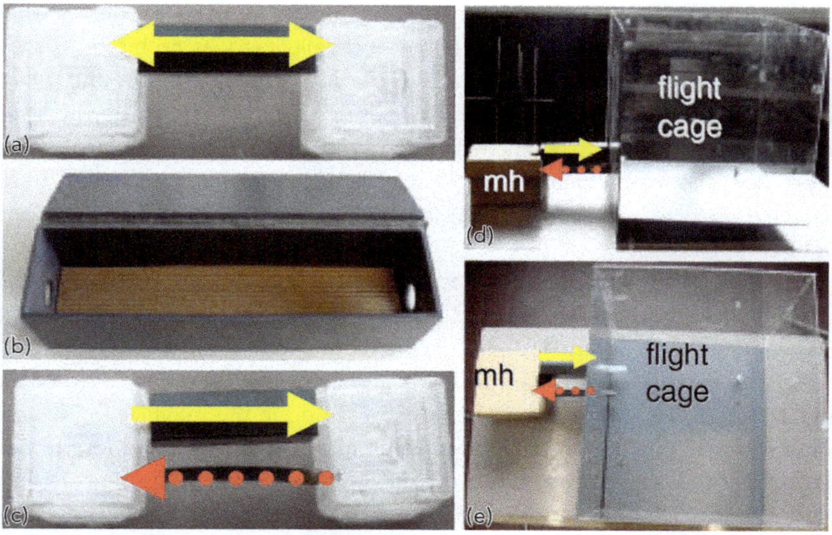

Figure 4 Overview of the experimental setup also named "miniature dispenser test": (a) the one-way miniature dispenser bioassay showing a plastic microcolony nest box connected to another microcolony nest box by a miniature dispenser (20 × 5 × 5 cm); (b) a detail of a miniature dispenser containing ribbed carbon paper on the bottom; (c) the two-way miniature-dispenser bioassay whereby an additional connection is made between the two microcolony nest boxes by using a 20 cm long plastic tube; (d), (e) the flight-cage experiment, showing a minihive (mh) connected to a flight cage (50 × 50 × 50 cm) with (d) the side view and (e) the top view. The solid arrow indicates the route, with the miniature dispenser containing the formulated powder product, that the bumblebee workers follow to go from the microcolony nest box or minihive to the other next box or flight cage; the dotted arrow indicates the route, without the miniature dispenser and free of powder formulated product, that the bumblebee workers take to return from the food compartment to the nest box/minihive. Based on Mommaerts et al. (2012b).

First, the products were placed in a one-way-passage miniature dispenser that connected the microcolony with the food compartment. Second, potential hazards were evaluated in a two-way-passage miniature-dispenser bioassay, a design similar to the one-way-passage miniature-dispenser bioassay, except that the workers returned to the nest with their food via another route free of the product under investigation. Third, the two-way-passage miniature-dispenser bioassay was extended with a flight cage to evaluate the impact of the hazards by a flight.

Mommaerts et al. (2012b) used this "miniature dispenser test" technique to investigate the compatibility of five model products, consisting of one microbial biofungicide, one chemical fungicide, and two carrier powders (Prestop-Mix, Signum, kaolin, wheat flour, and cellulose), with bumblebees of *B. terrestris*. The laboratory one-way miniature-dispenser bioassay (Tier-2) demonstrated that chemical and biological fungicides and the carrier/diluent kaolin caused a worker mortality of >70% after 5 weeks of exposure, while the loss of bumblebees with wheat flour and cellulose was no higher than in the blank control (i.e. empty miniature dispenser) (<25%). The laboratory two-way miniature-dispenser bioassay comprised separated passageways and demonstrated that only kaolin was toxic (89 ± 11%). In addition, a negative effect was observed against reproduction/colony development when nests were exposed to kaolin ($P < 0.05$) in the two-way miniature-dispenser and flight-cage bioassays.

Then a flight-cage bioassay was set up with a minihive, consisting of 50 workers and brood, and connected via a miniature dispenser to a flight cage (50 × 50 × 50 cm) (Figure 4, d and e). The flight cage consisted of 3 mm thick Plexiglass (transparent for the front, side, and top plates, and dark grey for the end plate). Upon connection, pollen was provided in the minihive, and sugar water in the flight cage. After 2 days, when foraging workers had found their way to the sugar source and back to the minihive, the miniature dispensers were filled with the different powder products. Based on the toxicity, it was decided to conduct further investigations of the hazards of kaolin, Prestop-Mix, Signum, and cellulose. For each powder product, one minihive was exposed, and this was repeated twice. Fresh pollen was provided in the minihives every 2 days. Using this flight-cage laboratory bioassay, no worker mortality was observed when workers were exposed over the course of 5 weeks to the four different products–kaolin, Prestop-Mix, Signum, and cellulose. However, it should be mentioned here that for kaolin that workers stayed in the flight cage after the first week of exposure. This typical behavior was indicative of the bumblebee workers avoiding contact with the kaolin powder. As a consequence, the sugar water transport in nests exposed to kaolin was negatively affected. The latter resulted in reduced worker survival of 77 ± 14% after 5 weeks, representing class T (harmful) in the IOBC classification. In contrast, this avoidance behavior was not observed with cellulose, Signum, and Prestop-Mix, and subsequently, no worker mortality was scored with these products after 5 weeks (corresponding to class N as harmless). In conclusion, the results with the miniature dispenser were confirmed in the flight-cage experiment. In addition, it can be concluded that the developed laboratory two-way miniature-dispenser bioassay gave a reliable prediction of the hazards associated with powder products. Additionally, the present data indicated the possibility of using cellulose and kaolin as respective negative and positive control carriers/diluents in future risk assessment experiments.

Finally, as stated by the "Microbials" working group of ICPPR, there is a need for standardization in risk assessment assays when using biological products based on microorganisms. For instance, the use of a microorganism as a positive control will require development and may present unreasonable challenges. It was also noted that the autoclaved test substance does not provide a suitable attenuated control. In addition, this expert group expressed numerous concerns with the current approaches (as developed for chemical products) and suggested alternatives for the detection of microorganisms, such as molecular PCR analysis.

3.3 Tier-3 risk assessment

As a good example, Ceuppens et al. (2015b) conducted risk assessment experiments with mini-hives of *B. terrestris*. They examined the effects under a higher level of complexity, i.e. on queen-right, free-flying bumblebee hives in flight cages in the greenhouse. Per flight cage (1.5 m x 1.2 m x 1.2 m), one queen-right mini-hive (1 queen, 25 workers and brood) was placed (Figure 5a). The mini-hives were provided with sufficient pollen, while the sugar water solution was placed *ad libitum* in a container at a distance of 1 m from the entrance of the mini-hives with artificial flowers. In this way, the workers needed to fly over a distance of at least 1 m in order to obtain sugar water, which is in contrast to walking a distance of 20 cm in the laboratory "foraging behavior test" as described above (Mommaerts et al., 2010c). In this setup, queen and worker survival was monitored for 2 weeks, and the sugar water consumption was used as a proxy of the foraging activity. Lambda-cyhalothrin was used as a model pyrethroid insecticide, and these authors investigated lethal and sublethal effects by different dilutions, ranging from 1/10 to 1/100 of its MFRC (37.5 ppm). With free-flying bumblebees, the toxic effects at 1/10 of the MFRC were more pronounced as compared to a classic Tier-2 laboratory test with a microcolony. Under these conditions, the bumblebee workers perform the complex task of foraging when the bumblebees must fly for their food. Moreover, all queens were dead.

More recently, Klein et al. (2022) conducted a ring test (consisting of nine laboratories from Germany, Spain, Switzerland, and UK) to design an outdoor (Tier-3) experimental setup for examining the effects of PPPs on *B. terrestris* colonies. The aim was to optimize the design of a tunnel (foraging) study in an outdoor setting and propose suitable controls and risk assessment endpoints. These bumblebee field trials used tunnels ranging from 40 to 108 m^2 depending on the laboratory, as shown in Figure 5b. Purple tansy (*Phacelia tanacetifolia*) and rapeseed (*Brassica napus*) were used as bumblebee-attractive crops. Tunnels each contained one bumblebee colony set up on one side of the tunnel, the hive entrance facing the crop; and hives were covered with a rainproof roof to provide shading and shelter from rain. The risk assessment includes a phase of exposure to the PPP of at least 10 days, where bumblebees foraged on the sprayed crop throughout its flowering period. Colonies were then transferred to a monitoring site where they continued to develop into the colonies' reproductive stage. Today, these

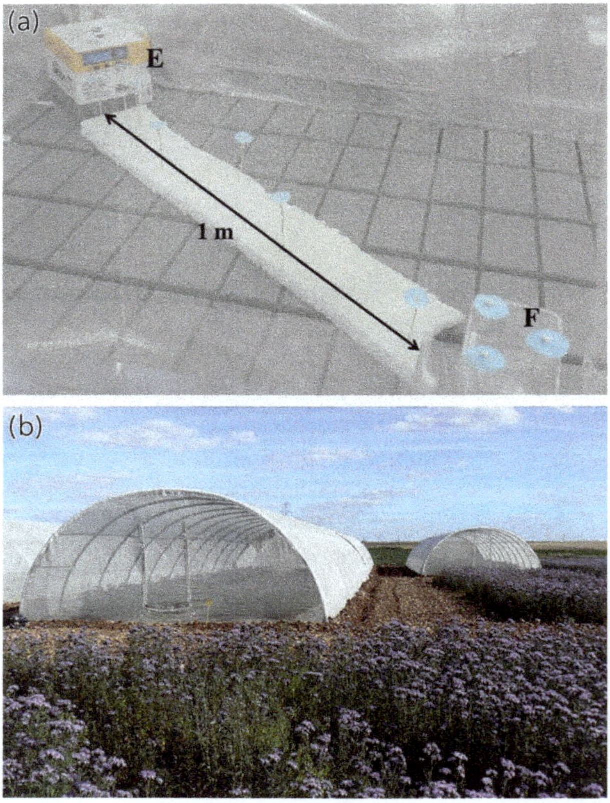

Figure 5 Risk assessment setup with free-flying bumblebees. (a) Setup with "minihives" of *B. terrestris* in flight cage. In separated flight cages (1.5 m x 1.2 m x 1.2 m), one queen-right mini-hive (1 queen, 25 workers, and brood) was placed per cage. The mini-hives were provided with sufficient pollen, while the sugar water solution was placed *ad libitum* in a container at a distance of 1 m from the entrance of the mini-hives with artificial flowers. Four mini-hives were fed with treated sugar water, while the control consisted of 4 mini-hives fed with untreated sugar water. Temperature and light conditions in the greenhouse followed a diurnal pattern with a minimum of 12°C and a maximum of 30°C. Queen and worker survival were monitored for 2 weeks, and the sugar water consumption was used as a proxy for foraging activity. Based on Ceuppens et al. (2015b). (b) Setup of field trails with "tunnels" ranging from 40 to 108 m^2 depending on laboratory, and with purple tansy (*Phacelia tanacetifolia*) and rapeseed (*Brassica napus*) used as forage. Each tunnel contains one bumblebee colony on one side of the tunnel, the hive entrance facing the crop, and the hive is covered with a rainproof roof to provide shading and shelter from rain. From: https://testapi.fr/

tunnel assays are done for different GLP trials with pollinators, such as GLP honeybee studies under tunnels, to determine the impact on honeybees and on honeybee colonies under semi-controlled conditions, as well as mortality, foraging, colony development, and brood development (OEPP

170, CEB 230, OECD 75); for GLP honeybee behavior studies to determine the impact on the behavior of adult honeybees on sprayed crops or seed treated crops (CEB 230); for GLP bumblebee studies to determine the impact on the mortality, foraging, and colony development on several crops (EFSA guidance document); and for GLP *Osmia* (mason bee) studies to determine the impact on mortality, foraging, mating, and brood development on flower crops (European Ring Test).

In the experimental design of Klein et al. (2022), reliable study parameters for foraging and colony success include flight activity (number of foragers exiting and entering the hive, visible pollen loads enabling more detailed observation) and queen production (combining emerged gynes, queen larvae, and queen pupae in the colony). Interestingly, Table 1 presents the minimum detectable differences (MDDs) for different parameters of the 2016 and 2017 ring tests, as reported by Klein et al. (2022). While the data from the above-mentioned ring test were generated using *B. terrestris*, the design should be suitable for other bumblebee species. However, adaptation to other bumblebee species may require that different crops and/or positive controls be used. Furthermore, more research is necessary to understand what influences queen production within semi-field and field setups; this is vital for reducing variability in queen production.

Finally, and important for future, relevant and effective risk assessment, based on this ring test, Klein et al. (2022) provided a series of recommendations for working methods, and a copy of these is presented here in Table 2.

Table 1 Minimum detectable differences (MDDs) for different parameters of the 2016 and 2017 ring tests for risk-based assessment of bioprotectants and PPPs at Tier-3 level, including free-flying bumblebees. From Klein et al. (2022)

Year	Parameter	*n*	Mean	Median	Minimum	Maximum
2016	Flight	3	34	27	24	50
	Adult mortality	5	42	40	34	52
	Colony weight development	6	66	71	37	87
	Queen production [a]	3	55	62	25	78
2017	Flight	10	29	30	21	44
	Adult mortality	10	61	59	29	114
	Colony weight development	10	33	34	8	60
	Queen production [a]	4	98	93	70	148

[a] Number of emerged gynes, queen pupae, and queen larvae (considering datasets with mean queen production of ≥10 only).
n = number of trials which were considered for the respective parameters.

Table 2 Recommendations for a possible semi-field study design for the risk-based assessment of bioprotectants and PPPs on bumblebees. From Klein et al. (2022) and including information from the "Microbials" working group on bioprotectants.

1. General design of the test method

The test consists of at least three treatment groups, the untreated Control C, the test item treatment Group T, and a toxic reference item (applied with the active ingredient dimethoate or in the case of insect growth regulators the active ingredient diflubenzuron). The application takes place as a spray application at least 2–4 days after the set-up of the bumblebee colonies in the tunnels.

As relevant parameters, brood development, flight activity, colony weight development, and reproduction performance [queen stages (defined as the sum of queen larvae, queen pupae, and emerged gynes)] are considered.

The adult bees and their larvae will be exposed to the nectar and pollen of the crop throughout the flowering period. After the end of exposure in the tunnels, the colonies are transferred to a monitoring location to assess the subsequent development of the colonies and the production of queens.

2. Test method

2.1. Test system/test organism

Species/variety	*Bombus terrestris* (Linnaeus, 1758) Recommendations might be applicable with adaptations also for other species, that is, *Bombus impatiens* (Cresson, 1863), *Bombus ignitus* (Smith, 1869), and so forth.
Source	Colonies from commercial suppliers.
Colonies	Normal queen-right colonies, each with approximately 10–40 bumblebee workers upon arrival at the test site, depending on the time of the study start and the crop (the colonies will be larger when they are introduced in the tunnels, because the colony growth between arrival at the test site and study start has to be considered). To ensure uniform colony development and low variability in the endpoints/parameters, the following approach is a possibility but is not mandatory: the colonies are kept indoors (laboratory, storage room, climatic chamber) at ambient temperature for approximately 1–2 weeks to monitor the development and exclude queens that were damaged by the transport. To reduce variability between the replicates colonies which are similar in size and composition and are developing at an equal pace should be selected for the test. Also, colonies should not have reached the "switch-point" (produced males or queen larvae) before the study start. Bumblebees are supplied with auxiliary food (sugar solution and pollen) before the set-up of the colonies in the tunnels. During the exposure and monitoring phase, the sugar solution supply is closed and pollen supply stopped.
Condition of the colonies, development of the bumblebee brood	At least once before set-up in the tunnels (initial brood assessments). Once at the end of the monitoring phase (final brood assessment).

(Continued)

Table 2 (*Continued*)

Crop	For the test, *Phacelia tanacetifolia* or *Brassica napus* as bee-attractive crops, or other bee attractive and relevant crops, can be used.
2.2. Test design	
Study type	Semi-field study
Tunnel size	Minimum 40 m^2, better 60 m^2 (~maximum one worker per m^2 at introduction of colonies in the tunnels)
Number of treatment groups	≥1 test item treatment, one control (negative control), one toxic reference (positive control)
Toxic reference item	In case of classic PPP, dimethoate (800 g a.s./ha); in case of insect growth regulators (IGRs), diflubenzuron (216 g a.i./ha); in case of bioprotectants, a representative standard microorganism species for testing can include one species of bacteria and one species of fungus (e.g. *Bacillus thuringiensis*, *Beauveria bassiana*, or *Metarhizium anisopliae*)
Replicates per treatment group	≥6 (for the toxic reference three replicates can be sufficient)
Test organism per treatment group	Six bumblebee colonies per treatment group (one per tunnel)
Bumblebee colony selection before test initiation	Approximately two times more colonies with approximately 10–40 workers should be ordered for initial brood assessment. The best and most uniform colonies can be selected. The following criteria should be taken into account for the final choice of the colonies: • Number of workers • Number of workers + pupae • Total number of brood stages • Development speed (increase/time) if assessed. The colonies should be as uniform as possible regarding these criteria.
Pre-exposure period	Set-up of bumblebee colonies in tunnels: 2–4 days before application
Set-up	Sun and rain protection of hives is recommended (i.e. hives placed on palettes, protection with styrofoam roofs) Water supply is generally not required (exception under very hot conditions, in greenhouses, etc.)
Application	Foliar application following good agricultural practice, timing during bee flight (flight activity at the colony entrance must be ≥2 bees exiting the colony entrance per 10 min before the test item is applied). During the application, the colonies must be covered (e.g. by plastic sheets) to prevent direct overspray of the colonies. For the test item, a different application scheme (i.e., application after or before flight, application before flowering, seed treatments, etc.) might be tested.

(*Continued*)

Table 2 (*Continued*)

Exposure phase	Duration of approximately 2 weeks (during the flowering period of the crop), as long as crop conditions are favorable (providing sufficient nectar and pollen)
Post-exposure period (monitoring site)	After the exposure phase in the tunnels, the bumblebee colonies are transferred to one monitoring site because different conditions on separate monitoring sites may substantially influence colony development. This site should be a non-cropped area with foraging resources providing a sufficient pollen and nectar supply to avoid starvation of the colonies; mass flowering crops should be avoided, and crop protection product exposure should be minimal. Colonies of the different treatment groups should be placed at least 20 m apart from each other. Hives should be well protected against rain and animal interference. The surrounding flowering vegetation is assessed once during the monitoring phase (listing of most abundant flowering species). Additional feeding (sugar solution, pollen) during the monitoring phase should be avoided.
2.3. Quality criteria	
Parameters	At least a significant reduction in one of the parameters in the toxic reference compared to the control.
2.4. Assessments and parameters	
Initial brood assessment	Adult and brood stages are determined before set-up of the colonies in the tunnels. For the initial brood evaluations, each hive is opened under red light and the following brood stages and observations are documented: • Presence of living foundress queen • Count of egg cells, larvae, and pupae • Count of dead larvae and pupae • Count of dead and alive workers • Weight of alive workers • Presence of diseases and predators The initial brood assessments are performed to choose bumblebee colonies that are healthy and similar in size and composition. It is recommended to perform the initial brood assessments with approximately twice as many colonies as needed to keep variability between the replicates at a minimum.
Flight activity	The observation of the flight activity takes place at the entrance of each bumblebee hive. At each assessment, the number of bumblebees entering and exiting the hive (both separately) will be counted during a defined time interval (at least 10 min). Flight assessments should be conducted in parallel in the control and the test item treatment(s). Assessments should be performed at least every 3–4 days (two times per week). Depending on the test item, daily assessments might be advisable in the first days after the application to be able to detect small, acute effects. While observing entering forager bumblebees, the number of foragers with pollen loads of the target crop should be documented to evaluate exposure.

(*Continued*)

Table 2 (*Continued*)

Mortality	Mortality of adult bumblebees and larvae is assessed inside the hives at least every 3–4 days (two times per week). Depending on the test item daily, assessments might be advisable in the first days after the application to be able to detect small, acute effects. At each assessment, dead adults and larvae are counted and removed from the hive. Foundress queen mortality is also noted.
Colony weight development	The weight of the bumblebee colonies is determined at least every 3–4 days (two times per week). Depending on the test item, daily assessments might be advisable in the first days after the application to be able to detect small, acute effects.
Switch-point	All colonies are monitored for the "switch point" (= detection of first queen larvae) every 3–4 days (two times per week, during the exposure and monitoring phase).
Queen production/ termination of the test	The bumblebee colonies are expected to switch to the reproduction phase (= the production of sexuals) at the end of the exposure phase or during the monitoring phase. The occurrence of queen larvae and pupae as well as the presence of males is documented. Approximately 7 days after the first queen pupae have been detected, queen excluders have to be installed, and emerged gynes are collected every 3–4 days (two times per week). If at least 30% of the estimated queen pupae (or at least 10 gynes) have emerged, the hive is deep-frozen (each hive individually, approximately 1 month after the first queen pupae have been detected). Colonies which do not reach the switch point are deep-frozen on the same day as the last control hive. If it is foreseeable that a hive will not reach the switch point, it can be deep-frozen earlier (e.g. toxic reference treated colonies; expert judgment and documentation necessary).
Queen production and final brood assessment	At the end of the study, bumblebee colonies will be killed by deep-freezing and the following brood stages and observations will be documented: • number of emerged gynes • weight of gynes individually • number of queen larvae and queen pupae.

4 Regulation of PPPs in Europe: Regulation (EC) No 1107/2009

In the EU, any 'plant protection product' or PPP that protects crops or other useful plants and is placed on the market needs authorization. As shown in Figure 6a, the legal framework is defined in the Regulation (EC) No 1107/2009 (EC, 2009) (https://webgate.ec.europa.eu/dyna2/pgd/). In essence, the authorization process is divided into the approval of the active substance (a.s.) and in the authorization of the PPP. A precondition for the authorization of a

PPP is an approval of the active substance by the EU Commission based on a risk assessment led by EFSA. A PPP containing an approved active substance is assessed and authorized by the Member States (MS). To enhance the efficiency of the authorization process, the EU is divided into three zones, the Northern, Central, and Southern zone (EC, 2009). The risk assessment of a PPP, including bioprotectants, is conducted by one Member State (zonal rapporteur Member State, zRMS) for the whole zone. In principle, MS of the same zone is forced to take over the results of the assessment and the decision of the zRMS. However, MS can make claims on national ecological or agricultural specificities to decide on dividing risk management options for their country.

Data requirements for assessing active substances are laid down in Regulation (EC) No 283/2013 (EC, 2013a) and for PPPs in Regulation (EC) No 284/2013 (EC, 2013b), respectively. Uniform principles for the evaluation and authorization of PPPs are laid down in the implementing Regulation (EU) No 546/2011 (EC, 2011a). The goal of the regulations is to achieve a high level of protection of human and animal health as well as the environment in all MS. The regulations are complemented by guidance documents produced by OECD, EPPO and EFSA to describe the methodological requirements for the risk assessment of active substances and PPP products.

What are the nuances of bioprotectants compared to classical chemical products? In principle, the Regulation (EC) No 1107/2009 (EC, 2009) differentiates between PPPs based on chemicals and microorganisms, where microorganism means any microbiological entity, including lower fungi and viruses, cellular or non-cellular, capable of replication or of transferring genetic material. The EU regulatory framework with Regulation (EC) No 1107/2009 for the risk assessment process for the (renewal of) approval of chemical and microbiological active substances is shown in Figure 6B. In detail, specific data requirements for the assessment of microorganisms are laid out for active substances in Part B of Regulation (EC) No 283/2013 (EC, 2013a), and for PPPs in Part B of Regulation (EC) No 284/2013 (EC, 2013b). Additionally, further categories of PPPs are defined in the Regulation (EC) No 1107/2009 (EC, 2009). Also, PPPs which are classified as low-risk products must fulfill strict criteria in particular low toxicological potential and low persistence in the environment; for details, see Article 22 Annex II of EC (2009). Furthermore, no specific risk mitigation measures should be required following the risk assessment for low-risk PPPs, and the PPP must be sufficiently effective. In practice, only a few active substances, including bioprotectants, are authorized in the EU as low-risk PPPs; for more details, see EU (2021). Another category of PPPs is basic substances. A basic substance is not predominantly used for plant protection purposes but nevertheless, it is useful in plant protection either directly or in a product consisting of the substance and a simple diluent. It does not have any inherent capacity to cause endocrine-disrupting, neurotoxic, or immunotoxic effects

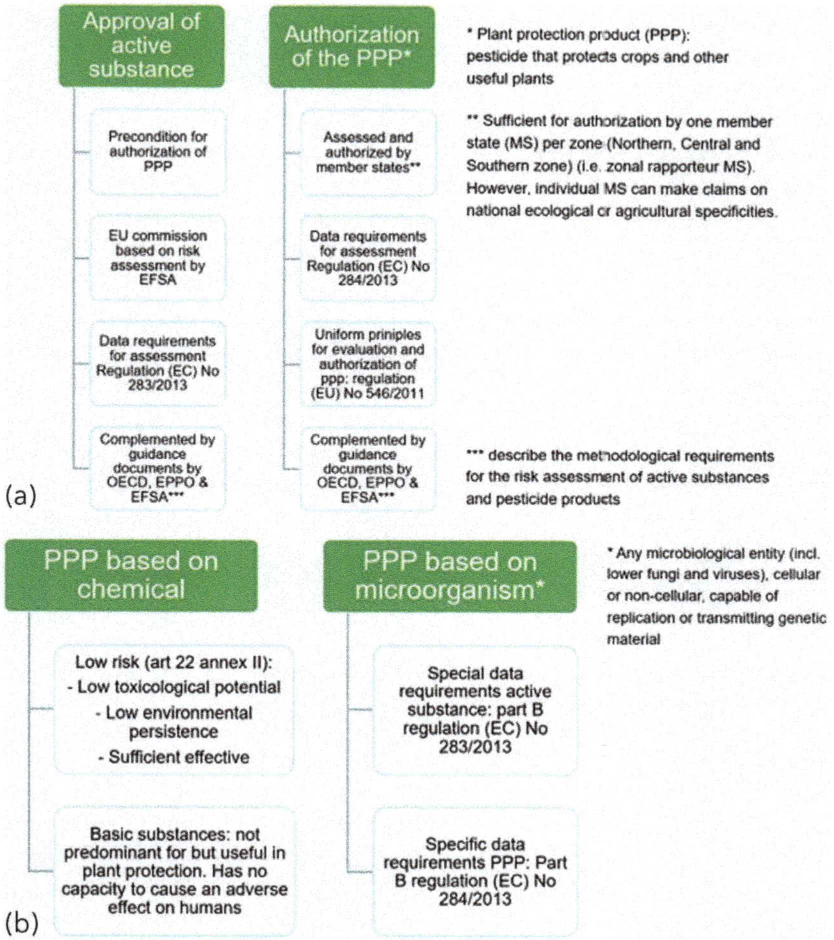

Figure 6 EU regulatory framework with the Regulation (EC) No 1107/2009: (a) for PPPs that protect crops and other useful plants, with the methodological requirements for the risk assessment of the active substance and the PPP product; (b) for the risk assessment process for the (renewal of) approval of chemical and microbiological active substances and for the authorization of chemical and microbiological PPPs. Both (a) and (b) are based on https://webgate.ec.europa.eu/dyna2/pgd/.

and it is not a substance of concern (Article 23, EC, 2009), which means it has no capacity to cause an adverse effect on humans (Article 3, EC, 2009). Currently, 23 basic substances are approved in the EU; good examples are beer, *Allium cepa* extract, and L-cysteine (EU, 2021).

The initial focus of the regulations was set on chemicals as active substances. However, Article 77 of Regulation (EC) No 1107/2019 (EC, 2009) offers the possibility that "the Commission may adopt or amend technical and other guidance documents such as explanatory notes or guidance documents on the content of the application concerning microorganisms, pheromones and biological products, for the implementation of this regulation. The Commission may ask the Authority to prepare or to contribute to such guidance documents." The rationale behind Article 77 is that data requirements for these product classes deviate from those for chemicals. Such specific adaptations to the data requirement are already implemented for important bioprotectants such as microorganisms, pheromones and botanicals (EC, 2011b, 2014a, 2014b, 2016). In summary, we understand that, while this Regulation (EC) No 1107/2019 provides a unified framework, bioprotectants often require tailored risk assessment approaches due to their unique modes of action and ecological profiles, particularly in protecting NTOs, like bumblebees. In this context, we also refer to recent developments, such as the working document on "Microbial Biocontrol Agents" and the EFSA guidance document on microorganisms, pheromones, and bioprotectants. These provide updated data requirements for registration and commercialization for bioprotectants and biological control agents, and also the risk assessment pathways for microbial products used in sustainable farming (Dietz-Pfeilstetter et al., 2021; Boro et al., 2022; Palmieri et al., 2022; Chaudhary et al., 2024).

5 Conclusion: Improving risk-based assessment of PPP effects on bumblebees

Both OECD and EFSA have established and continued to build upon guidelines for risk assessment of PPPs, including bioprotectants of chemical and biological origin, in pollinators, including bumblebees. This also included bioprotectants for use in biological control programs, to provide relevant, effective risk-based assessment. For example, EFSA has defined four specific risk cases for bees; these include acute contact risk, acute dietary risk, chronic dietary risk, and larvae dietary risk. OECD guidelines for bees include analogously defined risk cases, yet scientific data obtained since their adoption suggest that these guidelines are insufficient. For example, chronic dietary toxicity tests under OECD last 10 days. This has been demonstrated as insufficient for testing certain PPPs, such as the fungicide boscalid, which showed cumulative toxicity appearing in honeybees after 17–18 days of chronic dietary exposure (Simon-Delso et al., 2018), results which would not be evident using OECD's standard 10-day feeding method. These data may be similar in wild bee models like *B. terrestris* or *B. impatiens*, and this chapter remains an excellent example of how longer study durations can more accurately predict the effects of PPPs–of chemical, botanical, or microbiological origin–compared to the 10-day feeding method. This may be the case for Tier-2 and Tier-3 risk assessments as well.

In a holistic context, there are also factors at play in the environment of bumblebees. These include: aspects of the bumblebee microbiome (Meeus et al., 2013, 2015; Praet et al., 2015, 2016, 2018; Maebe et al., 2021; Cornet et al., 2022; Botero et al., 2024); spillover of pathogens (Meeus et al., 2011; Arbetman et al., 2013; Maharramov et al., 2013; De Sousa Pereira et al., 2021; Straub et al., 2022), bumblebee genetics with the whole genome sequencing of *B. terrestris* and *B. impatiens*; marker-assisted mass-breeding of *Bombus*; population structuring and genetic pauperization (Maebe et al., 2013, 2015a, 2015b, 2016, 2019, 2020; International bumblebee genomics consortium, 2015; Gerard et al., 2015); the effects of climate change and agricultural intensification (Goulson et al., 2015; Gerard et al., 2019; Maebe et al., 2021a, 2021b; Hart et al., 2022); and crop perfume and flower scents (Ceuppens et al., 2015a). Also, bumblebees can be used as entomovectors in a "flying doctors" system to enhance pollination and biocontrol (Mommaerts et al., 2010b; Mommaerts and Smagghe, 2011a; Temmermans et al., 2022), and associated compounds need to be tested under adequate exposure conditions. Mommaerts et al. (2012b) developed a miniature dispenser, and this bioassay system was used to investigate the compatibility of five model products, the biofungicide Prestop-Mix, Signum, kaolin, wheat flour, and cellulose, with the bumblebee *B. terrestris*. Further, Smagghe et al. (2013a) tested with this setup the compatible use of *Metarhizium anisopliae* in entomovectoring with *B. terrestris*. Testing the impact of microorganism-based bioprotectants, as currently used in different biological control and IPM programs, on the gut microbiome of beneficial organisms (such as pollinators, bumblebees, and also natural enemies) is a relevant topic that can be recommended in future risk assessment. With their capacity to forage in the environment, bumblebees can also be used as "flying noses" in assessing environmental quality and the potential presence of toxicants. For example, Parmentier et al. (2014) placed commercial hives of the bumblebee *B. terrestris*, containing one queen and 40 workers, in three different locations in the region of Ghent (Belgium), and the performance of these hives was followed during a 4-week period. In parallel, they determined the floral richness and diversity index in the chosen study sites. The sites consisted of a rich urban environment with patchy green areas as opposed to an urban environment with poor landscape metrics; a third rural study site showed intermediate positive landscape metrics. Interestingly, the results demonstrated that hive biomass and the number of workers increased significantly in the rich compared to the poor environment, providing a mechanism to discriminate between study sites. In addition, the bumblebee-collected pollen showed that flowering plants belonging to *Salix* spp. and *Rosaceae/Prunus* spp. were dominant food sources in all anthropogenic environments during early spring. Finally, the results were discussed in relation to the optimization of the experimental setup, as well as in relation to the use of commercial bumblebee hives in assessing local pollinator support within any given environment. Hence, environmental contaminants can be of different sources such as heavy metals, industry products, and waste materials as PFOS or nanobodies and microplastics (Mommaerts et al., 2011a, 2012a; Raza et al., 2024).

On the extrapolation of toxicity data and risk assessment of bioprotectants with other pollinators, Mommaerts and Smagghe (2011b) reported on the sensitivity of side-effects of a selection of PPPs, of chemical and biological origin, on bumblebees (*B. terrestris*) versus honeybees (*A. mellifera*). They tried a linear regression analysis with the LD_{50}-24h values after contact and oral exposure for different chemical insecticides (carbamates, pyrethroids, organophosphates, neonicotinoids) and microbial products (bioprotectants). Although no linear regression could be drawn, this analysis gives a first idea of bumblebee versus honeybee sensitivity for PPPs. It should be noted that the power of this analysis is limited because an LD_{50} was not available for each insecticide. However, based on available data for different IGRs, molting accelerating compounds (MACs) are safe for both bumblebees and honeybees (Thompson et al., 2005; Mommaerts et al., 2006b), whereas no correlation can be found for the other two classes (CSI and JHA). Indeed, for diflubenzuron (chitin synthesis inhibitors, CSIs) the LD_{50}-24 h for larvae showed that *B. terrestris* larvae are more sensitive than *A. mellifera* (LD_{50}-72 h) (Tasei, 2001). For the same compound, Mommaerts et al. (2006a) reported a total loss of *B. terrestris* reproduction, while Thompson et al. (2005) found only short-term effects on *A. mellifera* colonies. In contrast, for the juvenile hormone analog (JHA) fenoxycarb, *B. terrestris* larvae were less susceptible (LD_{50}-24h: >0.650 μg/larvae) than *A. mellifera* larvae (LD_{50}-48h: 0.013 μg/larvae) (Tasei, 2001). In conclusion, these data and others (van der Steen et al., 2008; Gradish et al., 2019; Hardstone and Scott, 2010; Kueh Tai et al., 2022) indeed confirm that it is not acceptable to extrapolate honeybee (*A. mellifera*) toxicology data across other bee species such as bumblebees (*Bombus* spp.), even though they belong to the same taxonomic family (Apidae).

Exposure to multiple products as PPPs, or simultaneous exposure to a chemical, botanical, or microbiological product and another stressor, is a key issue in the biosafety of bumblebee populations. Numerous studies have examined this topic in bumblebees (Raimets et al., 2018; Carnesecchi et al., 2019; Siviter et al., 2022), yet to our knowledge none have examined this within a Tier-2 or Tier-3 risk assessment setup. This represents a key area of interest for forwarding biosafety testing in bumblebees and other wild bees. Here, modeling of the ecological service of pollination by wild bees and pollination deficiency are important tools (Ariza et al., 2022; Gimenez-Garcia et al., 2023). While such investigations would require more treatment groups, and therefore more experimental resources, the data generated may prove vital to our understanding of how we can reduce the risks related to PPPs–of chemical and biological origin–to bumblebee populations. Also, the development of newer chemistries and biologicals, and insecticide groups with novel modes of action demand for adequate selection of relevant and effective endpoints, especially looking at bumblebee behavior and their foraging behavior important for pollination (Chittka 2022). Some good examples are newer compounds of chemical and microbiological origin that can be used in biological and IPM programs, botanicals and mineral oils, next-generation neonicotinoid insecticides (class 4), low-risk compounds (Besard et al., 2011; Mommaerts and Smagghe, 2011b; Smagghe et al., 2013b; Barbosa et al., 2015a, 2015b; Taning et al., 2019; Lei et al., 2023), mixtures and cocktails of products (Eeraerts et al., 2015; Nicholson et al., 2023), and the most recent class of RNA-interference (RNAi) mediated target suppressors (class 35) (Taning et al., 2021; Smagghe et al., 2024). The recent efforts of the "Microbials" working group of the ICPPR are also important in the development of relevant and effective risk assessment and standards, especially for microorganism-based products. Also, newer techniques in insect biotechnology and single cell sequencing are available (Willson et al., 2013; Askri et al., 2023; Chen et al., 2024). This is in service of the protection of the bumblebees' health and the conservation of their populations, and safeguarding their pollination services in agricultural crops and wild plant populations.

Finally, we bring a selection of suggestions for a specialized pathway for bioprotectants, regarding concerns over assessment data requirements and risk assessment. We believe that by incorporating these aspects, this will provide a more holistic view of the evolving EU regulatory landscape for a more modern agriculture with biocontrol and IPM programs and the use of bioprotectants, and its alignment with sustainability and biodiversity preservation goals under the "Green Deal" and "Farm to Fork" strategies (EU 2024a, 2024b; https://commission.europa.eu/strategy-and-policy/priorities-2019-2024/european-green-deal_en).

1. Environmental risk assessment (ERA): For bioprotectants, ERA must evaluate unique aspects like microbial persistence, secondary metabolites, and potential synergistic effects when used alongside other PPPs. As discussed in the chapter above, field-realistic conditions and extended exposure durations are critical to accurate risk evaluation.
2. Data requirements and harmonization: For proportional assessment, toxicological data requirements for bioprotectants, which often differ from those for chemical pesticides, must be adapted. Harmonizing these requirements with global standards (e.g. OECD), and incorporating tiered testing approaches, will enhance regulatory consistency.
3. Stakeholder collaboration: Research organizations are pivotal in generating robust risk assessment data. Including those actively aligned with bioprotectant-specific guidance ensures comprehensive coverage of best practices and innovations in this space.

4. Focus on NTOs: Given the chapter's detailed discussion on biosafety for bumblebees, including entomovectoring and pollination efficiency, linking these insights to regulatory frameworks for non-target species will solidify its practical relevance.

6 Acknowledgements

Guy Smagghe is grateful for the support he received in the past 25 years for starting and continuing his scientific research on bumblebees by the Special Research Council of Vrije Universiteit Brussel (OZR-VUB) and Ghent University (BOF-UGent), the Foundation of Scientific Research-Flanders (FWO) (Flanders, Belgium), the Agency for Innovation by Science and Technology (IWT) (Flanders, Belgium), Belgian Science Policy (BELSPO, "Belbees"), EU-core-organic (Bicopoll), EU-COST (Super-B), the International Commission for Plant-Pollinator Relationships (ICPPR), the Southwest University Thousand Talents program (SWU, China) and the Guizhou University High Talents program (GZU, China), and also for the many gifts of bumblebees by Biobest (Westerlo, Belgium).

7 References

Alford, D.V. (1975). *Bumblebees*. London, UK: Davis-Poynter, 352 p.

Arbetman, M., et al. (2013). Co-introduction and spillover of parasites by invasive *Bombus terrestris*. Bumblebee's chauffer hitchhike alien parasites when traveling to Patagonia, Argentina. *Biological Invasions*, 15, 489–494.

Ariza, D., et al. (2022). Linking remote sensing data to the estimation of pollination services in agroecosystems. *Ecological Applications*, 32, e2605.

Askri, D., et al. (2023). Parasite and pesticide impacts on the bumblebee (*Bombus terrestris*) haemolymph proteome. *International Journal of Molecular Sciences*, 24, 5384.

Barbosa, W.F., et al. (2015a). Lethal and sublethal effects of azadirachtin on the bumblebee *Bombus terrestris* (Hymenoptera: Apidae). *Ecotoxicology*, 24, 130–142.

Barbosa, W.F., et al. (2015b). Pesticides and reduced-risk insecticides, native bees and Pantropical stingless bees: pitfalls and perspectives. *Pest Management Science*, 71(8), 1049–1053.

Besard, L., et al. (2010). Compatibility of traditional and novel acaricides with bumblebees (*Bombus terrestris*): a first laboratory assessment of toxicity and sublethal effects. *Pest Management Science*, 66(7), 786–793.

Besard, L., et al. (2011). Laboratory assessment of lethal and sublethal side-effects supports a more benign profile of spinetoram compared to spinosad in the bumblebee *Bombus terrestris. Pest Management Science*, 67(5), 541–547.

Blacquière, T., et al. (2012). Neonicotinoids in bees: a review on concentrations, side-effects and risk-assessment. *Ecotoxicology*, 21, 973–992.

Boro, M., et al. (2022). Microorganisms in biological control strategies to manage microbial plant pathogens: a review. *Archives of Microbiology*, 204, 666.

Botero, J., et al. (2024). A comparative genomic analysis of *Fructobacillus evanidus* sp. nov. from bumble bees. *Systematic and Applied Microbiology*. https://doi.org/10.1016/j.syapm.2024.12650

Buchmann, S.L. (1983). Buzz pollination in angiosperms. In *Handbook of experimental pollination biology*, eds. Jones, C.E., Little, R.J. New York, NY, USA: Van Nostrand Reinhold Company, pp. 73–113.

CABI (2010). Compendium-Bombus terrestris (bumble bee). https://doi.org/10.1079/cabicompendium.91578.

CABI (2024). https://www.cabidigitallibrary.org/doi/10.1079/cabicompendium.91578.

Carnesecchi, E., et al. (2019). Investigating combined toxicity of binary mixtures in bees: meta-analysis of laboratory tests, modelling, mechanistic basis and implications for risk assessment. *Environment International*, 133, 105256.

Ceuppens, B., et al. (2015a). Characterization of volatiles in strawberry varieties 'Elsanta' and 'Sonata' and their effect on bumblebee flower visiting. *Arthropod-Plant Interactions*, 9(3), 281–287.

Ceuppens, B., et al. (2015b). Effects of dietary lambda-cyhalothrin exposure on bumblebee survival, reproduction, and foraging behavior in laboratory and greenhouse. *Journal of Pest Science*, 88(4), 777–783.

Chen, J., et al. (2024). Rotenone impairs brain glial energetics and locomotor behavior in bumblebees. *Science of the Total Environment*, 907, 167870.

Chaudhary, R., et al. (2024). Microbial bio-control agents: a comprehensive analysis on sustainable pest management in agriculture. *Journal of Agriculture and Food Research*, 18, 10142.

Chittka, L. (2022). *The mind of a bee*. Princeton, NJ: Princeton University Press, 272 p. ISBN-100691180474

Cornet, L., et al. (2022). Phylogenomic analyses of *Snodgrassella* isolates from honeybees and bumblebees reveals taxonomic and functional diversity. *mSystems*, 7, 3. https://doi.org/10.1128/msystems.01500-21

Dag, A. and Kammer, Y. (2001). Comparison between the effectiveness of honey bee (*Apis mellifera*) and bumblebee (*Bombus terrestris*) as pollinators of greenhouse sweet pepper (*Capsicum annuum*). *American Bee Journal*, *141*, 447–448.

De Sousa Pereira, K., et al. (2021). Managed bumble bees acquire parasites from their foraging environment: a case study on parasite spillback. *Journal of Invertebrate Pathology*, 182, 107583.

Dietz-Pfeilstetter, A., et al. (2021). Considerations and regulatory approaches in the USA and in the EU for dsRNA-based externally applied pesticides for plant protection. *Frontiers in Plant Sciences*, 12, 682387. https://doi.org/10.3389/fpls.2021.682387

Eeraerts, M., et al. (2015). Effects of imidacloprid in combination with λ-cyhalothrin on the model pollinator *Bombus terrestris* at different levels of complexity. *Julius Kuhn Archives*, 450, 219–224.

EC (2009). Regulation (EC) no 1107/2009 of the European parliament and of the council of 21 October 2009 concerning the placing of plant protection products on the market and repealing Council Directives 79/117/EEC and 91/414/EEC. *OJ L*, 309, 1–50.

EC (2011a). Commission Regulation (EU) No 546/2011 of 10 June 2011 implementing Regulation (EC) No 1107/2009 of the European Parliament and of the Council as regards uniform principles for evaluation and authorisation of plant protection products. *OJ L*, 155, 127–175.

EC (2011b). Guidance Document on the Assessment of New Substances Falling Into the Group of Straight Chain Lepidopteran Pheromones (SCLPs) Included in Annex I of Council Directive 91/414/EEC (SANCO/5272/2009 rev. 3 28 October 2011).

EC (2013a). Commission Regulation (EU) No 283/2013 of 1 March 2013 setting out the data requirements for active substances, in accordance with Regulation (EC) No 1107/2009 of the European Parliament and of the Council concerning the placing of plant protection products on the market. *OJ L*, 93, 1–84.

EC (2013b). Commission Regulation (EU) No 284/2013 of 1 March 2013 setting out the data requirements for plant protection products, in accordance with Regulation (EC) No 1107/2009 of the European Parliament and of the Council concerning the placing of plant protection products on the market. *OJ L*, 93, 85–152.

EC (2014a). Guidance document for the assessment of the equivalence of technical grade active ingredients for identical microbial strains or isolates approved under regulation (EC) no 1107/2009 (SANCO/12823/2012 –rev. 4,12 December 2014).

EC (2014b). Guidance document on botanical active substances used in plant protection products (SANCO/11470/2012– rev. 8 20 March 2014).

EC (2016). Guidance document on semiochemical active substances and plant protection products (SANTE/12815/2014 rev. 5.2 May 2016).

EU (2021). EU Pesticide Data Base. https://ec.europa.eu/food/plant/pesticides/eu/pesticides/database/active/substances/?event=search.as; https://food.ec.europa.eu/plants/pesticides_en

EU (2024a). https://agriculture.ec.europa.eu/sustainability/environmental-sustainability/low-input-farming/pesticides_en

EU (2024b). https://environment.ec.europa.eu/topics/nature-and-biodiversity/nature-restoration-law_en

Gerard, M., et al. (2015). Discrimination of haploid and diploid males of *Bombus terrestris* (Hymenoptera; Apidae) based on wing shape. *Apidologie*, 46, 644–653.

Gerard, M., et al. (2019). Shift in size of bumblebee queens over the last century. *Global Change Biology*, 26(3), 1185–1195.

Giménez-García, A. et al. (2023). Pollination supply models from a local to global scale. *Web Ecology*, 23, 99–129. https://doi.org/10.1002/ecy.3614

Goulson, D. (2003a). *Bumblebees: their behaviour and ecology*. Oxford, UK: OUP, 248 pp.

Goulson, D. (2003b). Conserving wild bees for crop pollination. Journal of Food Agriculture & Environment, 1(1), 142–144.

Goulson, D. (2010). *Bumblebees behaviour, ecology and conservation*. New York: Oxford University Press.

Goulson, D., et al. (2008). Decline and conservation of bumble bees. *Annual Review of Entomology*, 53, 191–208.

Goulson, D., et al. (2015). Bee declines driven by combined stress from parasites, pesticides, and lack of flowers. *Science*, 27(347), 1255957.

Gradish, A.E., et al. (2019). Comparison of Pesticide Exposure in Honey Bees (Hymenoptera: Apidae) and Bumble Bees (Hymenoptera: Apidae): Implications for Risk Assessments. *Environmental Entomology*, 48, 12–21.

Hanewald, N., et al. (2013). Optimizing laboratory toxicity test methods for bumblebees (*Bombus terrestis* L.). Presented by BASF SE on the SETAC Conference in Glasgow 2013.

Hardstone, M.C. and Scott, J.G. (2010). Is *Apis* more sensitive to insecticides than other insects? *Pest Management Science*, 66(11), 1171–1180.

Hart, A.F., et al. (2022). Signals of adaptation to agricultural stress in the genomes of two European bumblebees. *Frontiers in Genetics*, 13, 993416.

Heinrich, B. (1979). *Bumblebee economics*. Cambridge, MA: Harvard University Press, 246 pp.

Hughes, A.C. and Grumbine, R.E. (2023). The Kunming-Montreal global biodiversity framework: what it does and does not do, and how to improve it. *Frontiers in Environmental Science*, 11. https://doi.org/10.3389/fenvs.2023.1281536

International Bumblebee Genomics Consortium (2015). The genomes of two key bumblebee species with primitive eusocial organisation. *Genome Biology*, 16, 6.

IPBES (2016). The assessment report of the intergovernmental science-policy platform on biodiversity and ecosystem services on pollinators, pollination and food production. In *Secretariat of the intergovernmental science-policy platform on biodiversity and ecosystem services*, eds. Potts, S.G., et al. Bonn, Germany. 552 pages. https://doi.org/10.5281/zenodo.3402856

Klein, O., et al. (2022). Results of ring-testing of a semifield study design to investigate potential impacts of crop protection products on bumblebees (Hymenoptera, Apidae) and a proposal of a potential test design. *Environmental Toxicology & Chemistry*, 41, 2548–2564.

Kueh Tai, F., et al. (2022). Honey bee toxicological responses do not accurately predict environmental risk of imidacloprid to a solitary ground-nesting bee species. *Science of the Total Environment*, 839, 156398.

Lei, H., et al. (2023). Progress in pesticide toxicology research on bumblebees. *Asian Journal of Ecotoxicology*, 18(4), 102–114. https://doi.org/10.7524/AJE.1673-5897.20230210002

Maebe, K., et al. (2013). Scientific note on microsatellite DNA analyses revealing diploid and haploid drones in bumblebee mass-breeding. *Apidologie*, 45(2), 189–191.

Maebe, K., et al. (2015a). Quantitative trait loci for light sensitivity, body weight, body size, and morphological eye parameters in the bumblebee, *Bombus terrestris*. *PLoS One*, 10(4), e0125011.

Maebe, M., et al. (2015b). Microsatellite analysis of museum specimens reveals historical differences in genetic diversity between declining and more stable *Bombus* species. *PLoS One*, 10(6), e0127870.

Maebe, K., et al. (2016). A century of temporal stability of genetic diversity in wild bumblebees. *Scientific Reports*, 6, 38289.

Maebe, K., et al. (2019). Pattern of population structuring between Belgian and Estonian bumblebees. *Scientific Reports*, 9(1), 9651.

Maebe, K., et al. (2020). Level of genetic diversity in European bumblebees is not determined by local species abundance. *Frontiers in Genetics*, 10, 1262.

Maebe, K., et al. (2021a). Impact of intraspecific variation on measurements of thermal tolerance in bumble bees. *Journal of Thermal Biology*, 99, 103002.

Maebe, K., et al. (2021b). Bumblebee resilience to climate change, through plastic and adaptive responses. *Global Change Biology*, 27, 4223–4237.

Meeus, I., et al. (2011). International trade, invasive parasites and pollinator biodiversity loss. *Conservation Biology*, 25(4), 662–671.

Meeus, I., et al. (2013). Assessment of mutualism between *Bombus terrestris* and its microbiota by use of microcolonies. *Apidologie*, 44, 708–719.

Meeus, I., et al. (2015). 16S rRNA amplicon sequencing demonstrates that indoor-reared bumblebees (*Bombus terrestris*) harbor a core subset of bacteria normally associated with the wild host. *PLoS One*, 10(4), e0125152.

Maebe, K., et al. (2021). The holobiont as a key to the adaptation and conservation of wild bees in the Anthropocene. *Frontiers in Ecology and Evolution*, 9. https://doi.org/10.3389/fevo.2021.781470

Maharramov, J., et al. (2013). Genetic variability of the neogregarine *Apicystis bombi*, an etiological agent of an emergent bumblebee disease. *PLoS One*, 8(12), e81475.

Mommaerts, V., et al. (2006a). Hazards and uptake of chitin synthesis inhibitors in bumblebees *Bombus terrestris*. *Pest Management Science*, 62, 752–758.

Mommaerts, V., et al. (2006b). Bumblebees can be used in combination with juvenile hormone analogues and ecdysone agonists. *Ecotoxicology*, 15, 513–521.

Mommaerts, V., et al. (2008). *Trichoderma*-based biological control agents are compatible with the pollinator *Bombus terrestris*: a laboratory study. *Biological Control*, 46, 463–466.

Mommaerts, V., et al. (2009). Assessment of compatibility of microbiological control agents with the pollinator *Bombus terrestris*. *Pest Management Science*, 65, 949–955.

Mommaerts, V., et al. (2010a). Impact of *Bacillus thuringiensis* strains on survival, reproduction and foraging behavior in bumblebees (*Bombus terrestris*). *Pest Management Science*, 66(5), 520–525.

Mommaerts, V., et al. (2010b). Development of a new dispenser for bumblebees and evaluation to disseminate microbiological control agents in strawberry in the greenhouse. *Pest Management Science*, 66(11), 1199–1207.

Mommaerts, V., et al. (2010c). Risk assessment for side-effects of neonicotinoids against bumblebees with and without impairing foraging behaviour. *Ecotoxicology*, 19, 207–215.

Mommaerts, V., et al. (2010d). Bumblebees and neonicotinoids: a bioassay to evaluate sublethal effects on the foraging behaviour. *Proceedings of the Netherlands Entomological Society Meeting*, 21, 9–16.

Mommaerts, V. and Smagghe, G. (2011a). Entomovectoring in plant protection. *Arthropod-Plant Interactions*, 5, 81–95.

Mommaerts, V. and Smagghe, G. (2011b). Side-effects of pesticides on the pollinator *Bombus*: an overview. In *Pesticides in the Modern World – Pests Control and Pesticides Exposure and Toxicity Assessment*, Stoytcheva, M., ed. Croatia: InTech Publishing, pp. 507–552. ISBN 978-953-307-457-3.

Mommaerts, V., et al. (2011a). Impact of a perfluorinated organic compound PFOS on the terrestrial pollinator *Bombus terrestris* (Insecta, Hymenoptera). *Ecotoxicology*, 20(2), 447–456.

Mommaerts, V., et al. (2011b). *Bombus terrestris* as pollinator-and-vector to suppress *Botrytis cinerea* in greenhouse strawberry. *Pest Management Science*, 67(9), 1069–1075.

Mommaerts, V., et al. (2012a). Assessment of side-effects by Ludox TMA silica nanoparticles following a dietary exposure on the bumblebee *Bombus terrestris*. *Nanotoxicology*, 6(5), 554–561.

Mommaerts, V., et al. (2012b). Miniature dispenser-based bioassay to evaluate the compatibility of powder formulations used in an entomovectoring approach. *Pest Management Science*, 68, 922–927.

Nicholson, C.C., et al. (2023). Pesticide use negatively affects bumble bees across European landscapes. *Nature*. https://doi.org/10.1038/s41586-023-06773-3

OECD (1998a). *OECD guideline for testing of chemicals, No. 213: Honeybees, acute oral toxicity test*. Paris: Organisation for Economic Cooperation and Development.

OECD (1998b). *OECD guideline for testing of chemicals, No.214: Honeybees, acute contact toxicity test*. Paris: Organisation for Economic Cooperation and Development.

OECD (2017a). *OECD guideline for testing of chemicals, No. 246: Bumblebee, acute contact toxicity test*. Paris: Organisation for Economic Cooperation and Development.

OECD (2017b). *OECD guideline for testing of chemicals, No. 247: Bumblebee, acute oral toxicity test*. Paris: Organisation for Economic Cooperation and Development.

Palmieri, D., et al. (2022). Advances and perspectives in the use of biocontrol agents against fungal plant diseases. *Horticulturae*, 8, 577.

Parmentier, L., et al. (2014). Commercial bumblebee hives to assess an anthropogenic environment for pollinator support: a case study in the region of Ghent (Belgium). *Environmental Monitoring and Assessment*, 186, 2357–2367.

Praet, J., et al. (2015). *Bifidobacterium commune* sp. nov. isolated from the bumble bee gut. *Antonie Van Leeuwenhoek International Journal of General and Molecular Microbiology*, 107(5), 1307–1313.

Praet, J., et al. (2016). *Apibacter mensalis* sp. nov.: a rare member of the bumble bee gut microbiota. *International Journal of Systematic and Evolutionary Microbiology*, 66, 1645–1651.

Praet, J., et al. (2018). Large-scale cultivation of the bumblebee gut microbiota reveals an underestimated bacterial species diversity capable of pathogen inhibition. *Environmental Microbiology*, 20(1), 214–227.

Raimets, R., et al. (2018) Synergistic interactions between a variety of insecticides and an ergosterol biosynthesis inhibitor fungicide in dietary exposures of bumble bees (L.). *Pest Management Science*, 74, 541–546.

Rasmont, P. (1983). Catalogue of bumble bees of the West Palearctic region. (Catalogue commenté des bourdons de la région ouest-paléarctique (Hymenoptera, Apoidea, Apidae).) *Notes Fauniques de Gembloux*, 7, 71.

Rasmont, P. (1984). Bumble bees of the genus Bombus sensu stricto in western and central Europe (Hymenoptera, Apidae). *Spixiana*, 7(2), 135–160.

Raza, M.F., et al. (2024). Honeybees for monitoring of microplastic pollution: a narrative review on the sources, fates, effects and future constructing predictive models for environmental microplastic distribution and accumulation patterns. *Environmental Science and Pollution Research*, under review.

Simon-Delso, N., et al. (2018). Time-to-death approach to reveal chronic and cumulative toxicity of a fungicide for honeybees not revealed with the standard ten-day test. *Scientific Reports*, 8, 7241.

Siviter, H., et al. (2022). A combined LD50 for agrochemicals and pathogens in bumblebees (*Bombus terrestris* [Hymenoptera: Apidae]). *Environmental Entomology*, 51, 378–384.

Smagghe, G., et al. (2013a). Assessment of *Metarhizium anisopliae* in entomovectoring with bumblebees *Bombus terrestris*. *Journal of Economic Entomology*, 106, 277–282.

Smagghe, G., et al. (2013b). Dietary chlorantraniliprole suppresses reproduction in worker bumblebees. *Pest Management Science*, 69, 787–791.

Smagghe, G., et al. (2024). *RNAi in arthropods: powerful tool for basic research and product for entrepreneurs*. Dordrecht, the Netherlands: Springer-Verlag, 843 p.

Straub, L., et al. (2022). Do pesticide and pathogen interactions drive wild bee declines? *International Journal of Parasitology: Parasites and Wildlife*, 18, 232–243.

Taning, C.N.T., et al. (2019). With or without foraging for food, field-realistic concentrations of sulfoxaflor are equally toxic to bumblebees (*Bombus terrestris*). *Entomologia Generalis*, 39(2), 151–155.

Taning, C.N.T., et al. (2021). A sequence complementarity-based approach for evaluating off-target transcript knockdown in *Bombus terrestris*, following ingestion of pest-specific dsRNA. *Journal of Pest Science*, 94(2), 487–503.

Tasei, J.N. (2001). Effects of insect growth regulators on honey bees and non-*Apis* bees: a review. *Apidologie*, 32(6), 527–545.

Temmermans, J., et al. (2022). Entomovectoring/apivectoring: using pollinators to spread biocontrol agents. In *Promoting pollination and pollinators in farming*, eds. Chan, S. W. and Kevan, P. Cambridge: Burleigh Dodds Science Publishing, pp. 1–25.

Thompson, H.M. (2005). Assessing the exposure and toxicity of pesticides to bumblebees (*Bombus* sp.). *Apidologie*, 32(4), 305–321.

van der Steen, J.J.M., et al. (1996). *Methods to determine the acute oral and contact LD50 of pesticides for bumble bees* (*Bombus terrestris* L.). Proceedings ICPBR 6th Symposium on the Hazard of Pesticides to Bees 1996 Braunschweig, Germany

van der Steen, J.J.M., et al. (2008). *Can pesticide acute toxicity for bumblebees be derived from honeybees LD50-values?* Proceedings 10th International Symposium Hazard of Pesticides to Bees, Bucharest, Romania October 2008.

Velthuis, H.H.W. and van Doorn, A (2006). A century of advances in bumble bee domestication and the economic and environmental aspects of its commercialization for pollination. *Apidologie*, 37, 421–451.

Williams, P.H., et al. (2014). *Bumble Bees of North America: An identification guide.* Princeton: Princeton University Press.

Willson, D.E., et al. (2013). Use of primary cultures of Kenyon cells from bumblebee brains to assess imidacloprid side-effects. *Archives of Insect Biochemistry and Physiology*, 84, 43–56.

Wolf, S. and Moritz, R.F.A. (2008). Foraging distance in *Bombus terrestris* L. (Hymenoptera: Apidae). *Apidologie*, 39, 419–427.

Chapter 3

Developing a global regulatory regime for bioprotectants

Imme Gerke, IDRG Plant Protection, Germany

1 Introduction

Pest organisms do not think of themselves as pests. Insects, mites, nematodes, pathogens, weeds, rodents, birds, and others exploit human efforts to grow food, feed, and fibre crops, and they can thrive on these crops before and after harvest. Our monocultures, even if it is just a handful of apple trees, are a paradise for an insect that measures the world by the distance to the next apple or the next tree. As long as the youngest and smallest larvae can reach food and shelter, as well as a place to develop and to mate, pests will continue to compete with humans for the food that the nature provides (1).

Farmers have tried to protect their crops as long as they have farmed. From prehistoric times, farmers have sought to remove anything that appeared to damage the plants they planted. Large enough insects, diseased leaves/plants, and weeds were removed manually, while pests such as birds and rodents were chased away. These approaches worked as long as our farming practices were relatively rudimentary and small in scale. When our farming skills increased and

http://dx.doi.org/10.19103/AS.2024.0142.23

we developed technologies that allowed us to plough, plant, water, and harvest larger surfaces, we both created a monoculture in which pests could proliferate and a scale of operations that outstripped the capacity of traditional, manual methods to keep pests under control. As pests started to outcompete our capacity to deal with them, our fight to try to control them intensified, resulting in the first attempts at chemical control. As early as 4500 years ago (2500 BC), the Sumerians were using sulphur dusting of crops to control insect pests (1). The ancient Chinese and Egyptians used plant extracts with the same goal. In 300 BC, the Chinese were using natural enemies such as ants to control insect pests in citrus.

In the beginning, chemical and biological pest management developed side by side with farmers trying things out and learning from each other. Eventually, chemical substances that had been developed for completely different purposes took over the market and grew rapidly. Within only a relatively short period of time after industrialization in the nineteenth and twentieth centuries, companies started to provide farmers with a new generation of synthetic chemicals that, in many cases, had originally been developed for military purposes in the two World Wars. When these wars ended, these highly toxic materials were adapted for use in pest control in agriculture (2). Initially, this new generation of pesticides was widely welcomed, particularly as many products had very rapid and broad effects against a wide range of pests.

2 Pesticide regulation: evolution and challenges

At the start of the twentieth century, governments began to regulate pesticides, initially with the goal to protect farmers from being sold ineffective or fraudulent products. At the time the concern of regulators was not human or environmental safety but to detect and punish criminals who cheated farmers out of their income by selling them fraudulent 'miracle' products claiming that they would control all pest organisms but which actually controlled none at all (3).

However, long before Rachel Carson wrote her now famous book *Silent Spring* in 1962 (2), the toxic effects of chemical pesticides were becoming apparent. Farmers, their workforce, their families, and their livestock all started to show symptoms of intoxication as well as other symptoms such as damage to eyes and skin rashes. It took a while for farmers and doctors to connect these symptoms to the products that had been used to protect crops. When the link was understood, additional regulations were put in place requiring companies not only to demonstrate the efficacy of their products but their safety as well. Governments started to assess the toxicity of products before they were allowed to be manufactured and distributed and set standards for product labels that warned farmers of the dangers of the products they handled and explained how these products could be used safely (4).

Over time it became clear that it was not only people who were at risk but that pesticides were also damaging the wider environment (soil, water, air, and non-target organisms). Scientific studies began to describe the toxic effects of chemicals from industry and agriculture on the environment, including on particular species. In an effort to respond to growing public concerns, governments began to include the assessments of environmental hazards and risks in regulatory approval schemes for pesticides.

As this account suggests, the scope of pesticide regulation has evolved in a sometimes piecemeal way over time. An additional complication is that different countries and regions (such as the EU) have pursued different approaches and priorities at different times. Depending on the priorities of a particular government, regulations were developed either to test the chemical/physical/biological properties of a product, its efficacy on the pests to be controlled, or its effect on human health and the environment. This has led to today's situation where each has its own regulatory requirements for approving products for each relevant crop/pest combination individually. If we assume 10 000 crops grown worldwide, multiplied by a very conservative assumption of 10 pests per crop and by at least three different product types with a different mode of action for pest control, we reach a staggering number of 300 000 potential combinations. If we multiply that by about 200 bodies responsible for pesticide regulation in the world, we reach over 60 000 000 combinations. Since many crops are traded globally, exporters and importers of agricultural commodities are faced with a potentially impossible level of complexity and the risk of crops being rejected and even lost because they fail a particular set of national or regional regulations. As an example, since individual regulatory bodies initially worked in isolation from each other, they calculated different safe pesticide residue limits for the same crop/product combination. When the limit is higher in the exporting country than in the importing country, the shipment will be rejected if too high amounts of residues are detected. As advances in detection technologies has continued to increase the speed and precision of detection, this problem has only become more challenging.

A further complication in this already complex landscape is the sometimes very different perspectives and priorities of the different parties are involved in pesticide regulation. This can be seen, e.g., in the contrasting perspectives of regulators, pesticide producers, and farmers. In part, this can be understood in terms of the different approaches of fundamental, applied, and regulatory science (Fig. 1). When a new phenomenon is discovered, scientists working in fundamental science, typically within an academic environment, will set up experimental designs that gradually isolate the key factors until the core trigger has been found. The results do not matter as long as they reflect reality. Applied science, however, tries to make use of discoveries from fundamental science and to ensure replicable results until a new product or its effects can be trusted

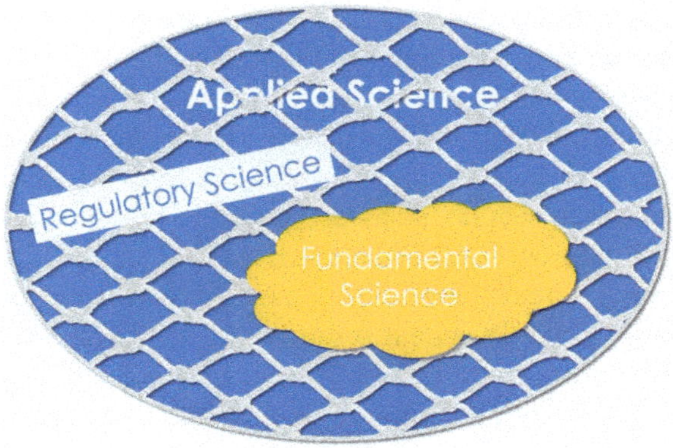

Figure 1 Comparing fundamental, applied, and regulatory science.

and potential customers can be convinced. The result of this kind of research has to be effective and affordable for the customer. Only then does a company have a profitable product. In contrast, regulatory science adds the dimension of risk. Regulators need to know (1) how efficacious and reliable a product is (farmer protection), (2) whether the product poses a threat to farmers, workers, consumers, and others (human health), and (3) if the product poses a risk to the environment (environmental health). These studies and the comparison of the resulting data to standard values differentiate regulatory science from fundamental and applied science (5).

Fundamental Science (yellow) studies and describes phenomena that were previously not known or not understood. Its only purpose is understanding the world around us.

Applied Science (blue) uses the findings of fundamental science to develop products, technologies, or processes in the interest of people. Its purpose is profitability to human needs and to the economy.

Regulatory Science (white net) explores the reliability of new products in terms of efficacy and human and environmental safety. It is guided by laws and regulations. Its purpose is to set and monitor limits to risks that can be considered acceptable (5).

As this picture suggests, regulatory evaluators are typically scientists who assess the safety and efficacy of products such as pesticides. They do so by reviewing scientific data generated by companies who specialize in a product, the crops it protects, and the pests it controls. Following this assessment of the accuracy and robustness of this data, evaluators then compare their findings to approved standards that have been set to limit risks to an acceptable level. To meet regulatory standards, the evidence that companies must supply to regulators needs to follow very detailed processes developed by regulators.

For pesticide companies, regulations, regulators, and regulatory science represent a very significant hurdle and cost in new product development.

Many companies will start with developing and testing product efficacy under laboratory, greenhouse, and field conditions. They then need to invest in scaling production and securing appropriate distribution and marketing resources to secure sales to ensure a successful and profitable product. It can be a major challenge to then meet a complex set of different and sometimes conflicting regulatory requirements. This can best be addressed by regulators developing standardized and harmonized data requirements that companies can then follow.

Finally, it is important to recognize that farmers are critically dependent on income typically from a few key cash crops harvested at particular times in the year, which needs to offset their daily costs. Successful and cost-effective pest control is key to protecting and enhancing these critical cash crop yields on which farmers rely. This means that, whilst they recognize the vital importance of regulation to protect human and environmental health, farmers will also prioritize the continued availability of effective crop protection solutions, including chemical pesticides where other solutions are not seen as sufficient.

3 Harmonizing approval of new crop protection products globally

The increase in international trade in agricultural produce after the Second World War, and the development of increasingly complex, globalized agri-food supply chains (with the potential to source food from almost anywhere in the world), highlighted a growing need to harmonize data requirements and assessment methods in pesticide regulation between different countries and trading blocs (such as the EU). An important early initiative occurred in 1994, under the leadership of the OECD (Organization for Economic Cooperation and Development), when government representatives, companies, farmers, and non-governmental organizations (NGOs) from 38 countries came together to discuss what should be done to remove regulatory differences between them (6).

From the start, there was the realization that synthetic chemical products and biological products needed to be treated differently. From as early as 1996, it was agreed that biological products were likely to be inherently safer (7). After long and intensive discussions, biological products were defined to include microbials, macrobials, pheromones/semiochemicals, plant extracts, and certain minerals. They were considered to be generally safer than chemicals because the active (biological) ingredients had typically evolved along with humans and other life forms in the environment for millions of years. It was assumed that co-evolution indicated a degree of proof of compatibility. However, it was also recognized that biological products were not risk free, e.g., with live microbial products potentially also able to affect non-target species and potentially to reproduce in the environment to unsafe levels.

A starting point was to develop and send out a questionnaire on the different data requirements from governmental regulators in individual countries, with the responses from all participating countries summarized in a single document (8). This was used to develop a comprehensive list of key data requirements for microbial active ingredients that would meet the requirements of individual regulators (7). A similar list was also developed for chemical products (9). Another output of this move to a more harmonized approach was the development of internationally accepted standards of Good Laboratory Practice (GLP) governing efficacy and safety tests conducted in laboratories, growth chambers, or greenhouses (10).

An early example of the benefit of this harmonized approach is related to the use of neem oil which was approved as a biopesticide in many jurisdictions. However, when Australian regulators found that Azadirachtin, a key component of neem oil, had some toxic effects in humans, recognition of the validity of the Australian findings led to a rapid re-evaluation of neem oil in many other countries until further safety testing and updated guidance on its use allowed its re-authorization as a biopesticide (11).

A key distinction in these guidelines is the separate list of data requirements for individual active ingredients and formulated end-use products. The approach is important because a single active ingredient can be used in several end-use products that are formulated to best fit specific crops, pests, growing methods, or application technologies. In both the chemical and microbial lists of data requirements (Appendixes 1 and 2), active ingredients and end-use products are subdivided into different categories:

- administrative information on the products, their manufacturers, and their formulators;
- physical, chemical, and/or biological characteristics of the active ingredients and formulants;
- production and quality control methods;
- mammalian toxicology;
- environmental toxicology and fate;
- value (efficacy, phytotoxicity, environmental benefits, and commercial benefits to farmers); and
- packaging requirements/issues.

The most crucial of the data required are from studies on an end-use product because the concern of the regulator is the total risk of the product rather than just of the active ingredients. The list for chemical products comprises 782 individual data requirements. The list for microbial biopesticides comprises only 148 data requirements. The difference is a consequence of the relative safety of biological products compared to chemical products.

In addition to harmonizing data requirements, one of the first actions that regulators took under the leadership of OECD in 1998 was to develop common formats for core data packages and all attachments (subsequently revised in 2005) (12). Today, there are templates for companies to use for chemical, microbial, and pheromone products (12). Simultaneously, OECD also developed templates for regulators to use that match those for companies for the purposes of scientific evaluations, risk assessments, and decision-taking processes. In 2009, when the EU established its core regulation governing authorization, sale and use of plant protection products (PPPs) (EC 1107/2009) (13), it created a format very similar to that of OECD. The main difference between the OECD version and the EU version is the fact that, in the EU version, the templates for companies and those for regulators have been merged. Following the new EU format, the FAO created a format that especially addressed regulators in developing countries (14). The suggested formats for the FAO data packages are almost identical to both the OECD and EU formats. Companies are therefore well advised to create their data packages in OECD format and then make necessary small adjustments when they decide to individually submit to a country with a slightly different format.

As late as 2000, data packages still had to be provided to regulators in hardcopy with specific numbers of copies for different chapters in the package. Since then, electronic data submissions have been allowed in form of DVDs by some countries, subsequently updated to zip folders. Today, very few countries expect to receive a hardcopy data package. The approach still exists as a backup to an electronic file but has become very rare. Instead, it is safe to build the data package electronically with the help of the relevant OECD template to save it in a zip folder and then either to burn it onto a DVD or send it through an online government portal.

There are numerous computer programs available to create locked zip folders. Many of them require the recipient of such a folder to access the same software to open the folder and get access to the relevant PDF, Word, and Excel files. The OECD has therefore created relevant software that can be uploaded from the OECD website free of charge (15).

When applying for a Global Joint Review (14) (discussed below), the company prepares a full data package according to OECD data requirements, assembles all information using the OECD template for the right product type, creates a zip folder of all documents, and then sends the folder to the regulatory authorities in all participating countries. All countries should receive the same number of copies. Countries that wish to receive the folder through their online portal should receive this in addition to the DVD copies. Countries that require data packages to be submitted in paper format should receive these as well for the same reason as the use of the portals.

Where English is not the official language, a regulator might request an additional copy of the data package in the national language. It is typically not used during the registration process but has to be made available to governments for reasons of transparency. The public can demand access to certain parts of the data package (non-CBI (Confidential Business Information)) and government is then obliged to provide this information in the national language. It is reasonable to expect that over time this requirement will disappear but it was still in place in 2023 for some countries. A reasonable approach to the translation into a national language is the use of automatic translation software followed by proofreading by an expert.

4 The current regulatory approval process

Though the terminology might differ slightly, the regulatory process is almost identical across OECD countries, following a commonly accepted chronological and scientific framework.

4.1 Pre-submission consultation

Most regulators offer pre-submission consultation services that companies can use to inquire about the data they have to submit when they apply for the authorization of their products. Although the data requirements are mostly self-explanatory, companies might have questions about where to locate efficacy trials, which study protocol to apply or the number of repetitions needed for a certain study. Some countries operate on a base of fee-for-service while others do not charge.

For a pre-submission consultation, companies provide basic information that regulators need to have to generate any advice. Questions include the following but are not limited to:

- Is the product chemical or biological?
- Which function does the product have (insecticide, fungicide, and herbicide)?
- Which pests can the product control?
- Which crops (sites) can be protected?
- Are these commercial or domestic uses?
- Are these field, forest, orchard, greenhouse, or inhouse uses?
- Is it a dry or a liquid formulation?
- Is it suggested as a reduced risk product?

And finally, the regulator will ask if the company wishes to proceed with a joint review or a workshare with regulators in other countries. If the company chooses a cooperation with other countries, the proposed countries should be listed.

When the data requirements have been clarified with all countries, the company proceeds to compiling all information that the regulators require and then submits the full package simultaneously to all regulators involved. It is important that all regulators receive identical information. Cooperation is not possible otherwise.

4.2 Review

When the regulators have received the full data package, the following steps take place. The terminology might differ between countries, but the process is the same (Fig. 1). For the purpose of this chapter, we have chosen the Canadian approach to labelling and description of different levels:

Level A: At reception, the data package receives a unique reference number, and the company is informed of the reception by the authority.

Level B: An administrator compares the list of data requirements that were provided to the company at the pre-submission stage with the chapters that the company has provided in their data package. At this point, the content of the different chapters is not assessed. However, if any of the requested chapters is missing, the data package (dossier) will be put on hold and the company will be informed with a letter of deficiency. The company then has a specific amount of time to provide the missing documents. These can contain a scientific study or a scientific rationale including all cited literature. A rationale can be the request for not having to conduct a study or an extrapolation of data from existing studies.

At the end of level B, the company will be informed that the data package is complete and will now be transferred to the scientific evaluators. The administrator informs the evaluators of their deadlines for the review of the different chapters that they are responsible for.

Level C: Evaluators now verify if the information that was provided to them is sufficient to assess the content of the respective chapters in terms of the questions they must respond to. This information can be a scientific study, a review of scientific literature or a scientific rationale that explains why it is technically or scientifically not possible or necessary to conduct a study.

If the evaluator finds the information provided to be insufficient or inadequate, the company will be informed and requested to complete the missing information.

Level D: With all information being available to them, evaluators assess the technical or scientific content for its correctness, they cross-check methods, numbers, statistics, references, and conclusions, and then draft their opinion according to the norms that they must comply with.

The first evaluator in each division conducts the primary review and then forwards his draft opinion to a colleague who conducts peer-review I. After peer-review I the file either goes back to the primary evaluator for peer-review II or to a third person. The reality depends on the number of trained evaluators available to the authorities. The work of new evaluators is typically peer-reviewed by two experienced colleagues.

At the end of level D, evaluators compare the draft label that has been provided by the company with the findings in their reviews and assessments. If they find incompatibilities, they first inform their colleagues in other departments. If the issue can be resolved, the company will be informed of proposed changes to their label. If the issue can't be resolved, the company will be asked to propose a solution.

Level E: When all incompatibilities have been resolved, the evaluators from the different departments come together with the next higher level of scientific experts and present their findings in the reviews of their respective data requirements (chapters). The evaluators of phys/chem/bio, of mammalian toxicity, environmental toxicity, environmental fate, efficacy, and crop tolerance defend their findings and aim at reaching compatible decisions in terms of application rates and frequency, safety phrases, and warning statements.

This step is critical. For example, if the efficacy team finds that the data they reviewed show a great effect on the pest organisms but only at a rate that is higher than what their colleagues in the health or environmental departments consider safe, both have to find a shared solution for the label.

If all departments are in agreement, the company is informed, and the regulator proceeds to the legal components of the authorization process.

Level F: The legal division receives the finalized label and verifies if all legal requirements are met. These include wording, location on the label, size of lettering, registration number, address of company, emergency contact information, and promotional imagery.

Level G: The registration report and the label are provided to the department responsible for legislative affairs in the respective country.

Level H: Has been deleted.

Level I: When the legal process is completed, the authority is informed and prepares the registration certificate for the company. The product can now be legally distributed to users and applied according to label directions attached to each package sold.

In some cases, regulators have chosen to cooperate across national borders (Global Joint Reviews) (16). In this case, each participating country conducts level A and B individually. Cooperation begins at level C and continues to level E.

Primary reviews are split across the participating countries and peer-reviews are exchanged between countries according to the available human resources of the different authorities. A country with 300 evaluators can provide more input than a country with only three evaluators. Countries with limited resources often aim to develop expertise in specific data requirements and rely on the work of their colleagues in other countries for the remaining requirements. Most important is that, at level E, regulators from different countries reach decisions that are at least compatible if they are not identical.

Levels F to I are again national activities that each country delivers individually.

For the majority of countries, the key limitation in the speed of the regulatory approval process is the availability of relevant scientific expertise which is why the regulatory processes typically take longer than official timelines permit. Well-resourced regulators such as Canada, US, Germany, France, China, or Australia have 300+ scientists and can additionally use external consultants for specific steps in the process or for specific data requirements. Some of these regulators charge high prices and these registration fees can become a barrier for a company considering a particular country as a potential market.

Developing countries often have a much lower regulatory workforce which can act as a barrier to the speed and reliability of the product-approval process. However, despite fewer resources, these regulators are still measured against the same international standards as much better-resourced regulators in developed economies. In an effort to meet these standards, regulators have found different solutions. In some regions of the world, regulators from different countries have developed a system in which they come together several times per year to jointly review submitted data packages, conduct the necessary risk assessments, and take regulatory decisions (Levels C–E). Based on these decisions, each country then approves the resulting labels nationally (Levels F–I). Some countries have developed unofficial alliances with each other or with stronger regulators in which they may provisionally accept the review outcome of a partner country in addition to the full data package and then cross-check certain points in the full package rather than conducting a full review themselves. A similar approach has been adopted by regulators independent of the size of their workforce where companies can submit the review outcomes of other regulators if their product is already approved in other countries.

4.3 The process of regulatory approval in the Europe Union

As discussed earlier, in 2009, the EU established a core regulation governing authorization, sale, and use of PPPs (EC 1107/2009) (13). The EU approach differs from the OECD in a number of ways. Rather than reviewing and assessing new active substances and new products simultaneously, the EU has put in place a system in which active substances have to be approved before the review of a product can begin. One country (rapporteur member state (RMS)) conducts the review of the active ingredient and proposes a regulatory decision to the relevant EU authority (the European Food Safety Authority – EFSA). All other countries receive copies of the data package and the findings of the RMS and are given a specific amount of time to request changes. As often occurs, the resulting scientific debate can be protracted and, as a result, the decision-taking process at EFSA level tends to be longer than the recommended timelines.

The approval of formulated end-use products is even more difficult in the EU than that of active substances. For regulatory purposes, the EU has been divided into three zones (Northern, Central, and Southern). If the product is to be distributed and used in all three zones, the company has to submit the corresponding data package for the end-use product in at least three countries, one for each zone (zonal rapporteur member state (zRMS)), has to fulfil the potential national requirements of different regulatory authorities, and has to pay the relevant registration fees.

After the first country in each zone has approved the product, the company then has to apply for mutual recognition in all other countries within the zone one by one. Countries can choose to reduce the number of uses and rates compared to the first zRMS. If they do, they must justify their decision which means that, rather than simply accepting the labelling and other requirements of the neighbouring country, they have to conduct yet another review. Since member states have their own arrangements, it is not surprising to find that the fees companies have to pay for mutual recognition can range from as low as €125 up to €33 000. These fees and the relative complexity of the registration process can be a barrier to small and medium companies. There is a growing recognition that more harmonization may be required to balance efficacy and safety with an effective approval process that allows useful new products to replace outdated chemistry with safer and more sustainable biological products.

5 Global Joint Reviews

A further OECD initiative has been to encourage cooperation between counties in conducting joint assessments. The first joint review of a new microbial product took place from 1999 to 2000 with the participation of two countries (Canada and US) Both countries received identical data packages (using the OECD 1998 format requirements) and shared the work of reviewing and assessing the information provided. It only look 10 months to complete the evidence review, risk assessments, and final decision for regulatory approval of a microbial product, VirosoftCP4, developed and produced by BIOTEPP.

By 2017, 44 chemical active ingredients had been reviewed and approved in numerous products for the use on hundreds of crop/pest combinations using the joint review process with another 11 data packages submitted. Fifteen countries from five continents had participated, though representation from African countries was absent, indicating the importance of pooling resources between poorer countries to allow participation.

The following countries have participated in all or some of the Global Joint Reviews:

- Asia: China, Japan, South Korea, and the Philippines
- Australia and New Zealand
- Europe: United Kingdom, Ireland, France, Germany, and Austria
- North America: Canada, USA, and Mexico
- South America: Brazil

For Global Joint Reviews, both companies (registrants) and regulatory authorities (regulators) have used the standard templates developed jointly under the leadership of OECD (15). Based on this approach, the FAO (Food & Agriculture Organisation) has developed a summary approach (14) to support developing countries and emerging economies in their cooperation with neighbouring countries within their regions and with regulators in other countries. In 2000, e.g., this material was used by INSAH (Institut du Sahel) to further advance existing cooperation among regulators in eight West African countries (17). In 2018, another nine countries joined the organization to create COAHP (Committee Ouest Africain d'Harmonisation des Pesticides) (18). About 470 products have been approved since 2021.

Companies that have submitted their products for Global Joint Reviews have mainly been the multinational manufacturers of chemical herbicides, fungicides, and insecticides who stood to gain most from faster access to global markets, though dealing with regulators in several countries simultaneously had it own challenges. Major companies that have participated in Global Joint Reviews include ALBAUGH, BASF, BAYER-CROPSCIENCE, DOW, DUPONT, ISHIHARA-SAYGYO, MONSANTO, NIPPON-SODA, NUFARM, PBI-GORDON, SUMITOMO, SUTERRA, SYNGENTA, and UPL.

The following 40 active ingredients have been through Global Joint Reviews either as new active ingredients, new formulations, new uses, or new sources: Ametoctradin, Bicyclopyrone, Chlorantraniliprole, Clodinafop-propargyl, Cyantraniliprole, Cyflumetofen, Dicamba, Difenoconazole, Ethaboxam, Ethiprole, Fludioxonil, Fluensulfone, Fluopyram, Flupyradifurone, Fluxapyroxad, Fosetyl-Al, Glyphosate, Isofetamid, Kasugamycin, Lambda-cyhalothrin, Mandestrobin, Mesotrione, Oxathiopiprolin, Penflufen, Penthiopyrad, Picoxystrobin, Proquinazid, Pyrasulfotole, Pyriofenone, Pyroxsulam, S-metolachlor, S-1563, Saflufenacil, Sedaxane, Solatenol, Spinetoram, Spirotetramat, Sulfoxaflor, Thiencarbazone, USF 0728 325 SC, and XDE-729.

6 Involving farmers in the regulatory approval process

The previous discussion has highlighted the ways companies developing new crop protection products interact with regulators. However, a critical third element are farmers as users of these products. Both pesticide manufacturers and regulators need to take account of their needs and views. Parallel to the developments among regulators and registrants, farmers in several countries have pushed to have a voice on which crop protection products they needed and should therefore be developed, authorized, and distributed. There have been various attempts to integrate farmers better into the regulatory approval process.

6.1 US-IR4

Starting in 1963, the US-IR4 Program (Interregional Project 4) (19) began under the coordination of Rutgers University. IR4 collects farmers' views on their crop protection needs, coordinates communication with companies manufacturing crop protection products and manages communication with the principal US regulatory body, the Environmental Protection Agency (EPA) (20). Today, IR4's headquarters are in Raleigh, North Carolina. From there, activities are delegated to five separate research stations across the country: Northeast Region, North Central Region, Western Region, Southern Region, and the USDA-ARS (21).

The process is as follows:

1. Farmers list their crop protection needs in the IR4 database (22).
2. IR4 organizes an annual three-day meeting to select up to 100 crop/pest/product combinations to be processed (23).
3. IR4 obtains cooperation agreements from companies developing appropriate products (24).
4. IR4 obtains from EPA the list of data that need to be provided and assessed.
5. IR4 reaches agreement with company on which data will be provided by the company and which data will be generated and provided by IR4 scientists.
6. IR4 forwards its study reports to the company.
7. Company submits all required information from both sources, IR4 and the company, to the EPA.
8. The EPA conducts reviews and risk assessment, takes regulatory decisions and informs the company and IR4.
9. If approved, the company makes its product available to farmers.
10. The list of farmer needs that are or have been supported by the different programs is available on the IR4 website (23). The support of biopesticides and organic agriculture are listed in a separate database on the IR4 website (25).

6.2 Canada

Other countries have subsequently followed the US example. In 2002, Canada established the Pest Management Centre (PMC) (26) within Agriculture and Agrifood Canada (AAFC), the main government agency supporting the agricultural sector in Canada. The PMC coordinates trials of new products across Canada as well as communication/cooperation with regulators and industry. In contrast to IR4 in the US, the PMC does not deal with farmers directly. Instead, farmers in each of the ten Canadian provinces consult with their provincial minor use coordinator who then briefs the PMC. Many trials are conducted in cooperation with the IR4 so that studies that were conducted in one country can be used for product approval in the other.

6.3 Europe

In 2015, the EU followed the North American example and founded the Minor Use Coordination Facility (MUCF) (27) located in Paris, France. The group collects and communicates the views of farmers across Europe and coordinates the research that needs to be conducted to achieve product approval in different European countries. The MUCF also coordinates European participation with the USA, Canada, Australia, and OECD.

6.4 Organisation for Economic Cooperation and Development

In 2018, the OECD created the Minor Use Foundation (28). The Minor Use Foundation collects views on crop protection priorities from farmers around the world (28), coordinates research activities across continents, communicates with companies and regulators, and cooperates with IR4 in the US (19), PMC in Canada (26), MUCF in Europe (28), and Department for Agriculture, Fisheries and Forestry (29) in Australia.

A separate development has been the commercial development of the Homologa Database which includes not only crop, pest, and product names but also the names of the registration holders (companies approved to sell particular crop protection products) in each country sorted by active ingredient, crop, and pest (30).

7 Conclusion

Figure 2 shows the full global process for regulatory approval of new crop protection process:

- from farmers/processors/traders identifying their crop protection needs (green);

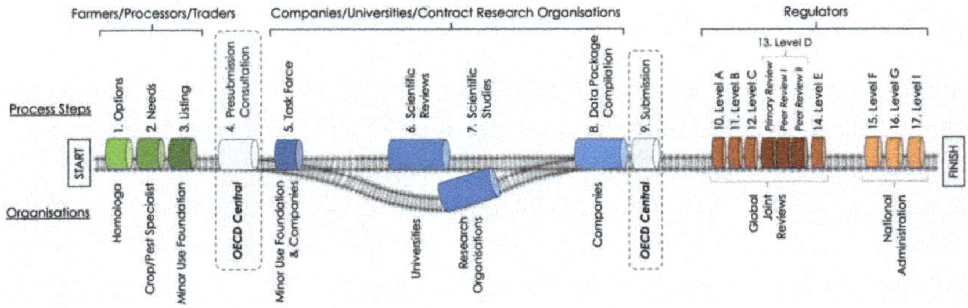

Figure 2 From product identification by farmers to product approval by regulators

- to companies/universities/contract research organizations compiling global data packages (blue); and
- and regulators conducting reviews and risk assessments and taking regulatory decisions (orange).

The only two blank steps that have not yet been fully harmonized and for which no shared tools exist yet are Step 4 (Presubmission consultation) and Step 9 (Submission). It is hoped that the OECD will act to address these gaps.

Farmers and their supporting crop and pest specialists need to know which products are available to their colleagues in other countries so that they can request access to these products themselves (step 1). When famers and their support staff have identified a specific product based on its function, they can upload this need in the database that is maintained by the Minor Use Foundation (step 2) (31). From here, companies can select the needs they feel able to address directly and the needs for which they would require support from the research community. They can indicate this and their willingness to be the authorization holder with all its legal implications in each country in the same database, following the example set by the IR4 (24).

Today, companies must still apply for pre-submission advice (step 3) to each regulator individually. Countries have forms, processes, and fees that should be integrated so that they can then be managed by an organization such as the OECD. It is realistic to envision that rather than sending out requests to several regulators in a single email message such a message could go to the OECD office in Paris instead. A good example for this approach is COAHP in Africa (18) discussed earlier where companies that seek authorization of their product in one of the 18 member states apply for a pre-submission consultation online (32) for all countries simultaneously. This could be the basis for a global data package.

When all points have been fully addressed in a data package (e.g. in form of a dedicated scientific study, existing reports, a scientific review of the literature or a report from a previous registration), the information can be formatted using in the correct OECD format (12). The full data package, scientific discussions, and scientific documentation will then be saved in a zip folder using software developed by OECD (15). Currently, this zip folder must be submitted to the regulators in different countries through a national portal or in the form of DVDs. This approach can easily be harmonized by the OECD creating a central portal into which the applicants upload the zip folder that contains their data package and from which individual regulatory authorities can download the zip folder for review, risk assessment, and regulatory decision.

The same OECD portal could be used by regulators to upload the reviews and risk assessments that they have agreed to deliver as part of a Global Joint Review. In the first round of reviews, regulators who have conducted a primary review would upload their documents. Other regulators would use these primary reviews for their peer-review I and then for peer-review II. For each step, the associated files would move from one subfolder in the cloud to the next.

At level E, countries would jointly prepare draft registration reports (DRR) and then move to level F on a national base. All following administrative steps would continue nationally. The process would be completed the same way as it is done today by national authorities completing and sending the registration certificate and the approved label to the applying companies.

In the past, data packages for new chemical pest control products relied mainly on new studies and were highly confidential. In contrast, data packages for biological products such as microbial active ingredients, pheromones, semiochemicals, and plant extracts rely mainly on existing studies and information from published scientific literature. Today synthetic chemicals still require more new studies given that they are typically new active ingredients but companies also draw more on existing scientific information whenever a new active ingredient belongs to a closely related group of chemicals.

There is an ongoing need to continue to overcome purely national or regional perspectives in the regulatory approval process. After all, environmental conditions that may affect a product's efficacy, safety, and environmental impact in a particular way in one region will have the same or very similar effects in another region with similar climatic conditions. This requires more cross-border cooperation and coordination. In addition, there needs to be better cooperation between and among farmers, companies, and regulators. All the necessary tools to achieve this (scientific, technical, and logistical) are in place, particularly in a modern digital environment. The only major hurdle to overcome is the will to embrace cooperative over competitive thinking. This will be essential if the sector is to play its role in achieving the UN's Sustainable Development Goals (SDGs) (33).

8 Appendix 1: Organisation for Economic Cooperation and Development data requirements for chemical products

ACTIVE INGREDIENT/SUBSTANCE/CONSTITUENT

1 Identity of the Active Substance/Plant Protection Product

- 1.1 Applicant (name, address, contact, telephone, and telefax numbers)
- 1.2 Manufacturer(s) (name, address, contact, telephone, and telefax numbers)
- 1.3 ISO common name proposed or accepted, and synonyms
- 1.4 Chemical Name as in Annex I to Directive 67/548/EEC, if not included in that Annex, in accordance with IUPAC and CA, nomenclature
- 1.5 Manufacturer's codes, names, and patent status
  - 1.5.1 Manufacturer's code number(s), for the active substance and formulations, materials concerned, countries in which used, and periods for which used
  - 1.5.2 Trade Name(s)
  - 1.5.3 Patent Status
- 1.6 Existing CAS, CIPAC, EINECS, and ELINCS numbers
- 1.7 Molecular formula, molecular mass, and structural formula
- 1.8 Method of Manufacture
  - 1.8.1 Method of manufacture (pathways, by-products and impurities) for each plant, whether or not relevant to a pilot plant
  - 1.8.2 Description of starting materials
- 1.9 Specification of purity of the active substance
  - 1.9.1 Minimum and/or nominal content (g/kg) of pure active substance (excluding inactive isomers), whether or not relevant to a pilot plant
    - 1.9.1.1 Minimum content (g/kg) of pure active substance (excluding inactive isomers), whether or not relevant to a pilot plant
    - 1.9.1.2 Nominal content (g/kg) of pure active substance (excluding inactive isomers), whether or not relevant to a pilot plant

	1.9.2	Certified limits of the active substances
	1.9.3	Control Product Specification Form or Confidential Statement of Formula
1.10	Identity, content, and structural formula of isomers, impurities, and additives	
	1.10.1	Inactive isomers For each isomer: • IUPAC and CA names • ISO common name proposed or accepted • CAS, CIPAC, EINECS, and ELINCS numbers • molecular and structural formula • molecular mass • ratio of the content of isomers/diastereo-isomers • maximum content in g/kg • whether or not relevant to a pilot plant
	1.10.2	Impurities and additives: • IUPAC and CA names • ISO common name proposed or accepted • CAS, CIPAC, EINECS, and ELINCS numbers • molecular and structural formula • molecular mass • maximum content in g/kg • whether or not relevant to a pilot plant • in the case of additives, their function and trade names • in the case of impurities and by-products of particular environmental concern, details of the analytical methods • guidance in identifying impurities of toxicological concern
1.11	Batch analysis data	
	1.11.1	Analytical profile of batches
	1.11.2	Results of analyses of batches produced in laboratory or pilot scale production systems and used in toxicological testing
1.12	Other/special studies	

(Continued)

ACTIVE INGREDIENT/SUBSTANCE/CONSTITUENT
2 Physical and Chemical Properties of the Active Substance
2.1 Melting point and boiling point
2.1.1 Melting point, freezing point, or solidification point of purified active substance
2.1.2 Boiling point of purified active substance
2.1.3 Temperature at which decomposition or sublimation occurs
2.2 Relative density of purified active substance
2.3 Vapour pressure and volatility
2.3.1 Vapour pressure of purified active substance
2.3.2 Henry's law constant
2.4 Appearance
2.4.1 Description of the physical state and colour of both the purified active substance and the active substance as manufactured (or technical grade active substance)
2.4.2 Description of the odour of the purified active substance and active substance as manufactured
2.5 Spectra and molecular extinction at relevant wavelengths
2.5.1 Spectra, a table of signal characteristics and molecular extinction at relevant wavelengths for purified active substance
2.5.1.1 UV/VIS
2.5.1.2 IR
2.5.1.3 NMR
2.5.1.4 MS
2.5.1.5 Wavelengths at which UV/VIS molecular extinction occurs, where appropriate, to include a wavelength at the highest absorption value above 290 nm
2.5.1.6 Optical purity

2.5.2 Spectra for impurities

- 2.5.2.1 UV/VIS
- 2.5.2.2 IR
- 2.5.2.3 NMR
- 2.5.2.4 MS

2.6 Solubility of purified active substance in water:

- determined in the neutral range
- determined in the acidic range (pH 4–6)
- determined in the alkaline range (pH 8–10)

2.7 Solubility in organic solvents at 15–25°C

2.8 Partition coefficient

- 2.8.1 n-octanol/water partition coefficient
- 2.8.2 Effect of pH (4–10) on the noctanol/water partition coefficient

2.9 Stability in water, hydrolysis rate, photochemical degradation, quantum yield and identity of breakdown products, dissociation constant

2.9.1 Hydrolysis rate of purified active substance at pH values 4, 7, and 9 under sterile conditions, in the absence of light:

- identity of hydrolysis products
- rate constant observed
- estimated DT50 value

2.9.2 Direct phototransformation of purified active substance in water using artificial light (simulating sunlight and excluding wavelengths $\lambda < 290$ nm) under sterile conditions to include the following:

- photochemical half-life
- mass balance to account for 90% of the applied radioactivity
- identity of breakdown products

2.9.3 Quantum yield of direct phototransformation

2.9.4 Calculated theoretical lifetime in the top layer of aqueous systems and the real lifetime of the active substance

2.9.5 Dissociation in water of purified active substance:

- dissociation constant(s) (pKa values)
- identity of dissociated species formed
- dissociation constant(s) (pKa values) of the active principle

(Continued)

ACTIVE INGREDIENT/SUBSTANCE/CONSTITUENT			
	2.10	Estimated photochemical oxidative degradation	
	2.11	Flammability including autoflammability	
		2.11.1	Flammability of the active substance as manufactured
		2.11.2	Auto-flammability of the active substance as manufactured
	2.12	Flash point of the active substance as manufactured	
	2.13	Explosive properties of the active substance as manufactured	
	2.14	Surface tension of the active substance as manufactured	
	2.15	Oxidizing properties of the active substance as manufactured	
	2.16	PH	
	2.17	Stability	
		2.17.1	Storage stability
		2.17.2	Stability (temperature, metals)
	2.18	Other/special studies	
3	Further Information on the Active Substance (Function, Mode of Action, Handling)		
	3.1	Function, e.g. fungicide	
	3.2	Effects on harmful organisms	
		3.2.1	Nature of the effects on harmful organisms, e.g. contact action
		3.2.2	Whether or not translocated in plants and if translocated whether such translocation is apoplastic, symplastic, or both
	3.3	Fields of use, e.g. forestry	
	3.4	Harmful organisms controlled and crops or products protected or treated	
		3.4.1	Details of existing and intended uses (crops, groups of crops, plants or plant products treated or protected)
		3.4.2	Details of harmful organisms against which protection is afforded
		3.4.3	Effects achieved, e.g. sprout suppression

3.5	Mode of action	
	3.5.1	Statement of the mode of action of the active substance in terms of biochemical and physiological mechanism(s) and biochemical pathway(s) involved
	3.5.2	Details of active metabolites and degradation products, cross referenced to the toxicological and residues data provided, to include the following: • IUPAC and CA names • ISO common name proposed or accepted • CAS, CIPAC, EINECS, and ELINCS numbers • molecular and structural formula • molecular mass
	3.5.3	Information relative to the formation of active metabolites and degradation products, to include the following: • the processes, mechanisms, and reactions involved • kinetic and other data concerning the rate of conversion and if known the rate limiting step • environmental and other factors effecting the rate and extent of conversion
3.6	Information on the possible occurrence of the development of resistance or cross-resistance	
3.7	A material safety data sheet for the active substance	
3.8	Procedures for destruction or decontamination	
	3.8.1	Pyrolytic behaviour of the active substance under controlled conditions at 800°C and the content of polyhalogenated dibenzo-pdioxins in the products of pyrolysis
	3.8.2	Detailed instructions for safe disposal
	3.8.3	Methods other than controlled incineration for the disposal of the active substance, contaminated packaging, and contaminated materials: • Detailed description of such methods • Data to establish their effectiveness and safety
3.9	Procedures for the decontamination of water in case of an accident	
3.10	Other/special studies	

(Continued)

ACTIVE INGREDIENT/SUBSTANCE/CONSTITUENT		
4	Analytical Methods and Validation	
4.1	Analytical standards and samples	
4.1.1	Analytical standards for pure active substance	
4.1.2	Samples of the active substance as manufactured	
4.1.3	Analytical standards for relevant metabolites and other components included in the residue definition	
4.1.4	Samples of reference substances for relevant impurities	
4.2	Methods for the analysis of the active substance as manufactured	
4.2.1	Description of analytical methods for the analysis of the active substance as manufactured. For each method submitted: • specificity • extent of interference by other substances present • explanation of interferences which contribute more than ± 3% of the total quantity determined For each method submitted, linearity over an appropriate range: • equation of the calibration line • correlation co-efficient • representative labelled documentation. e.g. chromatograms For each method submitted, accuracy: • pure active substance • impurities For each method submitted, repeatability (at least 5 determinations): • % relative standard deviation (RSD) • indication as to whether outliers identified have been discarded • reasons for the occurrence of outliers	
4.2.2	Applicability of existing CIPAC methods	
4.2.3	Description of analytical methods for the determination of impurities (non-active components arising from the manufacturing process or from the degradation during storage) which are of toxicological, ecotoxicological, or environmental concern or which are present in quantities ≥1 g/kg in the active substance as manufactured	

	For each method submitted: • specificity • extent of interference by other substances present • explanation of interferences which contribute more than ± 3% of the total quantity determined For each method submitted, linearity over an appropriate range: • equation of the calibration line • correlation co-efficient • representative labelled documentation, e.g. chromatograms For each method submitted, accuracy: • pure active substance • impurities For each method submitted, repeatability (at least 5 determinations): • % relative standard deviation (RSD) • indication as to whether outliers identified have been discarded • reasons for the occurrence of outliers
4.2.4	Description of analytical methods for the determination of additives (e.g. stabilizers) in the active substance as manufactured. For each method submitted: • specificity • extent of interference by other substances present • explanation of interferences which contribute more than ± 3% of the total quantity determined For each method submitted, linearity over an appropriate range: • equation of the calibration line • correlation co-efficient • representative labelled documentation e.g. chromatograms For each method submitted, accuracy: • pure active substance • impurities For each method submitted, repeatability (at least 5 determinations): • % relative standard deviation (RSD) • indication as to whether outliers identified have been discarded • reasons for the occurrence of outliers

(Continued)

ACTIVE INGREDIENT/SUBSTANCE/CONSTITUENT	
	4.2.5 Enforcement analytical methodology
	4.2.6 Inter-laboratory analytical methodology validation
	4.2.7 Storage stability of working solutions in analytical methodology
4.3	Description of analytical methods for the determination of residues (all components included in the residue definition proposed (see point 6) to enable compliance with MRLs to be determined or to determine dislogeable residues For each method and representative matrix: • Specificity (using a confirmatory method, if appropriate) • Repeatability • Validation - independent laboratory • Limit of determination • Individual and mean recovery, overall standard deviation and relative standard deviation at each fortification level
4.4	Description of methods for analysis of soil for parent compound and metabolites of toxicological, ecotoxicological or environmental concern For each method: • Specificity (using a confirmatory method, if appropriate) • Repeatability • Limit of determination • Individual and mean recovery, overall standard deviation at each fortification level
4.5	Description of methods for analysis of water (drinking water, ground water and surface water) for parent compound and metabolites of toxicological, ecotoxicological or environmental concern For each method: • Specificity (using a confirmatory method, if appropriate) • Repeatability • Limit of determination • Individual and mean recovery, overall standard deviation at each fortification level
4.6	Method for determining pesticides in sediment. For each method: • Specificity (using a confirmatory method, if appropriate) • Repeatability • Limit of determination

4.7	Description of methods for analysis of air for active substance and metabolites, formed during or shortly after application, of toxicological concern. For each method: • Specificity (using a confirmatory method, if appropriate) • Repeatability • Limit of determination • Individual and mean recovery, overall standard deviation at each fortification level
4.8	Analytical methods for parent compound and toxicologically, ecotoxicologically or environmentally significant metabolites in body fluids and tissues. For each method: • Specificity (using a confirmatory method, if appropriate) • Repeatability • Limit of determination • Individual and mean recovery, overall standard deviation at each fortification level
4.9	Other/special studies
5	Toxicological and Toxicokinetic Studies on the Active Substance
5.1	Absorption, distribution, excretion, and metabolism in mammals
5.1.1	Toxicokinetic studies – single dose, oral route, in rats
5.1.2	Toxicokinetic studies – second single dose, oral route, in rats
5.1.3	Toxicokinetic studies – repeated dose, oral route, in rats
5.2	Acute toxicity
5.2.1	Acute oral toxicity
5.2.2	Acute percutaneous toxicity
5.2.3	Acute inhalation toxicity
5.2.4	Skin irritation
5.2.5	Eye irritation
5.2.6	Skin sensitization
5.2.7	Potentiation/interactions of multiple active substances or products

(Continued)

ACTIVE INGREDIENT/SUBSTANCE/CONSTITUENT		
5.3	Short-term toxicity	
	5.3.1	Oral 28-day toxicity
	5.3.2	Oral 90-day toxicity (rodents)
	5.3.3	Oral 90-day toxicity (dog)
	5.3.4	Oral 1-year toxicity (dog)
	5.3.5	28-day inhalation toxicity (rodents)
	5.3.6	90-day inhalation toxicity (rodents)
	5.3.7	Percutaneous 28-day toxicity (rodents)
	5.3.8	Percutaneous 90-day toxicity (rodents)
5.4	Genotoxicity	
	5.4.1	In vitro genotoxicity testing – bacterial assay for gene mutation
	5.4.2	In vitro genotoxicity testing – test for clastogenicity in mammalian cells
	5.4.3	In vitro genotoxicity testing – test for gene mutation in mammalian cells
	5.4.4	In vivo genotoxicity testing (somatic cells) – metaphase analysis in rodent bone marrow, or micronucleus test in rodents
	5.4.5	In vivo genotoxicity testing (somatic cells) – unscheduled DNA synthesis or a mouse spot test
	5.4.6	In vivo studies in germ cells
5.5	Long-term toxicity and carcinogenicity	
	5.5.1	Long-term (2 years) oral toxicity in the rat (can be a combined long-term and carcinogenicity study)
	5.5.2	Carcinogenicity study in the rat (can be a combined long-term and carcinogenicity study)
	5.5.3	Carcinogenicity study in the mouse
	5.5.4	Mechanism of action and supporting data
5.6	Reproductive toxicity	
	5.6.1	Two generation reproductive toxicity in the rat
	5.6.2	Separate male and female studies

	5.6.3	Three segment designs
	5.6.4	Dominant lethal assay for male fertility
	5.6.5	Cross-matings of treated males with untreated females and vice versa
	5.6.6	Effects on spermatogenesis
	5.6.7	Effects on oogenesis
	5.6.8	Sperm motility, mobility, and morphology
	5.6.9	Investigation of hormonal activity
	5.6.10	Teratogenicity test by the oral route in the rat
	5.6.11	Teratogenicity test by the oral route in the rabbit
5.7	Neurotoxicity	
	5.7.1	Acute neurotoxicity – rat
	5.7.2	Delayed neurotoxicity following acute exposure
	5.7.3	28-day delayed neurotoxicity
	5.7.4	Subchronic neurotoxicity – rat – 90 day
	5.7.5	Postnatal developmental neurotoxicity
5.8	Toxicity studies on metabolites	
5.9	Medical data	
	5.9.1	Report on medical surveillance on manufacturing plant personnel
	5.9.2	Report on clinical cases and poisoning incidents
	5.9.3	Observations on exposure of the general population and epidemiological studies
	5.9.4	Clinical signs and symptoms of poisoning and details of clinical tests
	5.9.5	First aid measures
	5.9.6	Therapeutic regimes
	5.9.7	Expected effects and duration of poisoning as a function of the type, level, and duration of exposure or ingestion

(Continued)

ACTIVE INGREDIENT/SUBSTANCE/CONSTITUENT
5.9.8 Expected effects and duration of poisoning as a function of varying time periods between exposure or ingestion and commencement of treatment
5.9.9 Dermal penetration
5.10 Other/special studies
5.11 Summary of mammalian toxicity and overall evaluation
6 Metabolism and Residues Data
6.1 Stability of residues
6.1.1 Stability of residues during storage of samples
6.1.2 Stability of residues in sample extracts
6.2 Metabolism, distribution, and expression of residues
6.2.1 In plants, in at least three crops representative of the different categories of crop (root vegetables; leafy crops; fruits; pulses and oilseed; cereals)
6.2.2 Poultry
6.2.3 Lactating ruminants (goat or cow)
6.2.4 Pigs
6.2.5 Nature of residue in fish
6.2.6 Chemical identity (emphasis on impurities of residual concern)
6.3 Residue trials (supervised field trials) for crops or plant products used as food or feed on which use is proposed or where residues from soil can be taken up
6.3.1 Crop 1 (e.g. wheat)
6.3.2 Crop 2 (e.g. oilseed rape)
6.3.3 Crop 3 (as required)
6.3.4 Crop 4 (etc.)

6.4	Livestock feeding studies	
	6.4.1	Poultry
	6.4.2	Lactating ruminants (goat or cow)
	6.4.3	Pigs
	6.4.4	Fish
6.5	Effects of industrial processing and/or household preparation (representative processing situations) on	
	6.5.1	The nature of residue
	6.5.2	Distribution of the residue in peel/pulp
	6.5.3	Residue levels – balance studies on a core set of representative processes
	6.5.4	Residue levels – follow-up studies to determine concentration or dilution factors
6.6	Residues in succeeding crops	
	6.6.1	Theoretical consideration of the nature and level of the residue
	6.6.2	Metabolism and distribution studies on representative crops
	6.6.3	Field trials on representative crops
6.7	Proposed residue definition and maximum residue levels	
	6.7.1	Proposed residue definition
	6.7.2	Proposed maximum residue levels (MRLs) and justification of the acceptability of the levels proposed, including details of statistical analyses used
6.8	Proposed pre-harvest intervals, re-entry intervals or withholding periods to minimize residues in crops, plants, plant products, treated areas or spaces and a justification for each proposal	
	6.8.1	Pre-harvest interval (in days) for each relevant crop
	6.8.2	Re-entry period (in days) for livestock, to areas to be grazed
	6.8.3	Re-entry period (in hours or days) for man to crops, buildings or spaces treated
	6.8.4	Withholding period (in days) for animal feedingstuffs
	6.8.5	Waiting period (in days) between last application and sowing or planting the crops to be protected

(*Continued*)

ACTIVE INGREDIENT/SUBSTANCE/CONSTITUENT
6.8.6 Waiting period (in days) between application and handling treated product
6.8.7 Waiting period (in days) between last application and sowing or planting succeeding crops
6.9 Estimation of the potential and actual exposure through diet and other means
6.9.1 TMDI calculations
6.9.2 NEDI calculations
6.9.3 NESTI calculations
6.10 Other/special studies
6.11 Summary and evaluation of residue behaviour / Reasonable grounds in support of the petition
6.11.1 Summary and evaluation of residue behaviour
6.11.2 Reasonable grounds in support of the petition
7 Fate and Behaviour in the Environment
7.1 Route of degradation in soil – laboratory studies
7.1.1 Aerobic degradation
7.1.2 Anaerobic degradation
7.1.3 Soil photolysis
7.2 Rate of degradation in soil(s) • laboratory studies
7.2.1 Aerobic degradation of the active substance in soils at 20°C
7.2.2 Aerobic degradation of the active substance in soil at 10°C
7.2.3 Aerobic degradation of relevant metabolites, degradation and reaction products in soils at 20°C
7.2.4 Anaerobic degradation of the active substance in soil
7.2.5 Anaerobic degradation of relevant metabolites, degradation and reaction products in soil

7.3 Field studies

- 7.3.1 Soil dissipation testing in a range of representative soils - (normally 4 soils)
- 7.3.2 Soil residue testing
- 7.3.3 Soil accumulation testing on relevant soils

7.4 Mobility studies

- 7.4.1 Adsorption and desorption of the active substance
- 7.4.2 Adsorption and desorption of all relevant metabolites, degradation and reaction products
- 7.4.3 Column leaching studies with the active substance
- 7.4.4 Column leaching studies with relevant metabolites, degradation and reaction products
- 7.4.5 Aged residue column leaching
- 7.4.6 Leaching (TLC)
- 7.4.7 Lysimeter studies
- 7.4.8 Field leaching studies
- 7.4.9 Volatility - laboratory study

7.5 Hydrolysis rate of relevant metabolites, degradation and reaction products at pH values 4, 7 and 9 under sterile conditions, in the absence of light

- identity of the hydrolysis products
- rate constant observed
- estimated DT50 value

7.6 Direct phototransformation of relevant metabolites, degradation and reaction products in water using, artificial light (simulating sunlight and excited wavelengths $\lambda < 290$ nm) under sterile conditions, to include

- photochemical half-life
- mass balance to account for 90% of the applied radioactivity
- identity of breakdown products
- quantum yield of direct phototransformation
- calculated theoretical lifetime in the top layer of aqueous systems and the real lifetime of the substance added

(*Continued*)

ACTIVE INGREDIENT/SUBSTANCE/CONSTITUENT
7.7 Ready biodegradability of the active substance
7.8 Degradation in aquatic systems
7.8.1 Aerobic biodegradation in aquatic systems, including identification of breakdown products and metabolites
7.8.2 Anaerobic biodegradation in aquatic systems, including identification of breakdown products and metabolites
7.8.3 Water/sediment studies
7.9 Degradation in the saturated zone of the active substance, metabolites, degradation and reaction products
7.10 Rate and route of degradation in air
7.11 Definition of the residue
7.12 Monitoring data concerning fate and behaviour of the active substance and of relevant metabolites, degradation and reaction products
7.13 Other/special studies
8 Ecotoxicological Studies on the Active Substance
8.1 Avian toxicity
8.1.1 Acute oral toxicity to a quail species (Japanese or Bobwhite), mallard duck, or other bird species
8.1.2 Avian dietary toxicity (5-day) test in a quail species or in a mallard duck
8.1.3 Avian dietary toxicity (5-day) test in a second unrelated species
8.1.4 Subchronic and reproductive toxicity to birds
8.2 Fish toxicity
8.2.1 Acute toxicity of the active substance to fish
8.2.1.1 Rainbow trout (Oncorhynchus mykiss) - Analytical data on concentrations in the test media
8.2.1.2 Warm water fish species - Analytical data on concentrations in the test media
8.2.1.3 Acute toxicity of metabolites, degradation or reaction products to the more sensitive of the fish species used to test the acute toxicity of the active substance - Analytical data on concentrations in the test media
8.2.2 Chronic toxicity to fish
8.2.3 Chronic toxicity (28 day exposure) to juvenile fish - growth and behaviour - Analytical data on concentrations in the test media

	8.2.4	Fish early life stage toxicity test - Analytical data on concentrations in the test media	
	8.2.5	Fish life cycle test Analytical data on concentrations in the test media	
	8.2.6	Bioconcentration potential in fish	
		8.2.6.1	Bioconcentration potential of the active substance in fish
		8.2.6.2	Bioconcentration potential of metabolites, degradation and reaction products
	8.2.7	Aquatic bioavailability/ biomagnification/depuration	
8.3	Toxicity to aquatic species other than fish and aquatic species field testing		
	8.3.1	Acute toxicity to aquatic invertebrates	
		8.3.1.1	Acute toxicity (24 and 48 hour) for Daphnia preferably (Daphnia magna) - Analytical data on concentrations in the test media
		8.3.1.2	Acute toxicity (24 and 48 hour) for representative species of aquatic insects - Analytical data on concentrations in the test media
		8.3.1.3	Acute toxicity (24 and 48 hour) for representative species of aquatic crustaceans (species unrelated to Daphnia) Analytical data on concentrations in the test media
		8.3.1.4	Acute toxicity (24 and 48 hour) for representative species of aquatic gastropod molluscs - Analytical data on concentrations in the test media
	8.3.2	Chronic toxicity to aquatic invertebrates	
		8.3.2.1	Chronic toxicity in Daphnia magna (21-day) - Analytical data on concentrations in the test media
		8.3.2.2	Chronic toxicity for representative species of aquatic insects - Analytical data on concentrations in the test media
		8.3.2.3	Chronic toxicity for representative species of aquatic gastropod molluscs - Analytical data on concentrations in the test media
	8.3.3	Aquatic field testing	
8.4	Effects on algal growth and growth rate (2 species) Analytical data on concentrations in the test media		
8.5	Effects on sediment dwelling organisms		
	8.5.1	Acute test - Analytical data on concentrations in the test media	
	8.5.2	Chronic test - Analytical data on concentrations in the test media	

(*Continued*)

ACTIVE INGREDIENT/SUBSTANCE/CONSTITUENT			
8.6	Effects on aquatic plants - Analytical data on concentrations in the test media		
8.7	Effects on bees		
	8.7.1	Acute oral toxicity	
	8.7.2	Acute contact toxicity	
	8.7.3	Toxicity of residues on foliage to honey bees	
	8.7.4	8.7.4 Bee brood feeding test	
8.8	Effects on non-target terrestrial arthropods		
	8.8.1	Effects on non-target terrestrial arthropods using artificial substrates	
		8.8.1.1	Parasitoid
		8.8.1.2	Predatory mites
		8.8.1.3	Ground dwelling predatory species (selected to be relevant to the intended uses of preparations)
		8.8.1.4	Foliage dwelling predatory species (selected to be relevant - to the intended uses of preparations)
	8.8.2	Effects on non-target terrestrial arthropods in extended laboratory/semi-field tests	
		8.8.2.1	Parasitoid
		8.8.2.2	Predatory mites
		8.8.2.3	Ground dwelling predatory species (selected to be relevant - to the intended uses of preparations)
		8.8.2.4	Foliage dwelling predatory species (selected to be relevant to the intended uses of preparations)
		8.8.2.5	Other terrestrial invertebrates
8.9	Effects on earthworms		
	8.9.1	Acute toxicity to earthworms	
	8.9.2	Sublethal effects on earthworms	

8.10	Effects on soil microbial activity	
	8.10.1	Nitrogen transformation
	8.10.2	Carbon mineralization
	8.10.3	Rates of recovery following treatment
8.11	Effects on marine and estuarine organisms	
	8.11.1	Marine or estuarine organisms acute toxicity LC50/EC50
	8.11.2	Marine/Estuarine fish • salinity challenge
8.12	Effects on terrestrial vascular plants	
8.13	Effects on terrestrial vertebrates other than birds / wild mammal toxicity	
8.14	Effects on other non-target organisms (flora and fauna) believed to be at risk	
	8.14.1	Summary of all available data from preliminary tests used to assess biological activity and dose range finding, which may provide information on other non-target species (flora and fauna)
	8.14.2	A critical assessment as to the relevance of the preliminary test data to potential impact on non-target species
8.15	Effects on biological methods for sewage treatment	
8.16	Other/special studies	
	8.16.1	Other/special studies - laboratory studies
	8.16.2	Other/special studies - field studies
8.17	Summary and evaluation of points 7 and 8.1 to 8.16	
9	Proposals including justification for the proposals for the classification and labelling of the active substance • hazard symbol(s) • indications of danger • risk phrases • safety phrases	

(Continued)

END-USE PRODUCT

1 Identity of the Plant Protection Product

- 1.1 Applicant (name, address, contact, telephone and telefax numbers)
- 1.2 Manufacturer of the preparation, manufacturer and purity of the active substance(s)
  - 1.2.1 Manufacturer(s) of the preparation (name, address, contact, telephone and telefax numbers)
  - 1.2.2 Manufacturer(s) of the activesubstance(s) (name, address, contact, telephone and telefax numbers)
  - 1.2.3 Statement of purity (and detailed information on impurities) of the active substance
- 1.3 Trade name or proposed trade name and manufacturers code number(s), for the preparation and similar preparations (differences to be specified)
- 1.4 Detailed quantitative and qualitative information on the composition of the preparation
  - 1.4.1 Contents of:
    - Technical active substance
    - Pure active substance
    - Formulants
  - 1.4.2 Certified limits of each component
  - 1.4.3 Names and codes identifying the active substance
    - 1.4.3.1 ISO common name proposed or accepted for active substances, and synonyms
    - 1.4.3.2 Existing CIPAC, EINECS and ELINCS numbers for the active substance
    - 1.4.3.3 Salt, ester, anion or cation present for each active substance
  - 1.4.4 For each formulant, or component in formulants
    - Chemical name as in Annex I to Directive 67/548/EEC, if not included in that Annex, in accordance with IUPAC and CA nomenclature
    - Structure or structural formula
    - Existing CAS, CIPAC, EINECS and ELINCS numbers
    - Trade name
    - Specification of the formulant
    - Function of each formulant

		1.4.5	Formulation process
		1.4.5.1	Description of formulation process
		1.4.5.2	Discussion of the formation of impurities of toxicological concern
	1.5		Type of preparation (formulation) and code
	1.6		Function (herbicide, insecticide, etc.)
	1.7		Other/special studies
2			Physical, Chemical and Technical Properties of the Plant Protection Product
	2.1		Description of the physical state of the preparation (formulation) and its colour and odour
	2.2		Explosivity and oxidizing properties
		2.2.1	Explosive properties of the preparation
		2.2.2	Oxidizing properties of the preparation
	2.3		Flash point and other indication of flammability or spontaneous ignition
		2.3.1	The flash point of the preparation
		2.3.2	The flammability of the preparation
		2.3.3	The auto-flammability of the preparation
	2.4		Acidity/alkalinity and if necessary pH value
		2.4.1	Acidity or alkalinity and pH value
		2.4.2	pH of a 1 % aqueous dilution, emulsion or dispersion
	2.5		Viscosity and surface tension
		2.5.1	Kinematic viscosity of the preparation
		2.5.2	Viscosity of the preparation and details of the test conditions
		2.5.3	Surface tension of the preparation
	2.6		Relative density and bulk density
		2.6.1	Relative density of the preparation
		2.6.2	Bulk or tap density of the preparation

(*Continued*)

END-USE PRODUCT			
2.7	Storage stability and shelf-life		
	2.7.1	Stability after storage for 14 days at 54° C	
	2.7.2	Stability after storage for other periods and/or temperatures	
	2.7.3	Minimum content after heat stability testing	
	2.7.4	Effect of low temperature on stability	
	2.7.5	Shelf life following storage at ambient temperature	
	2.7.6	Shelf life in months	
2.8	Technical characteristics of the plant protection product		
	2.8.1	Wettability	
	2.8.2	Persistent foaming	
	2.8.3	Suspensibility and suspension stability	
		2.3.8.1	Suspensibility
		2.3.8.2	Spontaneity of dispersion
	2.8.4	Dilution stability	
	2.8.5	Sieve test	
		2.8.5.1	Dry sieve test
		2.8.5.2	Wet sieve test
	2.8.6	Particle size distribution	
		2.8.6.1	Particle size distribution
		2.8.6.2	Nominal size range of granules
		2.8.6.3	Dust content
		2.8.6.4	Particle size of dust
		2.8.6.5	Friability and attrition characteristics of granules

		2.8.7	Emulsion characteristics	
			2.8.7.1	Emulsifiability
			2.8.7.2	Emulsion stability
			2.8.7.3	Re-emulsifiability
			2.8.7.4	Stability of dilute emulsions
			2.8.7.5	Stability of emulsions
		2.8.8	Flowability, pourability and dustability	
			2.8.8.1	Flowability
			2.8.8.2	Pourability (including rinsed residue)
			2.8.8.3	Dustability following accelerated storage
	2.9	Physical and chemical compatibility with other products		
		2.9.1	Physical compatibility of tank mixes	
		2.9.2	Chemical compatibility of tank mixes	
	2.10	Distribution and adherence to seeds		
		2.10.1	Distribution (seed treatment)	
		2.10.2	Adhesion (seed treatment)	
	2.11	Miscibility		
	2.12	Dielectric breakdown voltage		
	2.13	Corrosion characteristics		
	2.14	Container Material		
	2.15	Other/special studies		
	2.16	Summary and evaluation of points 2.1 to 2.15		
3	Data on Application			
	3.1	Fields of use e.g. forestry		

(*Continued*)

END-USE PRODUCT			
	3.2	Nature of the effects on harmful organisms e.g. contact action	
	3.3	Details of intended use	
		3.3.1	Details of existing and intended uses (crops, groups of crops, plants or plant products treated or protected)
		3.3.2	Details of harmful organisms against which protection is afforded
		3.3.3	Effects achieved e.g. sprout suppression
	3.4	Rate of application per unit treated (ha, m2, m3, tonne) treated, in terms of g or kg of preparation and active substance	
	3.5	Concentration of active substance in material used (e.g. diluted spray, baits, treated seed) in g/l, g/kg, mg/kg or g/tonne	
	3.6	Description of the method of application, type of equipment used and type and volume of diluent per unit of area or volume	
	3.7	Number and timing of applications and duration of protection	
		3.7.1	Maximum number of applications and their timing
		3.7.2	For each application, growth stages of the crop or plants to be protected
		3.7.3	For each application, development stages of the harmful organism concerned
		3.7.4	Duration of protection afforded by each application
		3.7.5	Duration of protection afforded by the maximum number of applications
	3.8	Necessary waiting periods or other precautions to avoid phytotoxic effects on succeeding crops	
		3.8.1	Minimum waiting periods or other precautions between last application and sowing or planting succeeding crops
		3.8.2	Limitations on choice of succeeding crops
		3.8.3	Description of damage to rotational crops
	3.9	Proposed instructions for use as printed, or to be printed, on labels	
	3.10	Other/special studies	
4	Further Information on the Plant Protection Product		
	4.1	Packaging and compatibility with the preparation	
		4.1.1	Description and specification of the packaging and materials used in packaging, size, capacity, size of openings, types of closure and seals

	4.1.2	Suitability of the packaging and closures • Strength • Leakproofness • Resistance to normal transport and handling
	4.1.3	Resistance of the packaging material to its contents
4.2	Procedures for cleaning application equipment	
	4.2.1	Procedures for cleaning application equipment and protective clothing
	4.2.2	Effectiveness of the cleaning procedures
4.3	Re-entry periods, necessary waiting periods or other precautions to protect man, livestock and the environment	
	4.3.1	Pre-harvest interval (in days) for each relevant crop
	4.3.2	Re-entry period (in days) for livestock, to areas to be grazed
	4.3.3	Re-entry period (in hours or days) for man to crops, buildings or spaces treated
	4.3.4	Withholding period (in days) for animal feedingstuffs
	4.3.5	Waiting period (in days) between application and handling treated products
	4.3.6	Waiting period (in days) between last application and sowing or planting succeeding crops
	4.3.7	Information on any specific agricultural, plant health or environmental conditions under which the preparation may or may not be used
4.4	Statement of the risks arising from the recommended methods, precautions and handling procedures to minimize those risks, relating to	
	4.4.1	Warehouse storage
	4.4.2	User level storage
	4.4.3	Transport
	4.4.4	Fire
	4.4.5	Protective clothing and equipment proposed for use in storage, transport or in the event of fire – Nature
	4.4.6	Protective clothing and equipment proposed for use in storage, transport or in the event of fire – characteristics
	4.4.7	Sufficient data to evaluate suitability and effectiveness of the protective clothing and equipment under realistic conditions of use

(*Continued*)

END-USE PRODUCT			
	4.4.8	Procedures to minimize the generation of waste	
	4.4.9	Information on combustion products likely to be generated in the event of fire	
4.5	Detailed procedures for the use in the event of an accident during transport, storage or use		
	4.5.1	Containment of spillages	
	4.5.2	Decontamination of areas, vehicles and buildings	
	4.5.3	Disposal of damaged packaging, adsorbents and other materials	
	4.5.4	Protection of emergency workers and bystanders	
	4.5.5	First aid measures	
4.6	Neutralization procedures (e.g. reaction with alkali to form less toxic compounds) for use in the event of accidental spillages		
	4.6.1	Details of proposed procedures for small quantities	
	4.6.2	Evaluation of products of neutralization (small quantities)	
	4.6.3	Procedures for disposal of neutralized waste (small quantities)	
	4.6.4	Details of proposed procedures for large quantities	
	4.6.5	Evaluation of products of neutralization (large quantities)	
	4.6.6	Procedures for disposal of neutralized waste (large quantities)	
4.7	Pyrolytic behaviour of the active substance under controlled conditions at 800°C and the content of polyhalogenated dibenzo-pdioxins in the products of pyrolysis		
4.8	Disposal procedures for the plant protection product		
	4.8.1	Detailed instructions for safe disposal of the plant protection product and its packaging	
	4.8.2	Methods other than controlled incineration for disposal • detailed description of such methods • data to establish their effectiveness and safety	
4.9	Other/special studies		

5 Methods of Analysis

5.1 Analytical standards and samples

	5.1.2	Analytical standards for pure active substance
	5.1.3	Samples of the active substance as manufactured
	5.1.4	Analytical standards for relevant metabolites and all other components included in the residue definition
	5.1.5	Samples of reference substances for relevant impurities
5.2	Methods for the analysis of plant protection products	
	5.2.1	Description of analytical methods for the determination of the active substance in plant protection products. For each method submitted • Specificity • Extent of interference by other substances present in the preparation • Explanation of interferences which contribute more than ± 3% of the total quantity determined For each method submitted, linearity over an appropriate range • Equation of the calibration line • Correlation co-efficient • Representative labelled documentation e.g. chromatograms For each method submitted, accuracy • Pure active substance • Impurities For each method submitted, repeatability (at least 5 determinations) • % relative standard deviation (RSD) • Indication as to whether outliers identified have been discarded • Reasons for the occurrence of outliers
	5.2.2	For preparations containing more than one active substance, a description of a method capable of determining each in the presence of the other. For each method submitted • Specificity • Extent of interference by other substances present in the preparation • Explanation of interferences which contribute more than ± 3% of the total quantity determined For each method submitted, linearity over an appropriate range • Equation of the calibration line • Correlation co-efficient • Representative labelled documentation e.g. chromatograms

(*Continued*)

END-USE PRODUCT	
	For each method submitted, accuracy • Pure active substance • Impurities For each method submitted, repeatability (at least 5 determinations) • % relative standard deviation (RSD) • Indication as to whether outliers identified have been discarded • Reasons for the occurrence of outliers
5.2.3	Applicability of existing CIPAC methods
5.2.4	Description of analytical methods for the determination of impurities (non-active components arising from the manufacturing process or from degradation during storage) which are of toxicological, ecotoxicological or environmental concern, in the preparation. For each method submitted • Specificity • Extent of interference by other substances present in the preparation • Explanation of interferences which contribute more than ± 3% of the total quantity determined For each method submitted, linearity over an appropriate range • Equation of the calibration line • Correlation co-efficient • Representative labelled documentation e.g. chromatograms For each method submitted, accuracy • Pure active substance • Impurities For each method submitted, repeatability (at least 5 determinations) • % relative standard deviation (RSD) • Indication as to whether outliers identified have been discarded • Reasons for the occurrence of outliers
5.2.5	Description of analytical methods for the determination of formulants or constituents of formulants in the plant protection product. For each method submitted • Specificity • Extent of interference by other substances present in the preparation • Explanation of interferences which contribute more than ± 3% of the total quantity determined

	• Correlation co-efficient • Representative labelled documentation e.g. chromatograms For each method submitted, accuracy • Pure active substance • Impurities For each method submitted, repeatability (at least 5 determinations) • % relative standard deviation (RSD) • Indication as to whether outliers identified have been discarded • Reasons for the occurrence of outliers
5.3	Description of analytical methods for the determination of residues; Storage stability of working solutions in analytical methodology
	5.3.1 Description of analytical methods for the determination of residues (all components included in the residue definition proposed (see point 8) to enable compliance with MRLs to be determined or to determine dislodgeable residues). For each method and representative matrix • Specificity (using a confirmatory method, if appropriate) • Repeatability • Validation • independent laboratory • Limit of determination • Individual and mean recovery, overall standard deviation and relative standard deviation at each fortification level
	5.3.2 Storage stability of working solutions in analytical methodology
5.4	Description of methods for analysis of soil for parent compound and metabolites of toxicological, ecotoxicological or environmental concern. For each method • Specificity (using a confirmatory method, if appropriate) • Repeatability • Limit of determination • Individual and mean recovery, overall standard deviation and relative standard deviation at each fortification level
5.5	Description of methods for analysis of sediment. For each method • Specificity (using a confirmatory method, if appropriate) • Repeatability • Limit of determination • Individual and mean recovery, overall standard deviation and relative standard deviation at each fortification level

(Continued)

END-USE PRODUCT
5.6 Description of methods for analysis of water (drinking water, ground water and surface water) for parent compound and metabolites of toxicological, ecotoxicological or environmental concern. For each method • Specificity (using a confirmatory method, if appropriate) • Repeatability • Limit of determination • Individual and mean recovery, overall standard deviation and relative standard deviation at each fortification level
5.7 Description of methods for analysis of air for active substance and metabolites, formed during or shortly after application, of toxicological concern. For each method • Specificity (using a confirmatory method, if appropriate) • Repeatability • Limit of determination • Individual and mean recovery, overall standard deviation and relative standard deviation at each fortification level
5.8 Analytical methods for parent compound and toxicologically, ecotoxicologically or environmentally significant metabolites in body fluids and tissues. For each method • Specificity (using a confirmatory method, if appropriate) • Repeatability • Limit of determination • Individual and mean recovery, overall standard deviation and relative standard deviation at each fortification level
5.9 Other/special studies
6 Efficacy Data and Information (including Value Data)
6.1 Efficacy data
6.1.1 Preliminary range-finding tests
6.1.2 Minimum effective dose tests
6.1.3 Efficacy tests
6.1.4 Effects on yield and quality
6.1.4.1 Impact on the quality of plants and plant products
6.1.4.2 Effects on the processing procedure
6.1.4.3 Effects on the yield of treated plants and plant products

	6.2		Adverse effects
		6.2.1	Phytotoxicity to host crop
		6.2.2	Adverse effects on health of host animals
		6.2.3	Adverse effects on site of application (discoloration, corrosion, etc.)
		6.2.4	Adverse effects on beneficial organisms
		6.2.5	Adverse effects on parts of plants used for propagating purposes (e.g. seeds, cuttings, runners)
		6.2.6	Impact on succeeding crops
		6.2.7	Impact on other plants including adjacent crops
		6.2.8	Information on the possible occurrence of the development of resistance or cross-resistance
	6.3		Economics
	6.4		Benefits
		6.4.1	Survey of alternative pest control measures (chemical and non-chemical)
		6.4.2	Compatibility with current management practices including IPM
		6.4.3	Contribution to risk reduction
	6.5		Other/special studies
	6.6		Summary and assessment of data according to points 6.1 to 6.5
	6.7		List of test facilities including the corresponding certificates
7	Toxicological Studies and Exposure Data and Information		
	7.1		Acute toxicity
		7.1.1	Acute oral toxicity
		7.1.2	Acute percutaneous (dermal) toxicity
		7.1.3	Acute inhalation toxicity to rats
		7.1.4	Skin irritation
		7.1.5	Eye irritation

(Continued)

END-USE PRODUCT		
	7.1.6	Skin sensitization
	7.1.7	Supplementary studies for combinations of plant protection products
7.2	Short-term toxicity studies	
7.3	Operator exposure	
	7.3.1	Estimation of operator exposure assuming personal protective equipment is not used
	7.3.2	Estimation of operator exposure assuming personal protective equipment is used
	7.3.3	Measurement of operator exposure - (Mixer/Loader/Applicator)
7.4	Bystander exposure	
	7.4.1	Estimation of bystander exposure assuming personal protective equipment is not used
	7.4.2	Measurement of bystander exposure
7.5	Worker exposure	
	7.5.1	Estimation of worker exposure assuming personal protective equipment is not used
	7.5.2	Estimation of worker exposure assuming personal protective equipment is used
	7.5.3	Estimation of worker exposure assuming personal protective equipment is used and using data generated on dislogeable residues under the proposed conditions of use
	7.5.4	Measurement of worker exposure
7.6	Dermal absorption	
	7.6.1	Dermal absorption, in vivo in the rat
	7.6.2	Comparative dermal absorption, in vitro using rat and human skin
7.7	Dislodgeable residues	
	7.7.1	Dislodgeable residues – foliar
	7.7.2	Dislodgeable residues – soil
	7.7.3	Dislodgeable residues - indoor surface re-volatilization
7.8	Epidemiology	

	7.9		Data on formulants
		7.9.1	Material safety data sheet for each formulant
		7.9.2	Available toxicological data for each formulant
	7.10		Domestic animal/livestock safety
	7.11		Other/special studies
8			Metabolism and Residues Data
	8.1		Stability of residues
		8.1.1	Stability of residues during storage of samples
		8.1.2	Stability of residues in sample extracts
	8.2		Supplementary studies on metabolism, distribution and expression of residues in plants or livestock
	8.3		Supplementary residue trials (supervised field trials) for crops or plant products used as food or feed on which use is proposed – if it is not possible to extrapolate from the data provided in the context of point 6.3 AI, e.g. special formulations, different application methods, additional crops
		8.3.1	Crop 1 (e.g. wheat)
		8.3.2	Crop 2 (e.g. oilseed rape)
		8.3.3	Crop 3
		8.3.4	Crop 4, etc.
	8.4		Supplementary livestock feeding studies - if it is not possible to extrapolate from the data provided in the context of point IIA 6.4, e.g. use on additional fodder crops is proposed, leading to an increased intake of residues by livestock
		8.4.1	Poultry
		8.4.2	Lactating ruminants (goat or cow)
		8.4.3	Pigs
		8.4.4	Nature of residue in fish

(*Continued*)

END-USE PRODUCT			
8.5	Supplementary studies on the effects of industrial processing and/or household preparation on residue levels - if it is not possible to extrapolate from the data provided in the context of point IIA 6.5 e.g. additional crops		
	8.5.1	Effects of industrial processing and/or household preparation (representative processing situations) on the nature of the residue	
	8.5.2	Distribution of the residue in peel/pulp	
	8.5.3	Balance studies on a core set of representative processes	
	8.5.4	Follow-up studies to determine concentration or dilution factors • Potable water • Irrigated crops	
		8.5.4.1	Follow-up studies to determine concentration or dilution factors
		8.5.4.2	Potable waters
		8.5.4.3	Irrigated crops
8.6	Supplementary studies for residues in representative succeeding crops		
8.7	Proposed residue definition and		
	8.7.1	Proposed residue definition	
	8.7.2	Proposed MRLs and justification of the acceptability of the levels proposed, including details of statistical analyses used	
8.8	Proposed pre-harvest intervals, re-entry intervals or withholding periods to minimize residues in crops, plants, plant products, treated areas or spaces and a justification for each proposal		
	8.8.1	Pre-harvest interval (in days) for each relevant crop	
	8.8.2	Re-entry period (in days) for livestock, to areas to be grazed	
	8.8.3	Re-entry period (in hours or days) for man to crops, buildings, or spaces treated	
	8.8.4	Withholding period (in days) for animal feeding stuffs	
	8.8.5	Waiting period (in days) between last application and sowing or planting the crop to be protected	
	8.8.6	Waiting period (in days) between application and handling treated products	
	8.8.7	Waiting period (in days) between last application and sowing or planting succeeding crops	

	8.9	Other/special studies	
	8.10	Estimation of the potential and actual exposure through diet and other means	
		8.10.1	TMDI calculations
		8.10.2	NEDI calculations
		8.10.3	NESTI calculations
	8.11	Summary and evaluation of residue behaviour	
9	Fate and Behaviour in the Environment		
	9.1	Rate of degradation in soil – if it is not possible to extrapolate from the data provided for the active substance and relevant metabolites, degradation and reaction products (e.g. slow release formulations)	
		9.1.1	Aerobic degradation of the preparation in soil
		9.1.2	Anaerobic degradation of the preparation in soil
	9.2	Field studies	
		9.2.1	Soil dissipation testing on a range of representative soils
		9.2.2	Soil residue testing
		9.2.3	Soil accumulation testing
		9.2.4	Aquatic (sediment) field dissipation
		9.2.5	Forestry field dissipation
	9.3	Mobility of the plant protection product in soil	
		9.3.1	Column leaching
		9.3.2	Lysimeter studies
		9.3.3	Field leaching studies
		9.3.4	Volatility – laboratory study
		9.3.5	Volatility – field study

(*Continued*)

END-USE PRODUCT		
9.4	Predicted environmental concentrations in soil (PECs) for the active substance at the highest rate of application proposed and relating to the maximum number and highest rates of application proposed, for each relevant soil tested	
	9.4.1	Initial PECs value
	9.4.2	Short-term PECs values – 24 hours, 2 and 4 days after last application
	9.4.3	Long-term PECs values – 7, 28, 50, and 100 days after last application
9.5	Predicted environmental concentrations in soil (PECs) for relevant metabolites, degradation and reaction products, at the highest rate of application proposed and relating to the maximum number and highest rates of application proposed, for each relevant soil tested	
	9.5.1	Initial PECs value
	9.5.2	Short-term PECs value – 24 hours, 2 and 4 days after last application
	9.5.3	Long-term PECs values – 7, 28, 50, and 100 days after last application
9.6	Predicted environmental concentrations in ground water (PECgw) at the highest rate of application proposed and relating to the maximum number and highest rates of application proposed	
	9.6.1	Active substance PECgw value
	9.6.2	Relevant metabolites, degradation and reaction products PECgw values
	9.6.3	Additional field testing
	9.6.4	Information on impact on water treatment procedures
9.7	Predicted environmental concentrations in surface water (PECsw) for the active substance at the highest rate of application proposed and relating to the maximum number and highest rates of application proposed, relevant to lakes, ponds, rivers, canals, streams, irrigation/drainage canals and drains	
	9.7.1	Initial PECsw value for static water bodies
	9.7.2	Initial PECsw value for slow moving water bodies
	9.7.3	Short-term PECsw values for static water bodies – 24 hours, 2 and 4 days after last application
	9.7.4	Short-term PECsw values for slow moving water bodies – 24 hours, 2 and 4 days after last application
	9.7.5	Long-term PECsw values for static water bodies – 7, 14, 21, 28, and 42 days after last application
	9.7.6	Long-term PECsw values for slow moving water bodies – 7, 14, 21, 28, and 42 days after last application

9.8	Predicted environmental concentrations in surface water (PECsw) for the relevant metabolites, degradation and reaction products at the highest rate of application proposed and relating to the maximum number and highest rates of application proposed, relevant to lakes, ponds, rivers, canals, streams, irrigation/drainage canals and drains		
	9.8.1	Initial PECsw value for static water bodies	
	9.8.2	Initial PECsw value for slow moving water bodies	
	9.8.3	Short-term PECsw values for static water bodies – 24 hours, 2 and 4 days after last application	
	9.8.4	Short-term PECsw values for slow moving water bodies – 24 hours, 2 and 4 days after last application	
	9.8.5	Long-term PECsw values for static water bodies – 7, 14, 21, 28, and 42 days after last application	
	9.8.6	Long-term PECsw values for slow moving water bodies – 7, 14, 21, 28, and 42 days after last application	
	9.8.7	Additional field testing	
9.9	Fate and behaviour in air		
	9.9.1	Spray droplet size spectrum – laboratory studies	
	9.9.2	Drift – field evaluation	
9.10	Other/special studies		
	9.10.1	Other/special studies – laboratory studies	
	9.10.2	Other/special studies – field studies	
10 Ecotoxicological Studies on the Plant Protection Product			
10.1	Effects on birds		
	10.1.1	Acute toxicity exposure ratio (TERA) for birds	
	10.1.2	Short-term toxicity exposure ratio (TERST) for birds	
	10.1.3	In the case of baits, the concentration of active substance in the bait in mg/kg	
	10.1.4	In the case of pellets, granules, prills, or treated seed	
		10.1.4.1	Amount of the active substance in or on each pellet, granule, prill, or treated seed
		10.1.4.2	Proportion of the LD50 for the active substance in 100 particles and per gram of particles
	10.1.5	In the case of pellets, granules, and prills, their size and shape	

(Continued)

END-USE PRODUCT			
	10.1.6		Acute oral toxicity of the preparation to the more sensitive of the species identified in tests with the active substance
	10.1.7		Supervised cage or field trials
	10.1.8		Acceptance of bait, granules or treated seeds by birds (palatability test)
	10.1.9		Effects of secondary poisoning
10.2	Effects on aquatic organisms		
	10.2.1	Toxicity exposure ratios for aquatic species	
		10.2.1.1	TERA for fish
		10.2.1.2	TERLT for fish
		10.2.1.3	TERA for Daphnia
		10.2.1.4	TERLT for Daphnia
		10.2.1.5	TERA for an aquatic insect species
		10.2.1.6	TERLT for an aquatic insect species
		10.2.1.7	TERA for an aquatic crustacean species
		10.2.1.8	TERLT for an aquatic crustacean species
		10.2.1.9	TERA for an aquatic gastropod mollusc species
		10.2.1.10	TERLT for an aquatic gastropod mollusc species
		10.2.1.11	TERLT for algae
	10.2.2	Acute toxicity (aquatic) of the preparation	
		10.2.2.1	Fish acute toxicity LC50, freshwater, cold-water species
		10.2.2.2	Acute toxicity (24 and 48 h) for Daphnia preferably Daphnia magna
		10.2.2.3	Effects on algal growth and growth rate
		10.2.2.4	Marine or estuarine organisms acute toxicity LC50/EC50
		10.2.2.5	Marine sediment invertebrates, acute toxicity LC50/EC50
	10.2.3	Microcosm or mesocosm study	

	10.2.5	Chronic fish toxicity data		
		10.2.5.1	Chronic toxicity (28 day exposure) to juvenile fish: Analytical data on concentrations in the test media	
		10.2.5.2	Fish early life stage toxicity test: Analytical data on concentrations in the test media	
		10.2.5.3	Fish life cycle test: Analytical data on concentrations in the test media	
	10.2.6	Chronic toxicity to aquatic invertebrates		
		10.2.6.1	Chronic toxicity in Daphnia magna (21-day): Analytical data on concentrations in the test media	
		10.2.6.2	Chronic toxicity for a representative species of aquatic insects: Analytical data on concentrations in the test media	
		10.2.6.3	Chronic toxicity for a representative species of aquatic gastropod molluscs: Analytical data on concentrations in the test media	
	10.2.7	Accumulation in aquatic nontarget organisms: Analytical data on concentrations in the test media		
10.3	Effects on terrestrial vertebrates other than birds			
	10.3.1	Toxicity exposure ratios for terrestrial vertebrates other than birds		
		10.3.1.1	Acute toxicity exposure ratio (TERA)	
		10.3.1.2	Short-term toxicity exposure ratio (TERST)	
		10.3.1.3	Long-term toxicity exposure ratio (TERLT)	
	10.3.2	Effects on terrestrial vertebrates other than birds, where the required information is not provided by testing in accordance with points IIA 5 and IIIA 7, and where exposure is likely		
		10.3.2.1	Acute oral toxicity of the preparation	
		10.3.2.2	Acceptance of bait, granules or treated seeds by terrestrial vertebrates (palatability test)	
		10.3.2.3	Effects of secondary poisoning	
	10.3.3	Supervised cage or field trials or other appropriate studies		
10.4	Effects on bees			
	10.4.1	Hazard Quotients for bees		
		10.4.1.1	Oral exposure QHO	
		10.4.1.2	Contact exposure QHC	
	10.4.2	Acute toxicity of the preparation to bees		

(*Continued*)

END-USE PRODUCT
10.4.2.1 Acute oral toxicity
10.4.2.2 Acute contact toxicity
10.4.3 Effects on bees of residues on crops
10.4.4 Cage tests
10.4.5 Field tests
10.4.6 Investigation of special effects
10.4.6.1 Larval toxicity
10.4.6.2 Long residual effects
10.4.6.3 Disorienting effects on bees
10.4.7 Tunnel testing to investigate effects of feeding on contaminated honey dew or flowers
10.5 Effects on arthropods other than bees
10.5.1 Effects on sensitive species already tested, using artificial substrates
10.5.2 Effects on non-target terrestrial arthropods in extended laboratory tests
10.5.3 Effects on non-target terrestrial arthropods in semi-field tests
10.5.4 Field tests on arthropod species
10.6 Effects on earthworms and other soil macro-organisms
10.6.1 Toxicity exposure ratios for earthworms, TERA and TERLT
10.6.2 Acute toxicity to earthworms
10.6.3 Sublethal effects on earthworms
10.6.4 Field tests (effects on earthworms)
10.6.5 Residue content of earthworms
10.6.6 Effects of other soil non-target macro-organisms
10.6.7 Effect on organic matter breakdown

10.7	Effects on soil microbial activity		
	10.7.1	Laboratory test to investigate impact on soil microbial activity	
	10.7.2	Further laboratory, glasshouse of field testing to investigate impact on soil microbial, activity	
10.8	Effects on non-target plants		
	10.8.1	Effects on non-target terrestrial plants	
		10.8.1.1	Seed germination
		10.8.1.2	Vegetative vigour
		10.8.1.3	Seedling emergence
		10.8.1.4	Terrestrial field testing
	10.8.2	Effects on non-target aquatic plants	
		10.8.2.1	Aquatic plant growth – Lemna
		10.8.2.2	Aquatic field testing
10.9	Effects on other non-target organisms (flora and fauna) belived to be at risk		
	10.9.1	Summary of available data from preliminary tests used to assess biological activity and dose range finding, which may provide information on other non-target species (flora and fauna)	
	10.9.2	A critical assessment as to the relevance of the preliminary test data to potential impact on non-target species	
10.10	Other/special studies		
	10.10.1	Other/special studies - laboratory studies	
	10.10.2	Other/special studies – field studies	
10.11	Summary and evaluation of points IIIA 9 and IIIA 10.1 to 10.10, together with a detailed and critical assessment of the data		
	10.11.1	Predicted distribution and fate in the environment and the time courses involved	
	10.11.2	Non-target species at risk and extent of potential exposure	
	10.11.3	Short and long term risks for non-target species, populations, communities and processes	
	10.11.4	Risk of fish kills and fatalities in large vertebrates or terrestrial predators	
	10.11.5	Precautions necessary to avoid or minimize contamination of the environment and for the protection of non-target species	

(*Continued*)

END-USE PRODUCT	
11	Further information
11.1	Information on authorizations in other countries
11.2	Information on established maximum residue limits (MRL) in other countries
11.3	Proposals including justification for the classification and labelling proposed: • Hazard symbol(s) • Indications of danger • Risk phrases • Safety phrases
11.4	Proposals for risk and safety phrases
11.5	Proposed label
11.6	Specimens of proposed packaging

9 Appendix 2 Organisation for Economic Cooperation and Development data requirements for microbial products

MPCA; Microbial Pest Control Agent

1 Identity of the Active Substance/Plant Protection Product

- 1.1 Applicant (name, address, contact, telephone, and telefax numbers)
- 1.2 Producer (name, address, contact, telephone, and telefax numbers)
  - 1.3.1 Scientific name of microorganism to species level or a level sufficient to show taxonomic relation to known microorganisms, especially pathogens:
    - accession no. of sample in a recognized culture collection
    - test procedures and criteria, using best available technology, to characterize the strain or serotype;
    - for mutant or genetically-modified strains, indicate all known differences between the modified microorganism and the parent wild strain(s)
    - include any trade names, common names, developmental code names
    - indigenous or non-indigenous at the species level to the intended area of application (required in the EU)
- 1.4 Composition of Technical Grade of MPCA/Active Substance
  - 1.4.1 Concentration of microorganism (and metabolite, if appropriate) in terms of g/kg, or g/L (for US and Canada also in % w/w) and cfu's/mL or appropriate potency units; include acceptable range for each term. Potency should be expressed in recognized units of potency or an appropriate expression of biological activity per unit weight/volume
    - 1.4.2.1 Composition of microbial material used for manufacture of end use products in terms of g/kg or g/L (for US and Canada also in % w/w) for each active ingredient including
      - the MPCA
      - additives (preservatives, stabilizers, diluents)
      - microbial impurities, classified/identified to a taxonomic level required by quality criteria to support the hygienic state of the production process
      - non-microbial impurities (eg. metabolic products, impurities in starting materials, fermentation residues, extraneous host residues). This information is not required if Technical Grade of MPCA is a hypothetical stage in a continuous production process of an end-use product.
    - 1.4.2.2 Composition in terms of % g/kg or g/L, w/w for each ingredient:
      The identity and maximum content of all microbial impurities must be reported, if possible and appropriate, expressed in appropriate units, as outlined in point 1.3.1 (in terms of cfu's/mL or appropriate expression of biological activity/viability)

(Continued)

MPCA; Microbial Pest Control Agent

1.4.3 Quality criteria for the production and storage of the MPCA, including: criteria for consistency and integrity of the master and working seed stock, typically, measures of biological activity and phenotypic or genotypic properties:

- acceptable range for content of MPCA, in appropriate terms;
- presence of human/ mammalian pathogens;
- presence or maximum accepted level of known mammalian toxins, if their presence is suspected at any stage in process, or if MPCA is closely related to a toxigenic human pathogen
- maximum accepted level for microbial impurities, using suitable indicators of an unhygienic process

1.4.4 Quality control data (measures of quality criteria) from three to five production batches, including storage stability data. If the Technical Grade of MPCA is a stage in a continuous production process of an enduse product, this information should be provided for the entire production process.

1.4.5.1 A theoretical discussion regarding the following:

- formation and/or presence of unintentional ingredients, including impurities of toxicological concern, likely to occur in the Technical Grade of the MPCA,
- the impact of these ingredients on product quality, and
- appropriate quality criteria.

1.4.5.2 Physical and chemical properties, if MPCA is produced as a manufacturing product that is stored prior to formulation of end-use products: physical state; density; viscosity or surface tension; explosivity, corrosive character, oxidizing properties; technical characteristics as appropriate.

1.4.5.3 International regulatory status of microorganism

1.4.5.4 Comprehensive Data Summary/Tier II summaries in OECD format: "Guidelines and Criteria for Industry for the Preparation and Presentation of Complete Dossiers and of Summary Dossiers for Plant Protection Products and their Microbial Active Substances in Support of Regulatory Decisions in OECD Countries, Appendix 7 and 8 / Tier II format required by Annex IIb of 91/414/EEC.

1.4.5.5 Sample of MPCA, analytical standard of metabolite (and reference substances for the relevant impurities – EU only): if requested

1.5.3 Patent Status

2	Properties of the Active Substance	
	2.1	Origin of the isolate; method of isolation; preservation and maintenance of strain during development; historical information on testing and use of the strain; history of use of closely related strains or species; describe any unusual morphological, physiological, pesticidal or resistance characteristics of MPCA which differ from classical description of species
	2.2	Natural occurrence of microorganism including geographic distribution, hosts, habitat, ecological niche, level of natural occurrence
	2.3	Description of target organism(s); information on mode of action, kind of antagonism to target host, infective/toxic dose, transmissibility
	2.4	Available information on host specificity; possible effects on species closely related to the target pest (Any experience of toxic effect of the active substance or its metabolic products on human or animals, of whether the organism is capable of colonizing or invading humans or animals and whether it is pathogenic shall be stated. Any experience of whether the active substance or its products may irritate skin, eyes or respiratory organs of humans or animals and whether it is allergenic in contact with skin or when inhaled – always required in the EU).
	2.5	Life cycle of micro-organism including various forms that may occur, differences in pathogenic/ toxigenic character of various forms, virulence, and survival time of resting stages, interactions with other species (vector, parasitism, competition)
	2.6	Among closely related species, provide available information on: • pathogenicity to plants, animals or humans • formation of toxic metabolites: structure, stability, conditions under which they are formed, mode of action.
	2.6.1	Physiological properties, especially effect of environmental parameters on growth, infectivity, dispersal, and colonization ability: temperature, pH, redox potential, humidity, light, nutritional requirements
	2.6.2	Description of any plasmids or other extra chromosomal genetic elements involved in pesticidal activity, pathogenicity, toxicity, etc.
	2.6.3	Genetic stability (mutation rate of traits related to the mode of action), factors affecting genetic stability; microorganism's capacity to transfer genetic information to another population
	2.6.4	Detailed discussion of relationship of microorganism to any known human dermatophyte (see point 5.2)
	2.6.5	Resistance/sensitivity to antibiotics/antimicrobial agents used in human or veterinary medicine
3	Further Information on the Active Substance (Function, Mode of Action, Handling)	
	3.1	Function, e.g. fungicide
	3.3	Fields of use, e.g. forestry

(*Continued*)

MPCA; Microbial Pest Control Agent

3.4.1 Details of existing and intended uses (crops, groups of crops, plants or plant products treated or protected)

3.4.2 Details of harmful organisms against which protection is afforded

3.4.3 Effects achieved, e.g. sprout suppression

3.5.1 Statement of the mode of action of the Microbial Pest Control Agent in terms of biochemical and physiological mechanism(s) and biochemical pathway(s) involved

3.5.2 Details of active metabolites (especially toxins) and degradation products, cross referenced to the toxicological and residues data provided, to include the following:

- IUPAC and CA names
- ISO common name proposed or accepted
- CAS, CIPAC, EINECS, and ELINCS numbers
- molecular and structural formula
- molecular mass

3.5.3 Information relative to the formation of active metabolites (especially toxins) and degradation products, to include the following:

- the processes, mechanisms, and reactions involved
- kinetic and other data concerning the rate of conversion and if known the rate limiting step
- environmental and other factors effecting the rate and extent of conversion

3.6 Information on the possible occurrence of the development of resistance or cross-resistance

3.7 A material safety data sheet for the Microbial Active Substance

3.8.1.2 Detailed instructions for safe disposal

3.9 Procedures for the decontamination of water in case of an accident

3.10 Other/special studies

3.1 Crops or products to be protected or treated

3.1 Measures to render microorganistm harmless, in case of an accident

4 Analytical Methods and Validation/MPCA

4.1.5 Method to preserve and maintain the master seed stock; criteria for an acceptable level of consistency and integrity of seed stock

4.2.8 Production process for Technical Grade of MPCA, describing techniques used to ensure a uniform product and procedures when hazardous contamination is detected in a batch. List starting and intermediate materials, with source and purity of each.

4.3 Quality control and post-registration monitoring methods:

- to detect, isolate, and enumerate the microorganism;
- to differentiate a mutant or genetically modified microorganism from the parent strain;
- to detect spontaneous change in major characteristics of microorganism;
- to define content of microorganism in appropriate terms (same as 1.4.1), incl. standardization, sensitivity, reproducibility, statistical validity, and representative data to validate the bioassay;
- to show control to a specified and acceptable level, of microbial impurities and of any other impurities of toxicological concern, including toxic metabolites, which are known or suspected to be present at any stage of the manufacturing process;
- to show presence of any human and mammalian pathogens.

4.4 Storage stability test, data and determination of shelf life, if MPCA is stored

4.5 Post-registration monitoring methods to determine and quantify residues of viable or non-viable micro-organism and metabolites (especially toxins) on food, feed, animal tissue, in soil, water or air, where relevant. (analytical methods for amount or activity of proteinaceous products – required in the EU, where relevant)

5 Toxicological and Toxicokinetic Studies on the Active Substance

5.1 Summary: potential of microbial pest control agent to be hazardous to humans with consideration of its pathogenic potential, its ability to infect and pattern of clearance, and its toxicological effects

5.2 Occupational health surveillance report on workers during production and testing of MPCA, including information on the following:;

- details on any occurrence of hypersensitivity and chronic sensitization; and
- any significant clinical findings related to exposure, with special attention to those whose susceptibility may be affected

5.3.9 Acute oral infectivity and toxicity

5.4 Acute intra tracheal/ inhalation infectivity and toxicity

5.5 Acute intravenous/ intra peritoneal infectivity

5.6 Cell culture study, for viruses and viroids or specific bacteria and protozoa with intracellular replication

(Continued)

MPCA; Microbial Pest Control Agent

- 5.7 Genotoxic potential, especially for fungi and actinomycetes: a discussion of the potential for genotoxin production based on the relationship of the microorganism to a genus/species known to produce genotoxins. If a related fungus/ actinomycete produces a genotoxin, either an appropriate and sensitive analytical test (e.g. HPLC) must be done to detect its presence in the MPCA (for Canada), or genotoxicity testing is required (for EC).
- 5.8 Toxicity studies on metabolites (especially toxins)
    - 5.8.1.1 Published reports of adverse effects, especially reports of clinical cases and followup studies; list databases and key words used in a literature search.
    - 5.8.1.2 Short-term toxicity, pathogenicity, infectivity (28-day minimum)
        - for EU only
    - 5.8.1.3 Proposed first aid measures and medical treatment
- 5.10 Other/special studies CR 5.3
- 5.1 Summary of mammalian toxicity and overall evaluation

6 Metabolism and Residues Data/MPCA

- 6.1.1 Rationale for waiver of residue data based on information showing that MPCA is not hazardous to mammals, i.e. lack of potential for a known mammalian toxin and negative result from the acute oral toxicity test.
- 6.1.2 Rationale for waiver based on a substantiated estimation that MPCA is unlikely to occur on treated food/feed stuffs in concentrations considerably higher than under natural conditions.
- 6.1.3 Summary of residue behaviour and overall evaluation

7 Fate and Behaviour in the Environment/MPCA

- 7.1 Sufficient information on the origin, properties, survival and residual metabolites of the microorganism to assess its fate and behaviour in the environment.Information provided in parts 2–6 may suffice.
- 7.2 Viability/population dynamics, persistence, multiplication, and mobility.
- 7.1 Other/special studies

8 Effects on non-target organisms/MPCA

- 8.1 Birds
- 8.2 Fish

	8.3	Aquatic invertebrates	
	8.4	Effects on algal growth and growth rate (2 species)	
	8.6	Effects on aquatic or terrestrial plants	
	8.7	Bees	
	8.8	Non-target terrestrial arthropods	
	8.9	Other terrestrial invertebrates	
		8.9.1	In EU effects on earthworms required
	8.10	In EU effects on non-target soil microorganisms required	
	8.1	Other/special studies	
9	Summary and evaluation of environmental impact: summarize all data relevant to environmental impact and assess environmental risk by the following: • addressing distribution and fate of MPCA • identifying non-target species at risk and the extent of their exposure • identifying precautions necessary to minimize environmental contamination and to protect nontarget species.		
	TITLE (MPCP; Microbial Pest Control Product)		
1	Identity of the Plant Protection Product		
	1.1	Applicant (name, address, contact, telephone, and telefax numbers)	
		1.2.1	Manufacturer(s) of the preparation (name, address, contact, telephone and telefax numbers)
		1.2.2	Producer of the microbial pest control agent(s) (name, address, contact, telephone and telefax numbers)

(*Continued*)

TITLE (MPCP; Microbial Pest Control Product)

1.3 Trade name or proposed trade name and manufacturers code number(s), for the preparation and similar preparations (differences to be specified)

1.5 Physical state of MPCP (GIFAP formulation type)

1.6 Function (herbicide, insecticide, etc.)

1.6.1 Biological function category and field of use category, using terms defined by each country, e.g. 'control of weeds' for 'forestry'

1.7.1 Concentration of MPCA in MPCP, measured in terms of g/kg or g/L of the MPCP (for US and Canada, also provide figures in %w/w) and in cfu's or other appropriate potency units; provide content of MPCA in Technical Grade of MPCA, in the same terms.
Also indicate:
- scientific name and strain/serotype of MPCA, its accession number in a recognized culture collection,
- development phase (eg. spore) of MPCA in MPCP

1.7.2 Composition in terms of g/kg or g/L and % w/w of each ingredient in MPCP, including:
- Technical Grade of MPCA
- each additive: include chemical name and structure; CAS and EEC numbers of components of additive if they exist or an appropriate specification; trade name; function in MPCP
- microbial impurities: taxonomic identification as required by quality criteria to support the hygienic state of the production process; express content of microbial impurities in appropriate units, e.g. cfu's/ml.
- non-microbial impurities (eg. metabolic products, impurities in starting materials, fermentation residues, extraneous host residues)

1.7.3 Quality criteria for the production and storage of the MPCP, including:
- acceptable range for content of MPCA, in appropriate terms
- presence of human or non-target animal pathogens
- presence or maximum accepted level of known mammalian toxins, if their presence is suspected at any stage in process, or if MPCA is closely related to a toxigenic human pathogen
- maximum accepted level for microbial impurities, using suitable indicators of contamination

1.7.4 Quality control data (measures of quality criteria) from 3 to 5 production batches, including product stored for duration of shelf life if it is metabolically active. If the Technical Grade of MPCA is a stage in a continuous production process of an end use product, this information should be provided for the entire production process.

1.7.5 A theoretical discussion regarding

- the formation and/or presence of unintentional ingredients, including impurities of toxicological concern, likely to occur in MPCP,
- the impact of these ingredients on product quality, and
- appropriate quality criteria.

For metabolically-active MPCP, consider degradation or metabolic production during storage.

2 Physical, Chemical, and Technical Properties of the Plant Protection Product

2.1 Appearance (colour, odour, physical state)

2.2 Storage stability and shelf-life for MPCP which must contain metabolically active MPCP, include QC data for hazardous contaminants originating from degradation or metabolic production during storage

2.3 Explosivity, oxidizing properties, flash point, flammability, spontaneous ignition, acidity, alkalinity, pH, viscosity, surface tension: as appropriate

2.4

- Technical characteristics as appropriate: wettability, persistent foaming, suspensibility, suspension stability, dry/wet sieve test, particle size distribution, content of dust/fines, emulsifiability, emulsion stability, flowability, pourability, dustability
- Density
- Summary and evaluation of data on properties of the MPCP

3 Data on Application

3.1 Pest to be controlled, crop to be protected, available information on mode of action (site of uptake, toxic/competitive effect), is microorganism transmitted or translocated to another part of plant?)

3.2 Available information on the development of resistance in target pest and appropriate mitigation strategy

3.3 Application rate in terms of mass/vol of MPCP per unit area/volume (e.g. kg/ha). Content of micro-organism in material used (diluted spray, bait, treated seed).

3.4 Application rate in terms of units of micro-organism per unit area/volume

3.5 Method of application (incl. type of equipment and volume of diluent)

3.6 Number and timing of applications, related to: host/pest phenology, duration of protection, application of other pesticides, preharvest interval

3.7 Precautions to avoid phytotoxic/ phytopathogenic effects on protected crop or on succeeding crops, if appropriate. Proposed instructions for use as printed, or to be printed, on labels.

(*Continued*)

TITLE (MPCP; Microbial Pest Control Product)

4 Further Information on the Plant Protection Product

- 4.1 Packaging: description
- 4.2 Specifications of packaging and measures of its suitability
- 4.3 Label instructions re: cleaning equipment and protective clothing
- 4.4 Procedures to clean equipment and protective clothing; measures of their effectiveness
- 4.5 Necessary waiting periods for re-entry; recommended protective measures to reduce occupational exposure
- 4.6 Label instructions Re: safe handling and storage
- 4.7 Recommendations Re: handling, storage, transport, fire: specify risks, specify procedures to minimize hazards and the generation of waste
- 4.8 Label instructions re: cleanup of spills
- 4.9 Detailed procedures in case of accident to: contain a spill, decontaminate an area or vehicle, dispose of adsorbents and packaging, protect workers and bystanders, first aid
- 4.10 Procedures for destruction/disposal of MPCP and its packaging (eg. detailed instructions for controlled incineration)

5 Methods of Analysis / MPCP

- 5.1 Quality control and post-registration monitoring methods:
  - to differentiate a mutant or genetically-modified microorganism from the parent strain.
  - to detect spontaneous change in major characteristics of microorganism.
  - to define content of microorganism in appropriate terms (same as 1.4.1), incl. standardization, sensitivity, reproducibility, statistical validity, and representative data to validate the bioassay.
  - to identify contaminant microorganisms in MPCP
  - to show control to a specified and acceptable level, of microbial impurities and of any other impurities of toxicological concern, including toxic metabolites, which are known or suspected to be present at any stage of the manufacturing process.
  - to show presence of any human and mammalian pathogens.
- 5.2 Storage stability test and determination of shelf life
- 5.3 Production process for MPCP, describing techniques used to ensure a uniform product and procedures when hazardous contamination is detected in a batch. List starting and intermediate materials, with source and purity of each. Method for determination of residues: required if information provided for MPCA in Part 4 is insufficient, for MPCP

6.2 Performance assessment: field studies

6.3 Toxic or pathogenic effects on the crop or host which is to be protected

6.4 Compatibility with products in authorized tank mixes and with other products that are applied under expected conditions of use. Recommended interval between application of MPCP and chemical pesticide, to avoid loss of efficacy

6.5 Contribution to risk reduction and integrated pest management strategies, for the targeted crop or resource

7 Toxicological Studies and Exposure Data and Information/MPCP

7.1.1 Acute oral toxicity

7.1.2 Acute percutaneous (dermal) toxicity

7.1.3 Acute inhalation toxicity to rats

7.1.4 Skin irritation

7.1.5 Eye irritation

7.1.6 Skin sensitization

7.2 Operator and bystander exposure: monitoring data

7.3 Operator and bystander exposure: reporting of hypersensitivity incidents before and after registration

7.4 Safety data sheet for each additive

7.5 Supplementary information on all data points in part 7: Effects on Human Health, if it is recommended that MPCP be tank-mixed with an adjuvant or another pest control product

7.6 Summary and evaluation of health effects

8 Rationale to waive residue studies on MPCP

9 Rationale to waive testing, based on adequacy of information provided for MPCA, to permit an assessment of the fate and behaviour of MPCA in the environment

10 Rationale to waive additional testing, based on adequacy of information provided for MPCA, to permit an assessement of the impact of the MPCP on non-target organisms

11 Summary and evaluation of environmental impact: summarize all data relevant to environmental impact and assess environmental risk by:
- addressing distribution and fate of MPCP
- identifying non-target species at risk and the extent of their exposure
- identifying precautions necessary to minimize environmental contamination and to protect non-target species.

10 References

1. Pimentel, D., *CRC Handbook of Pest Management in Agriculture*. CRC Press; 2nd edition (November 30, 1990).
2. Carson, R., *Silent Spring*. Mariner Books Classics; Anniversary edition (February 1, 2022).
3. https://extension.psu.edu/a-short-history-of-pest-management.
4. Siggins, A., et al., *Impact of Historical Legacy Pesticides on Achieving Legislative Goals in Europe. Science of the Total Environment* (2023), Volume 873, Elsevier Press.
5. https://www.efsa.europa.eu/en/events/13th-global-summit-regulatory-science-gsrs23#.
6. https://www.oecd.org/chemicalsafety/pesticides-biocides/pesticides-publications-chronological-order.htm.
7. Guidance for Registration Requirements for Microbial Pesticides, Series on Pesticides No. 18, ENV/JM/MONO(2003)5.
8. Data Requirements for Pesticide Registration in OECD Member Countries: Survey Results, Series on Pesticides No. 1, OECD/GD(94)47.
9. Report of the OECD Survey on Pesticide Regulatory Data Requirements Regarding Product Chemistry of Active Ingredients and End-Use Formulations, Series on Pesticides No. 92, ENV/JM/MONO(2017)13.
10. https://www.oecd.org/chemicalsafety/testing/good-laboratory-practiceglp.htm.
11. Koul, O., *Neem: Today and in the New Millennium*. Kluwer Academic Publishers Dordrecht (2004).
12. https://web-archive.oecd.org/fr/2014-11-19/81844-oecdguidancedocumentsforpesticideregistration.htm.
13. https://eur-lex.europa.eu/legal-content/EN/TXT/?uri=celex%3A32009R1107.
14. https://www.fao.org/pesticide-registration-toolkit/en/.
15. https://www.oecd.org/chemicalsafety/submission-transport-standard.
16. Guidance Document on the Planning and Implementation of Joint Reviews of Pesticides Series on Pesticides No. 60, ENV/JM/MONO(2011)11.
17. https://insah.cilss.int/.
18. https://insah.cilss.int/du-csp-au-coahp/.
19. https://www.ir4project.org.
20. https://www.epa.gov/pesticides.
21. https://www.ir4project.org/about-ir4/participants-in-the-process-2/ir-4-regions/.
22. https://www.ir4project.org/about-ir4/submit-a-request.
23. https://www.ir4project.org/events/2023-food-use-workshop.
24. https://ir4app.cals.ncsu.edu/Ir4FoodPub/fullSearch.
25. https://www.ir4project.org/bpos/biopesticide-organic-database.
26. https://agriculture.canada.ca/en/science/agriculture-and-agri-food-research-centres/pest-management-centre/minor-use-pesticides-pest-management-centre.
27. https://minoruses.eu.
28. https://minorusefoundation.org.
29. https://www.agriculture.gov.au/.
30. https://v6.homologa.com/en/.
31. https://minorusefoundation.org/submit-priority/.
32. https://sigepao.insah.org/index.cfm.

33. https://www.undp.org/sustainable-development-goals.
34. Böl, G., Increasing Engagement in Regulatory Science: Reflections from the Field of Risk Assessment (2020). *Science Technology & Human Values* 46(12).
35. https://www.oecd-ilibrary.org/environment/oecd-guidelines-for-the-testing-of-chemicals_72d77764-en.

Part 2

Advances in the use of bioprotection agents

Chapter 4

Harnessing Actinobacteria for crop protection: current advances and future perspectives

Louise Thatcher and Marta Gallart, CSIRO Agriculture and Food, Australia

1 Introduction

Microbial communities including bacteria, archaea, viruses and fungi are present in all environments, often associated with macroorganisms. Bacterial populations can colonise plant components and thrive in and on plant cells and express gene functions sometimes at a greater level than plant cells, constituting the so-called *secondary genome*. Plant-associated bacterial communities are important drivers for host growth, health and fitness (Berg et al., 2016; Lakshmanan et al., 2014) and function as a first line defence against phytopathogens (Berendsen et al., 2012; Ghosh and Jha, 2023). Recent technological advances in the identification and characterisation of beneficial plant-associated bacteria have opened a myriad of opportunities for their exploitation as microbial biological control agents (MBCAs) for crop protection. Comparative genome analyses in host-associated microbiomes suggest that bacterial assemblies have taxonomic and functional overlaps among plant organs, but microbiota specialise to spatial niches (Bai et al., 2015). Insights into

http://dx.doi.org/10.19103/AS.2024.0142.11

the native functional roles of MBCAs and their interactions within the microbial community across plant compartments such as rhizosphere, endosphere and the phyllosphere are essential for the selection and design of effective microbiome engineering solutions. However, such considerations have been significantly understudied, also for keystone taxa with far-reaching metabolic potential like Actinobacteria.

The *Actinomycetota* phylum – popularly referred to as Actinobacteria – constitute a phylum of Gram-positive bacteria with genetic material of generally high guanine-cytosine (GC) content, whose origins date back to ~2700 million years ago (Battistuzzi et al., 2004). Members of the phylum are metabolically and morphologically diverse, ranging from anaerobic unicellular to aerobic multicellular, filamentous and spore-forming organisms. The popularity of Actinobacteria is due to their significant contribution to human health; they are the producers of most discovered antibiotics in modern medicine (Quinn et al., 2020). Their high diversity of bioactive compounds and ability to form beneficial plant relationships also lend them to applications in crop protection (Amin et al., 2020; Barka et al., 2016; Dow et al., 2023; Silva et al., 2022). In natural habitats, Actinobacteria often have a saprophytic lifestyle of nutrient scavenging from recalcitrant compounds, providing unique functional contributions to different ecosystems. Actinobacteria are resilient and ubiquitously distributed in aquatic and terrestrial ecosystems but also aerial and extreme hostile ecosystems such as Antarctic and hot, arid environments (Castillo et al., 2002; Idris et al., 2017; Núñez-Montero et al., 2019). A significant number of Actinobacteria strains have been isolated from soil samples, consistent with their key role in soil processes (Zheng et al., 2019). Comparative studies based on amplicon sequencing (16S rRNA gene) suggest that Actinobacteria dominate the microbiome of terrestrial plants (Trivedi et al., 2020). The plant phyllosphere and rhizosphere microbial communities are frequently enriched in members of the Actinobacteria; constituting approximately 20% of the community composition. Actinobacteria are also well-known colonisers of the plant endosphere, mainly acting as root endophytes. The interest for actinobacteria-based biological products has increased in the recent years given their ability to colonise plant surfaces and inhabit the endosphere. While *Streptomyces* spp. endophytes have been popularly sought after as candidate MBCAs, non-*Streptomyces* spp. recovered from plant roots have started gaining attention given their promising attributes, including strains that belong to the genera *Microbispora*, *Micromonospora*, *Actinoplanes*, *Streptosporangium* and *Rhodococcus* (El-Tarabily, 2003a; Hong et al., 2016; Inahashi et al., 2011; Martínez-Hidalgo et al., 2015; O'Sullivan et al., 2021).

The emergence of next-generation high-throughput sequencing (NGS) has assisted the latest advances in microbiome research, contributing to fundamental efforts towards uncovering the complex multidimensional network of genetic, biochemical and metabolic relationships between plants and microbes. In the past decade, the development of metagenomic and metatranscriptomic tools led to an impressive progress in the characterisation of Actinobacteria taxonomic and functional attributes, including the role of unculturable strains (Bomar et al., 2011; Lewin et al., 2013). Recent efforts in bioprospection have uncovered new Actinobacteria bioactivities ranging from antibiotics, fungicides, insecticides and herbicides and have renewed interests in the development of Actinobacteria-based inoculants, mainly composed of microbial cells and cell-free microbial cultured extracts. In this chapter, we review the surge in Actinobacteria research and multi-*omics* resources available, compile case studies of Actinobacteria MBCAs that depict new bioactivities in crop protection and identify challenges in bioproduction and resistance durability affecting the commercialisation and deployment of Actinobacteria-derived products. Finally, we discuss emerging fields and future approaches in the research and development of Actinobacteria MBCAs.

2 Actinobacteria multi-*omics* resources

Among the diverse microbial communities, Actinobacteria have gained significant attention due to their extensive genetic and biochemical attributes. The genomes of *Actinobacteria* vary, with an average genome size of over 4 megabases (Mb), from less than 1 Mb in some *Tropheryma* species to over 12 Mb in *Actinocrispum* species. Genetic material in Actinobacteria is organised in single chromosomes, with the presence of linear mega-plasmids that range from 12 kB to 600 kB (Cornell et al., 2018; Sakaguchi, 1990). As of the latest available data in 2023, the Actinobacteria phylum represents 8% of the 240,000 publicly available genomes in 2023 (NCBI Datasets GenBank®). Approximately 50% of the 19,774 actinobacterial genome assemblies recovered in the last 10 years belong to human pathogens *Mycobacterium* spp. and microbiome gut colonisers *Bifidobacterium* spp. (Fig. 1). While most genomes have been obtained from the whole-genome sequencing (WGS) of pure isolate cultures, close to 3,500 were metagenome-assembled genomes reconstructed from complex environmental microbial communities.

Actinobacteria produce diverse secondary metabolites such as polyketides, terpenoids, alkaloids and flavonoids with a broad range of bioactivity (Dinesh et al., 2017). While secondary metabolites or specialised metabolites are not essential for cellular development, growth and reproduction of the Actinobacterial life cycle (Keller et al., 2005), contribute to their ecological adaptability and resilience and confer a significant competitive advantage over other microbes. For instance, complex metabolic regulations enable the break down and utilisation of organic compounds in forms unavailable for other microbes. This metabolic versatility makes them one of the most prolific producers of antibiotics, which

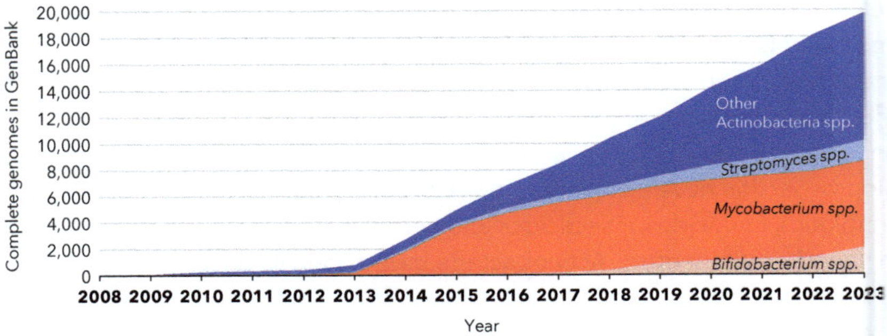

Figure 1 Number of genomes available of phylum Actinobacteria in GenBank. Assemblies included in the figure are dereplicated complete genomes obtained from WGS and whole-metagenome sequencing (WMS). Data were obtained from NCBI *datasets*, populated with GenBank sequences. Data accessed June 2023.

enable suppression of competitors for nutrient-rich sources (Viaene et al., 2016). To date, over 5,000 bioactive molecules have been isolated and characterised from Actinobacteria producer strains. In recent years, genome-guided discovery of bioactive compounds through genome mining has significantly contributed to the exponential rise of identified biosynthetic gene clusters (BGCs, Fig. 2). This includes software such as antiSMASH (Medema et al., 2011) which allows the identification, annotation and analysis of BGCs encoding the production of secondary metabolites in bacterial and fungal genomes.

The acquisition of highly accurate genomes and improvements in the prediction of BGCs revealed that Actinobacteria genomes encode for the biosynthesis of many uncharacterised metabolites (Traxler et al., 2013), suggesting a much higher metabolic potential than 5,000 compounds. Many of these are biotechnologically relevant compounds with important roles in medicine and agriculture, including antibiotics, antifungal, anticancer and antiviral compounds, plant growth regulators, enzymes, enzyme inhibitors, biosurfactants and natural dyes (Selim et al., 2021). Most of these compounds have been discovered from the *Streptomyces* spp. (Alam et al., 2022; de Lima Procópio et al., 2012). Rare Actinobacteria isolates with unique biochemical pathways have also proven to be a rich source of antibacterial agents, antiparasitics, antifungal agents, herbicides, pesticides, anticancer, immunosuppressors and enzymes (reviewed in Amin et al., 2020). For instance, *Micromonospora* spp. have a great capacity for antimicrobial production (Hifnawy et al., 2020). Members of this genus have unique morphological characteristics such as single spores attached to short substrate hyphae and carotenoid mycelial pigments, resulting in yellow, orange, red, purple, brown, or black colonies.

Comparative genomic studies indicated that factors such as multicellularity, endospore formation and genome size influence the richness and diversity of

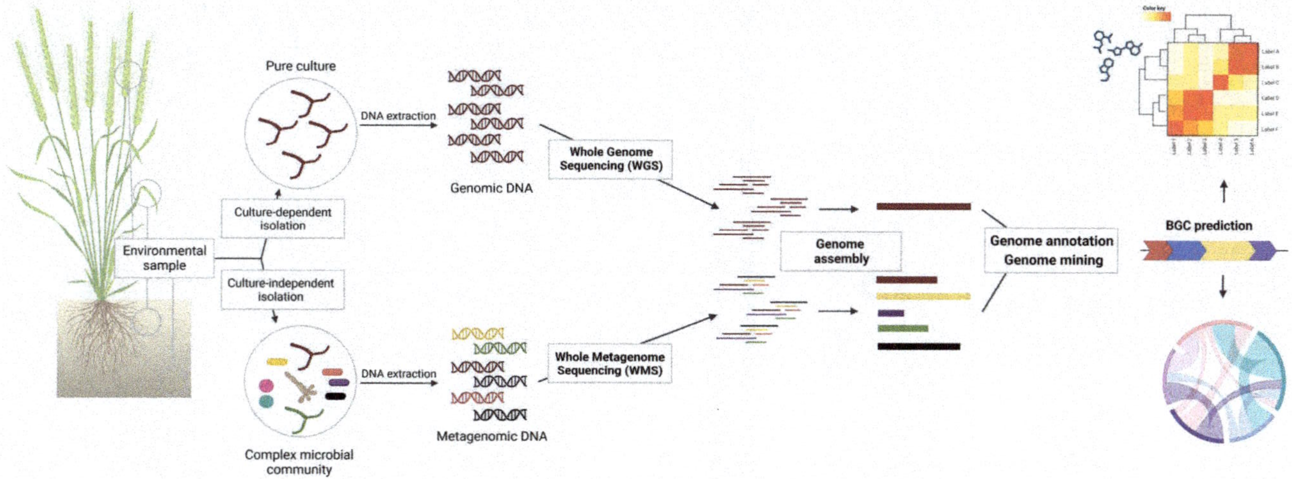

Figure 2 Genome-guided discovery of bioactive compounds. Summary of the genomics-based prediction of BGCs using genomics and metagenomics approaches. Assemblies of genomes coming from either single strains or complex microbial communities are annotated. Genome mining approaches such as antiSMASH allow the prediction of BGCs. Figure created with BioRender.com.

BGCs in bacterial genomes. Most BGCs in Actinobacteria reside in the single chromosome, although several studies have identified BGCs on predicted plasmids suggesting the horizontal transmission of genes (Novakova et al., 2013; Seshadri et al., 2022). The primary evolutionary events that determine gene content in Actinobacteria are largely unknown (Lewin et al., 2016). Gene acquisition in bacteria is either through both the origination of new genes and gene transference from other lineages through horizontal gene transfer (HGT). A recent meta-analysis of 824 Actinobacteria genomes demonstrated that, HGT is much more common than previously thought, suggesting Actinobacteria relies on BGCs acquired by HGT as a strategy to quickly adapt to new environments (Seshadri et al., 2022). For example, *Streptomyces* genomes have a very high proportion of HGT-derived BGC (41.5% of total BGCs), as well as *Micromonospora* (38%) and *Rhodococcus* (33%). The high level of BGCs acquired by plasmid-derived gene transference in Actinobacteria could be used as an opportunity to characterise candidate BGC targets that can potentially drive divergence of populations (Tran et al., 2019).

3 Actinobacteria ecosystem roles and modes of action for exploitation in crop protection

Actinobacteria have key ecological roles important to the soil food web. Actinobacteria play significant roles in biogeochemical processes, particularly in maintaining soil structure and recycling nutrients through the decomposition of organic matter (Lewin et al., 2016). They are keystone species in plant nutrition, as they provide plant-available nutrient forms essential for plant growth. Strains from the genera *Rhodococcus*, *Streptomyces*, *Nocardia* and *Corynebacterium* can produce cellulolytic enzymes that facilitate the breakdown of complex and recalcitrant plant polymers such as celluloses, hemicelluloses, lignin and chitin (reviewed in Anteneh and Franco (2019)), making them candidate biocatalysts for biofuel production. Actinobacteria play an instrumental part in soil nutrient cycles, for instance soil thermophilic Actinobacteria such as *Thermobifida*, *Saccharomonospora* and *Thermobispora* participate in composting processes (Lewin et al., 2016). Certain genera of Actinobacteria like *Streptomyces*, *Micromonospora* and *Nocardiopsis* can solubilise and mineralise phosphorous (P) by decreasing rhizosphere pH, promoting the conversion of insoluble P to plant-available soluble P forms (Boubekri et al., 2021; Sharma et al., 2013).

The coevolution of soil Actinobacteria and plants has translated into intimate mutualistic relationships that can influence plant health and growth in direct or indirect ways. Numerous studies have provided strong evidence that Actinobacteria species can antagonise phytopathogenic microorganisms directly by secreting antimicrobials or indirectly through stimulating plant immunity (reviewed in Dow et al. (2023) and Silva et al. (2022)). In the last few years, the increasing limitations on the access and efficacy of chemical pesticides has motivated the use of Actinobacteria inoculation as an alternative strategy of control management. An ample body of research have evaluated the efficacy of Actinobacteria antimicrobial-producer strains inhibiting growth of phytopathogens, both *in vitro* and *in planta* (review by Pacios-Michelena et al., 2021). For instance, soil-borne *Actinoplanes missouriensis* can parasitise fungal oospores of devastating root pathogens such as *Phytophthora* spp. and *Pythium* spp. through the production of cell wall degrading enzymes (Al-Rashdi et al., 2022; El-Tarabily, 2003b; Sutherland et al., 1984). Actinobacteria biocontrol agents also provide indirect crop protection through the stimulation of plant defence mechanisms or induced systemic resistance (ISR) that enhance plant immunity against pathogens (reviewed in Ebrahimi-Zarandi et al. (2022)). Similar to other biocontrol strains, priming of plant resistance mechanisms seems to be induced via both jasmonate and ethylene pathways and salicylic acid signalling pathway. However, only members of *Streptomyces* and *Micromonospora* have been so far reported to elicit ISR mechanisms in plants. For instance, root inoculation of tomato plants using a *Micromonospora* strain isolated from legume nodules significantly reduced aboveground effects of *Botrytis cinerea* through the induction of the jasmonate pathway (Martínez-Hidalgo et al., 2015).

4 Actinobacteria crop protection case studies

In the following sections, we present case studies that demonstrate the potential of Actinobacteria as effective agents in crop protection. The identification and isolation of plant-associated Actinobacteria in different ecosystems has motivated evaluation studies in crop protection to reduce crop losses to pests, weeds and diseases. Two main approaches have been employed: (1) screen existing Actinobacteria collections for bioactivity or (2) perform targeted isolations from healthy plants or rhizospheres or associated soils with the proviso for discovery of novel biocontrol microbes and metabolites against plant diseases (Brader et al., 2014; Collinge et al., 2022; Wang et al., 2022). Largely *Streptomyces* are identified, most likely due to their strong antibiosis activities, abundance in soils and in plant-associations, and the influence of isolation method or selection media used (Silva et al., 2022). Some studies have incorporated *Streptomyces* phages with high polyvalency in Actinobacteria isolations to selectively increase the abundance of non-*Streptomyces* strains in samples (Kamil et al., 2018). In this section, we provide case studies focused on the isolation of Actinobacteria from diverse environments and their evaluation for the control of tree, root, leaf and fruit diseases, plant pests and invasive weeds, highlighting the potential for *Streptomyces* and non-*Streptomyces* species as crop protection agents.

4.1 Control of a soil-borne root and crown pathogen of wheat

Cereal crops develop *Fusarium* crown rot disease, commonly associated with the fungal infection of *Fusarium pseudograminearum*, causing yield loss and the reduction of grain quality. So far, host genetics and chemical control have limited efficacy against the disease. In an attempt to utilise endophytic biological control agents as a disease management strategy, O'Sullivan et al. (2021) isolated 300 Actinobacteria strains from the roots of healthy, field-grown wheat plants. Of these 300 strains, 53 were selected based on *in vitro* pathogen screening and commercialisation criteria, including ease of culturing and production of spores. Using a dual-culture method consisting of measuring inhibition zones of *F. pseudograminearum* growth due to Actinobacteria diffusible antifungal metabolites, the study identified 11 Actinobacteria strains that strongly suppressed the growth of one or more *F. pseudograminearum* strains. The two strains showing greater inhibition, MH71 and MH243, were closely related *Streptomyces* species, while the other nine had genomic similarity with *Micromonospora*, *Microbispora* and *Rhodococcus* clades. Plant disease assays using wheat seeds coated with spores of strains MH71 and MH243 significantly reduced disease symptoms. While specific modes of action were not studied, fungal suppression was likely attributed to antifungal metabolites whose activity was detectable using *in vitro* assays and cell-free fermentation extracts (Belt et al., 2021; O'Sullivan et al., 2021).

4.2 Control of leaf blights on Eucalyptus plantations

Disease management in timber plantations poses many challenges, with chemical controls often unsuitable due to the sheer scale of areas to be treated. Microbial biological control agents offer a promising solution when they can colonise their plant host and persist in the environment. Himaman et al. (2016) isolated 477 Actinomycete strains from the roots and rhizosphere soil of Eucalyptus trees to screen potential MBCAs for control of leaf and shoot blight. Most isolates (439) were identified as members of the *Streptomyces* genus with less than 10% (38 isolates) identified as non-*Streptomyces* and predominantly members of the *Micromonspora* genera (17 isolates). *In vitro* disease assays based on dual-culture methods were used to screen the collection of Actinomycete isolates for antifungal activity against the causal fungal pathogens of Eucalyptus leaf and shoot blight, *Cryptosporiopsis eucalypti*, *Cylindrocladium* sp. and *Teratosphaeria destructans*. Himaman et al. (2016) identified 272, 118 and 241 isolates antagonistic to *Cryptosporiopsis eucalypti*, *Cylindrocladium* sp. and *T. destructans*, respectively, with 48 isolates showing fungal suppression towards all pathogens. Strain EUSKR2S82 showed the highest activity and was tentatively assigned as *Streptomyces ramulosus* based on 16S sequencing.

Interestingly, Eucalyptus root endophytes displayed greater antifungal activity against *Cryptosporiopsis eucalypti* than those isolated from rhizosphere soils, suggesting targeted isolation of endophytes is a useful strategy. Further characterisation of *Streptomyces* sp. strain EUSKR2S82 confirmed its ability to colonise plant roots as an endophyte and that pre-treatment of excised leaves with EUSKR2S82 spores before *Cylindrocladium* sp. inoculation resulted in significant inhibition of fungal growth compared to the untreated control.

4.3 Control of a fungal fruit disease of mango plantations

Mango dieback disease is a destructive disease caused by the fungus *Lasiodiplodia theobromae*, leading to significant yield loss and poor fruit quality. Dieback manifests as leaf drop and stem withering, as well as fruit lesions. Seeking potential biological control agents, Kamil et al. (2018) focused on environments surrounding non-diseased mango trees. The study isolated 53 Actinobacteria strains from rhizosphere soil samples collected from underneath healthy mango trees and conducted *in vitro* screens for antifungal activity against *Lasiodiplodia theobromae*. Kamil et al. (2018) identified 19 strains with antagonistic activity using two different approaches: dual-culture methods through measuring zones of *Lasiodiplodia theobromae* inhibition associated with Actinobacteria diffusible antifungal metabolites and through quantifying cell-wall degrading enzyme or chitinase enzyme activity. Twelve of the bioactive strains were *Streptomyces* and the remaining seven non-*Streptomyces* including *Actinoplanes*, *Actinomadura*, *Microbispora*, *Micromonospora*, *Nocardia*, *Rhodococcus* and *Streptosporangium* species. The *in vivo* evaluation of three of the strongest candidates in excised mango fruit recorded a significant reduction of lesion severity on fruits infected with *Lasiodiplodia theobromae*. The best species matches for the three bioactive strains were *Streptomyces samsunensis*, *Streptomyces cavourensis* and *Micromonospora tulbaghiae*. The three MBCAs were further evaluated in a greenhouse assay whereby young mango seedlings were wounded and artificially inoculated with Actinobacteria colonised agar plugs 1 week prior to agar plug inoculation with the pathogen. All MBCA applications reduced disease development, with *Streptomyces samsunensis* significantly reducing leaf defoliation and conidia counts of *Lasiodiplodia theobromae* in comparison to *S. cavourensis* and *Micromonospora tulbaghiae*. While the bioactive molecules were not identified in this study, the combination of *in vitro* screening techniques suggests the efficacy of these Actinobacteria strains *in planta* is likely attributed to the production of both antifungal metabolites and pathogen-inhibiting enzymes.

4.4 Control of nematode pests of trees and food crops

Nematodes are soil-borne pests affecting plant growth and health by feeding on roots and originating wounds that serve as entry points for fungal pathogens (Barker and Koenning, 1998). Liu et al. (2019) screened a collection of 5,000 soil-isolated Actinomycetes for nematicidal activity against the pine wood nematode *Bursaphelenchus xylophilus*, causal agent of Pine wilt disease. A mortality screening assay using juvenile and adult nematodes exposed to extracts from Actinomycetes liquid cultures identified 29 strains that caused more than 80% mortality after 24 h of exposure. Sixteen of the 29 strains were *Streptomyces* with the remaining strains distributed among the genera *Actinomadura*, *Arthrobacter*, *Kitasatospora*, *Mycobacterium*, *Nocardia*, *Nonomuraea*, *Nocardioides*, *Pilimelia* and *Streptosporangium*. Best species match for the highest mortality strain AN091965 was *Streptomyces spectabilis* able to produce a nematicidal compound identified as polyketide spectinabilin. Further, injection of spectinabilin in the form of crude extract from a *S. spectabilis* AN091965 ferment into pine seedlings inoculated with *Bursaphelenchus xylophilus* nematodes significantly reduced Pine wilt disease, increasing tree survival up to 80% compared with the 10% survival for controls. In a less demanding study by Ran et al. (2022), fermentation extracts from 10 Actinomycetes isolated from soils identified the *Micromonospora* species strain WH06 exhibited nematicidal activity against the root-knot nematode *Meloidogyne incognita*, a common nematode species worldwide with multiple hosts. Bioactivity was attributed to four isolated compounds: the alkaloid *N*-acetyltyramine, the carboxylic acid benzenepropanoic acid and the benzoic acid derivatives 3-hydroxybenzoic acid and 4-hydroxybenzoic acid.

4.5 Control of beetle pests on root and tuber crops

Colorado potato beetle is a destructive pest of potato crops. Its control relies heavily on the treatment with systemic insecticides, such as neonicotinoids, with an increased development of pest resistance to chemical control. In the study by Scott and colleagues (2023), the authors evaluated resistance risks in potato crops controlled for Colorado potato beetle using two spinosyn-based bioinsecticide products derived from the Actinobacteria *Saccharopolyspora spinosa*. Spinosyns are polyketide-derived macrolides effective against a broad range of insect pests including lepidopterans, thrips and aphids across a variety of crops (reviewed in Datta and Saxena (2001) and Sparks et al. (2021)). Spinosad is a commercial product composed of spinosyn A and spinosyn D, with increasing sales over the past 30 years since the product was registered (Sparks et al., 2021). A Canadian survey of Colorado potato beetle populations conducted between 2018 and 2022, compared resistance in the populations against neonicotinoids and spinosyns (Scott et al., 2023). The study found pest resistance against neonicotinoids ranged from 0% to 16%, while resistance against spinosyns ranged from 0% to 9%, with higher resistance development than expected in four out of the six potato-growing provinces. Resistance to spinosad was higher in organic potato fields, possibly due to a greater number of applications allowed per season compared to non-organic fields.

4.6 Weed control

Several Actinobacteria producers of specialised metabolites with bioherbicide or phytotoxic activity have been identified. The majority have been isolated from *Streptomyces* species and include for example nigericin, acivicin, gabaculin, neomycin, kanamycin, hygromycin and thiolactomycin (reviewed in Duke and Dayan (2011)). Well-characterised examples are bialaphos and phosphinothricin. Bialaphos is a natural tripeptide herbicide synthesized by *Streptomyces hygroscopicus* and *Streptomyces viridochromeogenes*. Once taken up by plants, the tripeptide is metabolized through hydrolytic cleavage, releasing L-phosphinothricin. L-phosphinothricin is an inhibitor of glutamine synthetase and is the actual phytotoxic molecule (reviewed in Takano and Dayan (2020)). Phosphinothricin, also known as glufosinate, is a successful commercial herbicide and is sold under several trade names (Duke and Dayan, 2011). It has been used for the control of invasive weeds for 30 years. Since first introduced as a non-selective herbicide in the United States and Canada in 1993–94, glufosinate has been used for weed control with limited reports of weed species evolving resistance against it (Takano and Dayan, 2020). It continues to be an important product for crop protection given the increasing development of glyphosate-resistant weeds and the use restriction of other herbicides with different mode of actions. With increasing demand for sustainable herbicide options, Prem Anand and Suthindhiran (2022) recently published methods for the rapid screening of Actinobacteria or their specialised metabolite with herbicide properties against the invasive weed *Chloris barbata*.

4.7 Plant pathogenic Actinobacteria

Some *Streptomyces* species are also known to be phytopathogens, including on relevant agricultural crops. A small number of phytotoxic specialised metabolites have been described as determinants for pathogenic *Streptomyces* phenotypes, causing disease on above- and below-ground plant structures. Common scab is a *Streptomyces*-induced disease with major economic impact in crop marketability, particularly in potato crops. More than ten phylogenetically diverse *Streptomyces* species are responsible for common scab, including *S. scabies*, *S. acidiscabies*, *S. turgidiscabies* and *S. ipomoeae*, and these can impact a broad range of hosts and plant tissues. Thaxtomin phytotoxins are key determinants of pathogenicity in *Streptomyces* strains causing scab lesions (Healy et al., 2000; Lawrence, 1990). Recent genomic studies identified that scab-producing *Streptomyces* can also produce bioactive specialised metabolites previously characterised in other plant pathogens. For example, *S. scabiei* can synthetise members of the coronafacoyl phytotoxin family, which display different biological activities in *Pseudomonas syringae* (Bender et al., 1999). Coronafacoyl phytotoxins, including coronatine, are non-essential for pathogenicity in *Pseudomonas syringae* but contribute to the virulence phenotype by enhancing the severity of disease symptoms against multiple crops (Bignell et al., 2018; Hyo-Shim et al., 2003). The coronatine toxin manipulates plant hormonal signalling pathways through jasmonic acid and salicylic acid signalling pathways to suppress plant immunity responses against *P. syringae* infection (Gimenez-Ibanez et al., 2016). Although the role of coronafacoyl phytotoxins in *S. scabiei* strains remains unclear, it is likely that their biosynthesis contributes to severity but not to development of scab lesions (Cheng et al., 2019).

5 Bioproduction

The above case studies demonstrate the broad potential of Actinobacteria in crop protection. However, challenges in the development of commercial Actinobacteria-based crop protection products exist, including the cost of bioproduction and scale-up, the downstream processing, the product formulation and the shelf-life. The slow reproduction rates of Actinobacteria, compared to other bacterial or fungal biological control agents, present a major challenge in cost-effective bioproduction. This often leads to prolonged fermentation processes. In addition, the complex Actinobacteria life cycle and metabolism requires multidisciplinary approaches and strain-specific optimisation. Further considerations related to the bioproduction system are related to the product type, which occur in form of Actinobacteria spores, their specialised metabolites or other molecules such as enzymes, and finding suitable and inexpensive feedstocks that facilitate bioproduction at high-scale volumes required for agriculture applications. It is important to note that most commercialised Actinobacteria-derived products are obtained from *Streptomyces* (reviewed in Ebrahimi-Zarandi et al. (2022) and Silva et al. (2022)), given the abundance of discovery and characterisation studies and the decades of knowledge from industrial pharmaceutical production. As non-*Streptomyces* Actinobacteria are being characterised and benchtop precision bioreactors become more affordable and scalable, the development of new Actinobacteria-based crop protection products is becoming increasingly realistic.

5.1 Upstream processing

The bioproduction of Actinobacteria-based crop protection products involves various stages of upstream and downstream processing. It is recommended that bioproduction processes initiate with the evaluation of media composition, undertaken using iterative media assessments on shake flasks (Fig. 3). Important consideration should also be prioritised on the impact of cheaper media or feedstocks on the yield, quality and performance of the final product and any negative impact on downstream processing (Balagurunathan et al., 2020; Ganeshan et al., 2021). Once media and initial physiochemical parameters are optimised in shake flasks, the optimisation of bioprocesses typically involves submerged fermentation or solid-state fermentation. If the desired end product is in form of Actinobacteria spores, then optimisation studies for spore production need to be conducted, typically involving the extension of fermentation time (van Keulen and Dyson, 2014; Silva et al., 2022). In most well-studied *Streptomyces* species, sporulation and the production of specialised metabolites are linked developmental processes, therefore findings from spore production studies are also valuable to optimise biopesticide titre and yield (van Keulen and Dyson, 2014).

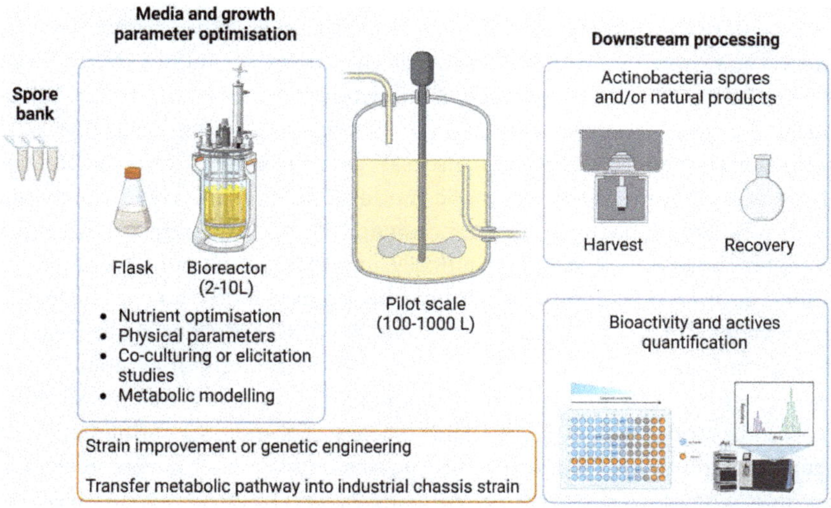

Figure 3 Workflow for upstream and downstream processing of Actinobacteria biomass production and recovery of specialised metabolites or other actives. Media and growth parameter optimisation is conducted in shaker flasks or small bioreactors before upscale to pilot production. Dependent on the process, downstream processing can be optimised at lab or pilot scale. Each stage should be validated for product recovery via biomass weight, active(s) quantification, or bioactivity screens. Improved strains or chassis strains can also be implemented in early stages of the production process. Figure created with BioRender.com.

5.2 Downstream processing

Actinobacteria typically produce biopesticides and other active compounds in relatively small quantities that fail to meet titre or yield requirements for commercial production. Various strategies can be employed to increase Actinobacteria production, such as nutrient optimisation, One strain many compounds (OSMAC) approaches, co-culturing or elicitation studies, metabolic modelling, strain improvement through mutagenesis or genetic engineering, or chassis production through transfer of the target biosynthetic pathway into an industrial chassis strain (Fig. 3) (reviewed in Adrio and Demain (2006), Dow et al. (2023), Hoskisson and Fernández-Martínez (2018) and Li et al. (2021)). An interesting example of Actinobacteria co-culturing with synergistic strains is the generation of the antibiotic keyicin, which is produced by *Micromonospora* sp. WMMB235 but only in the presence of *Rhodococcus* sp. WMMA185 (Adnani et al., 2017). The ability of keyicin to selectively inhibit Gram-positive bacteria suggests it may be effective for the control of plant-pathogenic Actinobacteria. A successful engineering approach is demonstrated with Actinobacteria *Saccharopolyspora spinosa*, source of spinosyns, which have been commercialised in various forms (reviewed in Li et al. (2021) and Silva et al. (2022)). Using omics-guided metabolic engineering and heterologous expression of spinosyns in the *Streptomyces* chassis strain *Streptomyces albus* J1074, Tan et al. (2017) increased the titre of spinosad by 1,000-fold after three rounds of rational engineering of the chassis. Another approach demonstrated by EL-Moslamy (2018) on the endophytic *Streptomyces coelicolor* strain E72 is applying statistical design methods to optimise microbial growth kinetics and production processes. Genome-scale metabolic models (GEMs) have been developed for several Actinobacteria to aid prediction of gene manipulation targets for overproduction of pharmaceutically important antibiotics (reviewed in Mohite et al. (2019)) and these should now be exploited for crop protection applications. These examples demonstrate the power of different approaches to elicit and optimise bioproduction from non-model Actinobacteria.

Harvesting and recovery of Actinobacteria products are critical steps to ensure high product purity and efficacy. Submerged fermentation of Actinobacteria tends to produce aggregates and mycelial pellets which can interfere with downstream processing (Silva et al., 2022). Processing may include biomass removal through centrifugation or filtration, harvesting of supernatant and product recovery or concentration through filtration, spray drying, or solvent extraction and formulation into wettable powder, liquid, or granular state (Jambhulkar et al., 2016). The final step of a bioproduction process is the product formulation to enhance delivery, viability, stability and shelf-life. As reviewed by Teixidó et al. (2022), the main objectives of formulation are to ensure product viability and efficacy at its target site and protection from environmental factors; to stabilise the product during shelf life, distribution and storage; and to assist in the handling and use of the product including compatibility with agricultural machinery. Product stability has implications in the durability of the Actinobacteria MBCA product and are discussed in the following section.

Actinobacteria-based formulations can be prepared in both solid and liquid forms, each with its strengths and shortcomings. While dry solid formulations facilitate longer storage times at higher temperatures with reduced risk of contamination, they involve high processing costs and might not be amenable to all MBCAs (reviewed in Teixidó et al. (2022)). Dehydration processes include freeze drying, spray drying, fluidized bed drying and fluidized bed-spray drying. Liquid formulations are generally cheaper but result in larger volumes of product with reduced protection from contamination and stability, requiring storage and distribution under refrigerated temperatures. Common options for liquid formulation forms are oil dispersion, soluble (liquid) concentrate, or suspension concentrate. Nanoparticles that encapsulate cells also offer potential success (reviewed in Compant et al. (2019) and Silva et al. (2022)). The shelf-life of a biocontrol product should be at least 6 months and preferably up to 2 years (Teixidó et al., 2020). For example, the Actinobacteria-based product Actinovate (*Streptomyces lydicus* strain WYEC 108) is formulated as a cell-based dry powder and has been registered for an extended shelf-life of 2 years.

5.3 Durability of Actinobacteria microbial biological control agents for crop protection

The rise in countries whose regulations mandate the reduction of chemical use has led to the investment into biological crop protection solutions and subsequent increase in biopesticide deployment (Helepciuc and Todor, 2021; Jacquet et al., 2022; Newitt et al., 2019; Sapbamrer et al., 2023; Silva et al., 2022; Syed Ab Rahman et al., 2018). However, the predicted increase in biopesticide deployment also increases the uncertainty and risk regarding the durability and the evolution of pathogen resistance to biopesticides. In natural habitats, Actinobacteria together with other members of the plant microbiome typically act through multiple modes of action and therefore are generally thought to be more durable as the pathogen or pest must simultaneously evolve multiple mechanisms to achieve resistance. Despite the vast amount of research into Actinobacteria crop protection ability and production of specialised metabolites, research into the durability of Actinobacteria for crop protection is largely deficient. There is an opportunity to learn from the successes and failures of other MBCAs and of chemical crop protection technologies, identify what research is missing and use this information to better deploy Actinobacteria-based solutions in ways that facilitates prolonging their durability. There are limited studies into durability of any fungal or bacterial biological control agents. To address this, we firstly need to consider what defines durability of a MBCA and what are the factors impacting durability. The durability of a crop protection control method can be defined as the persistence of its efficacy in space and time and the potential of the control to exert selective pressure against a target (Bardin et al., 2015). There are several factors that need to be considered in this aspect for Actinobacteria MBCAs and their specialised metabolites (Fig. 4). Using pathogen control as a representative case study, these factors are:

1. Stability and persistence
2. Pathogen lifestyle
3. Diversity/sensitivity of pathogen populations

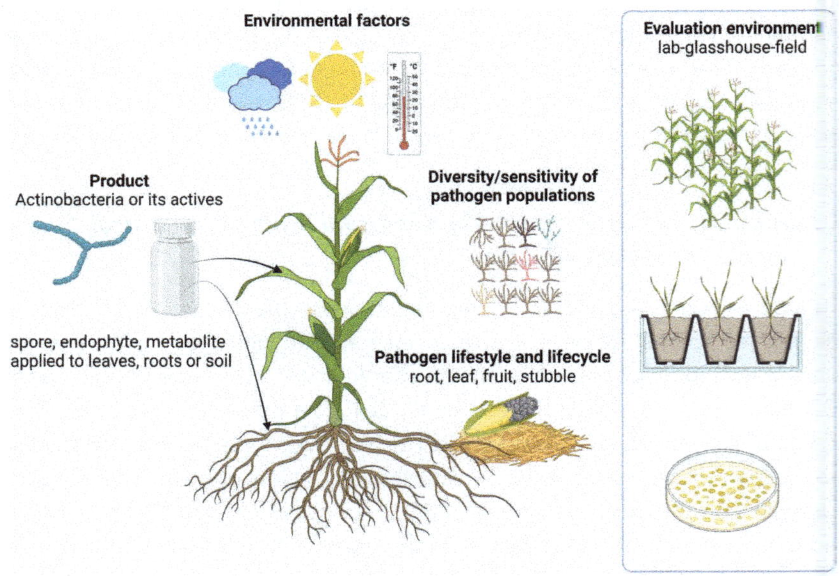

Figure 4 Factors influencing the durability of Actinobacteria MBCAs and their specialised metabolites. Product durability is influenced by its stability and persistence in each environment (soil, roots, within a plant host) and protection from natural elements. The risk of resistance evolution and loss of durability varies between pathogens, their lifestyle and lifecycle and is influenced by the diversity or sensitivity within the pathogen population. Due to the influence of the above factors, experimental results on Actinobacteria MBCA and specialised metabolite durability need to be interpreted in the context of their testing environment, from laboratory and glasshouse evaluation through to multi-environment field scale testing. Figure created with BioRender.com.

5.3.1 Stability and persistence

When considering stability, the physical form of the delivered product and the delivery formulation can have substantial influence. For example, will it be a liquid or solid formulation and whether the Actinobacteria microbe itself is the product and or its bioactive extracts. The robust, spore forming ability of Actinobacteria contributes to its stability in storage and delivery into the field (Barka et al., 2016; O'Sullivan et al., 2021). Suitable formulations are lacking for most MBCAs due to the diversity of these non-model microorganisms and the time and cost to individualise stable formulations for commercial use. Target delivery site (soil, seed, plant surface) and the timing of application are also important considerations for Actinobacteria MBCA stability and persistence.

The efficacy and stability of MBCAs for crop protection is often compared against their chemical counterparts where in many cases their efficacy under controlled conditions such as plant growth chambers or greenhouses does not equate to efficacy under field conditions. This can be due to poor formulation as well as numerous environmental factors affecting functional traits directly impacting the capacity of MBCA to produce specialised metabolites or enzymes (Fig. 4). Actinobacteria with endophytic abilities have the advantage of increased protection from UV radiation and of temperature, moisture, pH and nutrient variations as compared with rhizosphere associated MBCAs, and this trait can be used to select Actinobacteria more likely to succeed as MBCAs when delivered into the field (Silva et al., 2022). Strategically, Actinobacteria specialised metabolites with poor UV stability can be delivered as seed or soil treatments to avoid prolonged exposure to UV radiation. For host-beneficial Actinobacteria endophytes, it can be extrapolated that the host plant will maintain selective pressure to retain the endophyte and thus the ability of the Actinobacteria to persist in that environment. Compant et al. (2019) and Lugtenberg et al. (2016) suggest crops can lose beneficial endophytes when they do not benefit from their presence, or when chemical controls replace their activity or negatively impact their survival. Further, recurrent applications of the MBCA might be required to achieve sufficient microbial density that enables effective colonisation rates.

The slow-growing nature of Actinobacteria and competition amongst other rhizosphere or endosphere microbes is an important factor influencing their niche-colonisation and in-field efficacy and persistence. For example, there are multiple studies showing effective plant colonization under laboratory or controlled environments but fail to replicate this phenomenon under field conditions. Their slower reproduction rate compared to other microbes and the impact of host preference can play a major role in the speed and success of Actinobacteria colonisation (Merzaeva and Shirokikh, 2006; Silva et al., 2022) and are some of the main challenges faced by Actinobacteria-based products.

For Actinobacteria-derived extracts or specialised metabolites, field studies indicate their efficacy and persistence following delivery is typically limited to a few days. Even though their residues may be detected, antibiotics typically have a short duration of activity on plants growing under field conditions. For example, the *Streptomyces*-derived antibiotics streptomycin and oxytetracycline have been registered in several countries since the 1950s for the control of bacterial diseases such as *Erwinia amylovora* fire blight on pear and apple trees (reviewed in Ghorai et al. (2021) and Miller et al. (2022)). While streptomycin and oxytetracycline delivered to citrus seedlings via a root drench or stem injection were detectable in tissues for up to 35 days post treatment (Al-Rimawi et al., 2019), oxytetracycline sprayed on peach leaves rapidly degraded when exposed to natural sunlight, with a 92% loss in detection within four days (Christiano et al., 2010). The efficacy of *in planta* applications of streptomycin and oxytetracycline against *E. amylovora* was restricted to four days after treatment, despite the fact plants were growing under a screenhouse protected from rain and UV radiation (Stockwell et al., 2008).

While the efficacy of specialised metabolites significantly declines within days after application, it is still plausible that at sub-lethal doses they could exert selective pressure on the target pathogen and affect the durability of the control. Increased persistence of an MBCA in a farming system could also have negative consequences in the form of off-target effects on other beneficials within the farming system, residues on produce and longer windows of selective pressure on the target pathogen or pest. Winding et al. (2004) reviewed non-target effects of bacterial BCAs, including Actinobacteria, that are used to suppress root pathogenic fungi. They found non-target effects of MBCAs, but these were generally small and transient and generally limited to one crop growth season. They also highlighted the need to evaluate non-target effects of MBCA inoculums *in situ* versus direct introduction of an enzyme or antimicrobial compound. The former would result in a continuum of production while the later would have the advantage of limited long-term effects due to degradation and turnover. While these studies are currently absent, it will be important to study potential non-target effects of Actinobacteria MBCAs to better understand their ecological impacts.

Resistance risk is often a function of functional or metabolic diversity of the control agent and while many biologicals are predicted to have high metabolic diversity, very few have been experimentally evaluated. We know from chemical controls that repeated use of single-site pesticides with the same mode of action can lead to the evolution of resistance (Corkley et al., 2022). Given a product delivered as the Actinobacteria microbe may suppress pathogens or pests through several mechanisms or modes of action, potentially expressed successively, simultaneously, or synergistically, it could be hypothesized that MBCAs offer more effective and longer lasting control.

This hypothesis is, however, reliant on several factors. Firstly, the Actinobacteria needs to effectively colonize its host plant to an effective abundance and cannot be outcompeted by other plant-associated organisms or inhibited by environmental factors. Secondly, its mechanisms of crop protection such as antibiotic or volatile organic compound production or pathogen-inhibiting enzymes like cellulases or chitinases need to be expressed to a functional level. To overcome roadblocks at any of these steps, new Actinobacteria research focuses on the development of cell-free extract or spent ferment after the microbe is removed. In this respect, the product is defined by the activity or concentration of the active(s) which may be a fermentation cocktail of bioactives or a refined specialised metabolite (reviewed in Alam et al. (2022), Dow et al. (2023), Ebrahimi-Zarandi et al. (2022) and Ghorai et al. (2021)). As with a microbial delivered product, a fermentation cocktail would be predicted to have a reduced resistance risk compared to a refined specialised metabolite. It could also be hypothesised that since specialised metabolites are already produced within the ecosystem, pathogens and pests would have already encountered them, and the propensity of resistance already exists. The caveat being pathogens and pests would not have been exposed to the levels or extended duration of specialised metabolites in a crop protection agent in a farming system compared to at a natural ecosystem context.

To decipher hypothesises like the above, several studies have been undertaken to compare bacterial communities sprayed with or without antibiotics. One such case is the use of streptomycin. As described above, streptomycin is a well-studied Actinobacteria-derived specialised metabolite that has been in use for decades in crop protection and in the United States, along with oxytetracycline, accounts for most of the antibiotic use in plant agriculture. In one study, the application of streptomycin to an apple orchard did not influence the abundance nor diversity of major bacteria taxa in the soils from treated apple rows compared to water treated controls (Walsh et al., 2013). In a separate study, Yashiro and McManus (2012) compared phyllosphere bacterial communities of commercial apple orchards where streptomycin had been used for several years against communities from orchards where streptomycin had not been used. Interestingly, a higher percentage of culturable bacteria in the non-sprayed orchards were resistant to streptomycin (65%) compared to the sprayed orchards (50%). They concluded the use of streptomycin over many years did not lead to elevated levels of streptomycin resistance on the apple leaves. Contrasted against this finding was the detection of streptomycin resistant plant pathogens within 5-10 years of commercialisation of products containing the antibiotic (reviewed in Stockwell et al. (2008)).

5.3.2 Pathogen lifestyle

To evaluate and predict the resistance risks and durability of Actinobacteria and their specialised metabolites as new crop protection solutions, we need to determine if variability in their efficacy exists over diverse pathogens, if they exert differential selective pressure on their target and how might resistance develop. In this aspect, consideration to pathogen lifestyle and lifecycle are key factors. For pathogens, this includes necrotrophic, biotrophic or hemibiotrophic lifestyle which impacts the localisation, persistence, latency or non-disease expressing stage, and alternate hosts. Soil and stubble persisting pathogens or largely foliar pathogens also impacts the potential duration of exposure of the pathogen to the MBCA. For example, understanding when the pathogen is predicted to be on plant surfaces and exposed to specialised metabolite sprays could reduce the number of sprays for disease control and decrease the potential for resistance development. Lastly, pathogens with sexual life stages or a propensity for high mutation rates facilitates increased genetic diversification and the potential to develop resistance against MBCAs. Similarly, considerations need to be evaluated for pests and weeds.

Straightforward *in vitro* pathogen suppression assays can be used to quickly evaluate potential variation in efficacy of Actinobacteria MBCAs towards pathogen lifestyle by screening against pathogen diversity panels. For example, from a collection of 20 endophytic Actinobacteria representing *Streptomyces*, *Micromonospora*, *Microbispora* and *Rhodococcus* species isolated from wheat roots (O'Sullivan et al., 2021), screening for antifungal towards common root or foliar infecting fungal pathogens of canola was conducted. The foliar pathogens included the stubble borne and soil-borne pathogens *Sclerotinia sclerotiorum*, *Leptosphaeria maculans*, and the root-infecting pathogen *Rhizoctonia solani*. *In vitro* dual competition assays indicated some Actinobacteria isolates were equally effective against all pathogens, while others were most effective in suppressing single pathogens (Roper et al., 2018).

5.3.3 Diversity/sensitivity of pathogen populations

While the risk of evolution of resistance varies between pathogens, changes at population level are driven by mutations in non-adapted pathogen genotypes, changes in frequencies of pre-existing adapted genotypes, and introductions from external sources (Rimbaud et al., 2018). From the limited studies so far conducted for MBCAs, research has identified natural diversity in fungal pathogen populations existing towards sensitivity to some biologicals (reviewed in Bardin et al. (2015)). Research approaches include assessing natural variation in sensitivity towards a MBCA or specialised metabolite amongst isolates in pathogen collections, amongst natural populations from the field, or combining these with selective pressure experiments.

A few studies have assessed natural variation in sensitivity towards Actinobacteria MBCAs amongst isolates in pathogen collections. For example, via an *in vitro* dual-culture method O'Sullivan et al. (2021) screened 53 Actinobacteria isolates from a culture collection against three strains of the wheat fungal pathogen *F. graminearum*. Some Actinobacteria isolates conferred similar levels of suppression against all three *Fusarium* strains, while others were specific to one or two of the pathogen strains indicating natural variation in pathogen sensitivity to Actinobacteria with potentially diverse modes of action. In a similar but focused study, Boukaew et al. (2017) evaluated three *Streptomyces* strains against 41 strains of *Botrytis cinerea* differing in their date of isolation, geographic, plant host or origin. Sensitivity of *Botrytis cinerea* strains was assessed via dual-culture assays, *Streptomyces* sp. cell-free extracts and/or *Streptomyces* sp. volatile organic compounds, revealing some strains of *Botrytis cinerea* are less sensitive to the inhibitory effect of antimicrobials produced by *Streptomyces* MBCAs. Creating an MBCA-pathogen screening matrix, Otto-Hanson et al. (2013) screened 19 *Streptomyces* MBCAs against 15 plant pathogenic *Streptomyces* (*Streptomyces* scabies), and identified significant variation among the MBCAs to inhibit pathogen isolates and among pathogens in their susceptibility to inhibition. In another study, Sharma et al. (2022) identified natural variation in pathogen sensitivity to a *Streptomyces* antibiotic by exposing 11 strains of the bacterial blight pathogen *Xanthomonas axonopodis* pv. *punicae* to increasing doses of streptomycin. Studies such as these highlight that using even relatively small pathogen populations, variation in sensitivity to Actinobacteria MBCAs and their specialised metabolites exists. How this sensitivity would affect the field durability of commercialised Actinobacteria MBCA products remains largely unexplored.

Based on findings from evaluations of non-Actinobacteria commercialised products, it is plausible that the field efficacy and durability of Actinobacteria MBCAs and their specialised metabolites vary. For example, a commercial strain of the mycoparasitic fungus *Paraphaeosphaeria minitans* (formerly *Coniothyrium minitans*), was evaluated *in vitro* for efficacy against 75 *Sclerotinia sclerotiorum* strains isolated from diverse crop hosts, geographies and isolation times (Nicot et al., 2019). A wide range in *S. sclerotiorum* sensitivity was observed prompting the authors to hypothesise that increasing MBCA use and selection pressure could lead to a gradual increase in the frequency of less susceptible pathogen isolates.

An estimation of durability and potential risk can also be achieved through experimental evolution studies, with the production of successive generations of a pathogen under selection pressure and exposure to the MCBA or specialised metabolite. Due to the absence of studies using Actinobacteria, here we use the *Pseudomonas*-derived antibiotic pyrrolnitrin as a case study where similar methodology can be applied to future Actinobacteria-based research. In an experimental evolution study, Ajouz et al. (2010) observed *in vitro* adaptation of resistance to pyrrolnitrin in several isolates of the fungal pathogen *Botrytis cinerea*. The experiment started with ten successive generations of *Botrytis cinerea* produced in the presence of a sublethal dose of the antibiotic. For these ten generations, no major change in sensitivity to pyrrolnitrin was observed. In contrast, increasing the dose of pyrrolnitrin by over 10-fold in additional generations resulted in the rapid development of *Botrytis cinerea* variants with a high level of resistance to the antibiotic. However, a high fitness cost associated with pyrrolnitrin-resistance was observed with the *Botrytis cinerea* variants displaying reduced mycelial growth and sporulation and reduced *in planta* aggressiveness compared to the pyrrolnitrin-sensitive parental isolates.

5.4 Future research

The limited evidence on the durability of Actinobacteria MBCA highlights the need for additional research to assess the resistance risk levels of these new crop protection measures and to develop predictive tools and frameworks for proactive resistance management. This includes studies on optimal time and number of applications, optimal dose rates and the effects of combinatorial product applications and modes of action. It also emphasises the need for regular monitoring of pathogen, pest and weed populations for resistance as well as non-target effects, and also the assessment at scale in multiple environments and growing seasons. Screening Actinobacteria MBCAs against pathogen collections would also work to expedite the identification of underlying genetics and mechanisms potentially involved in resistance to MBCAs such as specialised metabolite excretion or metabolization (reviewed in Bardin et al. (2015)). Research should also focus on identifying types of MBCAs or specialised metabolites with lower risk of efficacy loss; i.e. modes of action that do not favour the selection of resistant isolates in natural populations. Further, interdisciplinary collaborations involving microbiologists, pathologists, ecologists, agricultural scientists, formulation chemists and growers are needed to comprehensively study the durability of Actinobacteria MBCAs and their specialised metabolites. Finally, as quoted from Corkley et al. (2022), 'Whichever crop protection measures are developed in the future, whether chemical, genetic, or biological, the same core principles of resistance management will apply: assume any control measure will exert a selection pressure for resistance, avoid over-reliance on any single measure, and use within an IPM [integrated pest management] strategy'.

6 Emerging fields and future approaches in research and development of Actinobacteria biological control agents

The need for new, effective and sustainable agricultural solutions to combat crop pests, pathogens and weeds and to provide relief to food security is growing more than ever. The spiking emergence of newly discovered Actinobacteria bioactivities ranging from antibiotics, fungicides, insecticides and herbicides have promoted a boom in Actinobacteria-based research. While there is a growing market for microbial-based solutions to improve crop productivity, challenges with stability, efficacy, cost of production and registration of these products remain (Trivedi et al., 2021). As advancements in genetic manipulation and genomic selection have accelerated our breeding of crops for disease or pest resistance, these approaches have parallel benefit to the *breeding* or domestication of non-model MBCAs such as Actinobacteria. These methodological advances could be used to enhance the growth rate and plant colonisation efficiency of Actinobacteria as well as the stability and potency of their bioactives. Although delivering a genetically engineered Actinobacteria MBCA directly into agricultural settings remains a distant goal, delivery into industrial settings for increased production and recovery of enzymes or specialised metabolites to be subsequently formulated into cell-free products is not. New advances in formulation and delivery of MBCAs and their specialised metabolites may offer the competitive edge to extend their stability and efficacy in field and facilitate their uptake into diverse agricultural settings.

There are further opportunities to build efficacy and durability into Actinobacteria biocontrol products through the selection of beneficial traits for the design of synthetic microbial communities or the development of microbiome interventions (Marín et al., 2021; Mendes et al., 2011). Knowledge on complex plant–Actinobacteria interactions that integrate multi-omics approaches, high-throughput culturing and computational and synthetic engineering advances are expanding our understanding on community diversity, resource competition (Wei et al., 2015) and interspecies and interkingdom interactions for the benefit of agriculture. Expanding our understanding of Actinobacteria-plant responses and how this impacts broader native communities and soil health will enhance the development of stable microbial products, microbiomes engineered towards durable traits, and the plant or microbial-derived signals that facilitate this. New *in situ* ways to target beneficial functional traits are becoming increasingly available, for example the use of metabolite-sensing biosensors incorporated plant hosts or MBCAs (reviewed in Dow et al. (2023)). Imaging mass spectrometry (IMS) approaches combined with high-resolution mass analysers are also facilitating the identification and quantification of small molecules and their spatial distribution *in situ* (Bourceau et al., 2023; Ho et al., 2017). This ranges from single cells to tissue sections and therefore offers great potential in the study of Actinobacteria specialised metabolite production and their role in microbe–microbe and plant–microbe interactions. For example, matrix-assisted laser desorption ionization–time of flight (MALDI-TOF) IMS identified metabolites produced by *Streptomyces* during interactions with other microbes, including biocontrols such as *Bacillus* species (reviewed in Li and Li (2022) and Sgobba et al. (2021)). In another application, Bourceau et al. applied MALDI mass spectrometry imaging to assign the spatial distribution of metabolites to microbiome at single cell resolution in animal tissues. This technique has great potential to assign modes of action in the tripartite MBCA–pathogen–crop interaction.

Finally, genomics and multi-omics approaches including genome-wide association studies and metagenome-wide association studies, combined with network analysis and genome-mining methods to predict metabolic pathways, have set up a strong foundation to model and engineer plant–microbiome interactions. Such technological advances have enabled the identification and characterization of genes and metabolites that shape plant–microbial interactions, thus increasing our understanding of how microbes like Actinobacteria adapt to plant ecological niches (e.g. Chevrette et al. (2022)). Characterisation of silent and cryptic BGCs, including unculturable Actinobacteria, through advanced microbial chassis engineering systems or cell-free expression systems may also facilitate the identification of Actinobacteria specialised metabolites with new modes of action to maintain plant health.

7 Conclusion

As highlighted in this review, Actinobacteria and their natural products could offer sustainable alternatives to synthetic products for crop protection. However, increased research funding, collaboration between academia and industry and policy support from governments are needed to overcome barriers preventing their commercial adoption. Research should focus on reducing the production costs of Actinobacteria-based crop protection solutions, improving product formulation, and selecting strains to increase stability and efficacy.

8 Where to look for further information

Good introductions to the subject of Actinobacteria and biological control are Bardin et al. (2015); Barka et al. (2016); Dow et al.(2023); Ganeshan et al. (2021);

Medema et al. (2023) and Silva et al. (2022). See in the References section for full details.

Key online materials for Actinobacteria research are The Society for Actinomycetes Japan (https://www.actino.jp/index-e.html) and ActinoBase (https://actinobase.org/). Actinobase, maintained by the Actinobacteria community, provides tools and protocols for researchers working on *Streptomyces* and other filamentous Actinobacteria.

There are a number of regular conferences on Actinobacteria related research with a major meeting held every two years by the International Society for the Biology of Actinomycetes (ISBA).

Centres of expertise include:

- The CSIRO Plant pathogens, pests and weeds research groups (https://www.csiro.au/en/research/plants/pathogens-pests-weeds).
- The John Innes Centre Advancing Plant Health (APH) programme (https://www.jic.ac.uk/research-impact/our-strategic-research-programmes/advancing-plant-health/).
- The INRAe biological control group (https://www6.paca.inrae.fr/pathologie_vegetale_eng/Our-research/MISTRAL/Biological-control).

9 References

Adnani, N., Chevrette, M. G., Adibhatla, S. N., Zhang, F., Yu, Q., Braun, D. R., Nelson, J., Simpkins, S. W., McDonald, B. R., Myers, C. L., Piotrowski, J. S., Thompson, C. J., Currie, C. R., Li, L., Rajski, S. R. and Bugni, T. S. (2017). Coculture of marine invertebrate-associated bacteria and interdisciplinary technologies enable biosynthesis and discovery of a new antibiotic, keyicin, *ACS Chemical Biology* 12(12), 3093–3102, doi: 10.1021/acschembio.7b00688.

Adrio, J. L. and Demain, A. L. (2006). Genetic improvement of processes yielding microbial products, *FEMS Microbiology Reviews* 30(2), 187–214, doi: 10.1111/j.1574-6976.2005.00009.x.

Ajouz, S., Nicot, P. C. and Bardin, M. (2010). Adaptation to pyrrolnitrin in *Botrytis cinerea* and cost of resistance, *Plant Pathology* 59(3), 556–566, doi: 10.1111/j.1365-3059.2009.02230.x.

Alam, K., Mazumder, A., Sikdar, S., Zhao, Y. M., Hao, J., Song, C., Wang, Y., Sarkar, R., Islam, S., Zhang, Y. and Li, A. (2022). Streptomyces: the biofactory of secondary metabolites, *Frontiers in Microbiology* 13, 968053, doi: 10.3389/fmicb.2022.968053.

Al-Rashdi, A., Al-Hinai, F. S., Al-Harrasi, M. M. A., Al-Sabahi, J. N., Al-Badi, R. S., Al-Mahmooli, I. H., Al-Sadi, A. M. and Velazhahan, R. (2022). The potential of endophytic bacteria from *Prosopis cineraria* for the control of *Pythium aphanidermatum*-induced damping-off in cucumber under saline water irrigation, *Journal of Plant Pathology* 105(1), 39–56, doi: 10.1007/s42161-022-01237-5.

Al-Rimawi, F., Hijaz, F., Nehela, Y., Batuman, O. and Killiny, N. (2019). Uptake, translocation, and stability of oxytetracycline and streptomycin in citrus plants, *Antibiotics* 8(4), 196, doi: 10.3390/antibiotics8040196.

Amin, D. H., Abdallah, N. A., Abolmaaty, A., Tolba, S. and Wellington, E. M. H. (2020). Microbiological and molecular insights on rare Actinobacteria harboring bioactive prospective, *Bulletin of the National Research Centre* 44(1), 5, doi: 10.1186/s42269-019-0266-8.

Anteneh, Y. S. and Franco, C. M. M. (2019). Whole cell Actinobacteria as biocatalysts, *Frontiers in Microbiology* 10, 77, doi: 10.3389/fmicb.2019.00077.

Bai, Y., Müller, D. B., Srinivas, G., Garrido-Oter, R., Potthoff, E., Rott, M., Dombrowski, N., Münch, P. C., Spaepen, S., Remus-Emsermann, M., Hüttel, B., McHardy, A. C., Vorholt, J. A. and Schulze-Lefert, P. (2015). Functional overlap of the *Arabidopsis* leaf and root microbiota, *Nature* 528(7582), 364–369, doi: 10.1038/nature16192.

Balagurunathan, R., Radhakrishnan, M., Shanmugasundaram, T., Gopikrishnan, V. and Jerrine, J. (2020). Production of bioproducts from Actinobacteria. In: *Protocols in Actinobacterial Research* (1st edn.), Springer, New York, pp. 113–128.

Bardin, M., Ajouz, S., Comby, M., Lopez-Ferber, M., Graillot, B., Siegwart, M. and Nicot, P. C. (2015). Is the efficacy of biological control against plant diseases likely to be more durable than that of chemical pesticides?, *Frontiers in Plant Science* 6, 566, doi: 10.3389/fpls.2015.00566.

Barka, E. A., Vatsa, P., Sanchez, L., Gaveau-Vaillant, N., Jacquard, C., Meier-Kolthoff, J. P., Klenk, H. P., Clément, C., Ouhdouch, Y. and van Wezel, G. P. (2016). Taxonomy, physiology, and natural products of Actinobacteria, *Microbiology and Molecular Biology Reviews* 80(1), 1–43, doi: 10.1128/MMBR.00019-15.

Barker, K. R. and Koenning, S. R. (1998). Developing sustainable systems for nematode management, *Annual Review of Phytopathology* 36(1), 165–205, doi: 10.1146/annurev.phyto.36.1.165.

Battistuzzi, F. U., Feijao, A. and Hedges, S. B. (2004). A genomic timescale of prokaryote evolution: insights into the origin of methanogenesis, phototrophy, and the colonization of land, *BMC Evolutionary Biology* 4(1), 44, doi: 10.1186/1471-2148-4-44.

Belt, K., Foley, R. C., O'Sullivan, C. A., Roper, M. M., Singh, K. B. and Thatcher, L. F. (2021). A plant stress-responsive bioreporter coupled with transcriptomic analysis allows rapid screening for biocontrols of necrotrophic fungal pathogens, *Frontiers in Molecular Biosciences* 8, 708530, doi: 10.3389/fmolb.2021.708530.

Bender, C. L., Alarcón-Chaidez, F. and Gross, D. C. (1999). *Pseudomonas syringae* phytotoxins: mode of action, regulation, and biosynthesis by peptide and polyketide synthetases, *Microbiology and Molecular Biology Reviews* 63(2), 266–292, doi: 10.1128/MMBR.63.2.266-292.1999.

Berendsen, R. L., Pieterse, C. M. J. and Bakker, P. A. H. M. (2012). The rhizosphere microbiome and plant health, *Trends in Plant Science* 17(8), 478–486, doi: 10.1016/j.tplants.2012.04.001.

Berg, G., Rybakova, D., Grube, M. and Köberl, M. (2016). The plant microbiome explored: implications for experimental botany, *Journal of Experimental Botany* 67(4), 995–1002, doi: 10.1093/jxb/erv466.

Bignell, D. R. D., Cheng, Z. and Bown, L. (2018). The coronafacoyl phytotoxins: structure, biosynthesis, regulation and biological activities, *International Journal of General and Molecular Microbiology* 111(5), 649–666, doi: 10.1007/s10482-017-1009-1.

Bomar, L., Maltz, M., Colston, S. and Graf, J. (2011). Directed culturing of microorganisms using metatranscriptomics, *mBio* 2(2), e00012–e00011, doi: 10.1128/mBio.00012-11.

Boubekri, K., Soumare, A., Mardad, I., Lyamlouli, K., Hafidi, M., Ouhdouch, Y. and Kouisni, L. (2021). The screening of potassium- and phosphate-solubilizing Actinobacteria and the assessment of their ability to promote wheat growth parameters, *Microorganisms* 9(3), 470, doi: 10.3390/microorganisms9030470.

Boukaew, S., Prasertsan, P., Troulet, C. and Bardin, M. (2017). Biological control of tomato gray mold caused by *Botrytis cinerea* by using *Streptomyces* spp., *BioControl* 62(6), 793–803, doi: 10.1007/s10526-017-9825-9.

Bourceau, P., Geier, B., Suerdieck, V., Bien, T., Soltwisch, J., Dreisewerd, K. and Liebeke, M. (2023). Visualization of metabolites and microbes at high spatial resolution using MALDI mass spectrometry imaging and in situ fluorescence labeling, *Nature Protocols* 18(10), 3050–3079, doi: 10.1038/s41596-023-00864-1.

Brader, G., Compant, S., Mitter, B., Trognitz, F. and Sessitsch, A. (2014). Metabolic potential of endophytic bacteria, *Current Opinion in Biotechnology* 27(100), 30–37, doi: 10.1016/j.copbio.2013.09.012.

Castillo, U. F., Strobel, G. A., Ford, E. J., Hess, W. M., Porter, H., Jensen, J. B., Albert, H., Robison, R., Condron, M. A. M., Teplow, D. B., Stevens, D. and Yaver, D. (2002). Munumbicins, wide-spectrum antibiotics produced by *Streptomyces* NRRL 30562, endophytic on *Kennedia nigriscans*, *Microbiology* 148(9), 2675–2685, doi: 10.1099/00221287-148-9-2675.

Cheng, Z., Bown, L., Piercey, B. and Bignell, D. R. D. (2019). Positive and negative regulation of the virulence-associated coronafacoyl phytotoxin in the potato common scab pathogen *Streptomyces scabies*, *Molecular Plant-Microbe Interactions* 32(10), 1348–1359.

Chevrette, M. G., Thomas, C. S., Hurley, A., Rosario-Meléndez, N., Sankaran, K., Tu, Y., Hall, A., Magesh, S. and Handelsman, J. (2022). Microbiome composition modulates secondary metabolism in a multispecies bacterial community, *Proceedings of the National Academy of Sciences of the United States of America* 119(42), e2212930119, doi: 10.1073/pnas.2212930119.

Christiano, R. S. C., Reilly, C. C., Miller, W. P. and Scherm, H. (2010). Oxytetracycline dynamics on peach leaves in relation to temperature, sunlight, and simulated rain, *Plant Disease* 94(10), 1213–1218, doi: 10.1094/PDIS-04-10-0282.

Collinge, D. B., Jensen, D. F., Rabiey, M., Sarrocco, S., Shaw, M. W. and Shaw, R. H. (2022). Biological control of plant diseases – what has been achieved and what is the direction?, *Plant Pathology* 71(5), 1024–1047, doi: 10.1111/ppa.13555.

Compant, S., Samad, A., Faist, H. and Sessitsch, A. (2019). A review on the plant microbiome: ecology, functions, and emerging trends in microbial application, *Journal of Advanced Research* 19, 29–37, doi: 10.1016/j.jare.2019.03.004.

Corkley, I., Fraaije, B. and Hawkins, N. (2022). Fungicide resistance management: maximizing the effective life of plant protection products, *Plant Pathology* 71(1), 150–169, doi: 10.1111/ppa.13467.

Cornell, C. R., Marasini, D. and Fakhr, M. K. (2018). Molecular characterization of plasmids harbored by Actinomycetes isolated from the great salt plains of Oklahoma using PFGE and next generation whole genome sequencing, *Frontiers in Microbiology* 9, 2282, doi: 10.3389/fmicb.2018.02282.

Datta, S. and Saxena, D. B. (2001). Pesticidal properties of parthenin (from *Parthenium hysterophorus*) and related compounds, *Pest Management Science* 57(1), 95–101, doi: 10.1002/1526-4998(200101)57:1<95::AID-PS248>3.0.CO;2-J.

Dinesh, R., Srinivasan, V., T E, S., Anandaraj, M. and Srambikkal, H. (2017). Endophytic Actinobacteria: diversity, secondary metabolism and mechanisms to unsilence biosynthetic gene clusters, *Critical Reviews in Microbiology* 43(5), 546–566, doi: 10.1080/1040841X.2016.1270895.

Dow, L., Gallart, M., Ramarajan, M., Law, S. R. and Thatcher, L. F. (2023). Streptomyces and their specialised metabolites for phytopathogen control – comparative in vitro and in planta metabolic approaches, *Frontiers in Plant Science* 14, 1151912, doi: 10.3389/fpls.2023.1151912.

Duke, S. O. and Dayan, F. E. (2011). Modes of action of microbially-produced phytotoxins, *Toxins* 3(8), 1038–1064, doi: 10.3390/toxins3081038.

Ebrahimi-Zarandi, M., Saberi Riseh, R. and Tarkka, M. T. (2022). Actinobacteria as effective biocontrol agents against plant pathogens, an overview on their role in eliciting plant defense, *Microorganisms* 10(9), 1739, doi: 10.3390/microorganisms10091739.

EL-Moslamy, S. H. (2018). Bioprocessing strategies for cost-effective large-scale biogenic synthesis of Nano-MgO from endophytic *Streptomyces coelicolor* strain E72 as an anti-multidrug-resistant pathogens agent, *Scientific Reports* 8(1), 3820, doi: 10.1038/s41598-018-22134-x.

El-Tarabily, K. A. (2003a). An endophytic chitinase-producing isolate of *Actinoplanes missouriensis*, with potential for biological control of root rot of lupin caused by *Plectosporium tabacinum*, *Australian Journal of Botany* 51(3), 257, doi: 10.1071/BT02107.

El-Tarabily, K. A. (2003b). An endophytic chitinase-producing isolate of *Actinoplanes missouriensis*, with potential for biological control of root rot of lupin caused by Plectosporium tabacinum, *Australian Journal of Botany* 51(3), 257, doi: 10.1071/BT02107.

Ganeshan, S., Kim, S. H. and Vujanovic, V. (2021). Scaling-up production of plant endophytes in bioreactors: concepts, challenges and perspectives, *Bioresources and Bioprocessing* 8(1), 63, doi: 10.1186/s40643-021-00417-y.

Ghorai, A. K., Patsa, R., Jash, S. and Dutta, S. (2021). Microbial secondary metabolites and their role in stress management of plants. In: Jogaiah, S. *Biocontrol Agents and Secondary Metabolites*, Elsevier, Cambridge. pp. 283–319.

Ghosh, S. and Jha, G. (2023). Editorial: utilization of microbiome to develop disease resistance in crop plants against phytopathogens, *Frontiers in Plant Science* 14, 1204896, doi: 10.3389/fpls.2023.1204896.

Gimenez-Ibanez, S., Chini, A. and Solano, R. (2016). How microbes twist jasmonate signaling around their little fingers, *Plants* 5(1), 9, doi: 10.3390/plants5010009.

Healy, F. G., Wach, M., Krasnoff, S. B., Gibson, D. M. and Loria, R. (2000). The txtAB genes of the plant pathogen *Streptomyces acidiscabies* encode a peptide synthetase required for phytotoxin thaxtomin a production and pathogenicity, *Molecular Microbiology* 38(4), 794–804, doi: 10.1046/j.1365-2958.2000.02170.x.

Helepciuc, F. E. and Todor, A. (2021). Evaluating the effectiveness of the EU's approach to the sustainable use of pesticides, *PLoS ONE* 16(9), e0256719, doi: 10.1371/journal.pone.0256719.

Hifnawy, M. S., Fouda, M. M., Sayed, A. M., Mohammed, R., Hassan, H. M., AbouZid, S. F., Rateb, M. E., Keller, A., Adamek, M., Ziemert, N. and Abdelmohsen, U. R. (2020). The genus *Micromonospora* as a model microorganism for bioactive natural product discovery, *RSC Advances* 10(35), 20939–20959, doi: 10.1039/D0RA04025H.

Himaman, W., Thamchaipenet, A., Pathom-aree, W. and Duangmal, K. (2016). Actinomycetes from Eucalyptus and their biological activities for controlling Eucalyptus leaf and shoot blight, *Microbiological Research* 188–189, 42–52, doi: 10.1016/j.micres.2016.04.011.

Ho, Y. N., Shu, L. J. and Yang, Y. L. (2017). Imaging mass spectrometry for metabolites: technical progress, multimodal imaging, and biological interactions, *Wiley Interdisciplinary Reviews. Systems Biology and Medicine* 9(5), doi: 10.1002/wsbm.1387.

Hong, C. E., Jeong, H., Jo, S. H., Jeong, J. C., Kwon, S. Y., An, D. and Park, J. M. (2016). A leaf-Inhabiting endophytic bacterium, *Rhodococcus* sp. KB6, enhances sweet potato resistance to black rot disease caused by *Ceratocystis fimbriata*, *Journal of Microbiology and Biotechnology* 26(3), 488–492, doi: 10.4014/jmb.1511.11039.

Hoskisson, P. A. and Fernández-Martínez, L. T. (2018). Regulation of specialised metabolites in Actinobacteria - expanding the paradigms, *Environmental Microbiology Reports* 10(3), 231–238, doi: 10.1111/1758-2229.12629.

Hyo-Shim, H., Koh, Y.-J., Hur, J.-S. and Jung, J.-S. (2003). Identification and characterization of coronatine-producing *Pseudomonas syringae* pv. actinidiae, *Journal of Microbiology and Biotechnology* 13(1), 110–118.

Idris, H., Goodfellow, M., Sanderson, R., Asenjo, J. A. and Bull, A. T. (2017). Actinobacterial rare biospheres and dark matter revealed in habitats of the Chilean Atacama Desert, *Scientific Reports* 7(1), 8373, doi: 10.1038/s41598-017-08937-4.

Inahashi, Y., Matsumoto, A., Ōmura, S. and Takahashi, Y. (2011). *Streptosporangium oxazolinicum* sp. nov., a novel endophytic actinomycete producing new antitrypanosomal antibiotics, spoxazomicins, *The Journal of Antibiotics* 64(4), 297–302, doi: 10.1038/ja.2011.18.

Jacquet, F., Jeuffroy, M.-H., Jouan, J., Le Cadre, E., Litrico, I., Malausa, T., Reboud, X. and Huyghe, C. (2022). Pesticide-free agriculture as a new paradigm for research, *Agronomy for Sustainable Development* 42(1), 8, doi: 10.1007/s13593-021-00742-8.

Jambhulkar, P. P., Sharma, P. and Yadav, R. (2016). Delivery systems for introduction of microbial inoculants in the field. In: Singh, D., Singh, H. and Prabha, R. *Microbial Inoculants in Sustainable Agricultural Productivity*, Springer India, New Delhi, pp. 199–218.

Kamil, F. H., Saeed, E. E., El-Tarabily, K. A. and AbuQamar, S. F. (2018). Biological control of mango dieback disease caused by *Lasiodiplodia theobromae* using streptomycete and non-streptomycete Actinobacteria in the United Arab Emirates, *Frontiers in Microbiology* 9, 829, doi: 10.3389/fmicb.2018.00829.

van Keulen, G. and Dyson, P. J. (2014). Production of specialized metabolites by *Streptomyces coelicolor* A3(2). In: Sariaslani, S. and Michael Gadd, G. *Advances in Applied Microbiology* (1st edn., vol. 89), Academic Press, Oxford, pp. 217–266.

Lakshmanan, V., Selvaraj, G. and Bais, H. P. (2014). Functional soil microbiome: belowground solutions to an aboveground problem, *Plant Physiology* 166(2), 689–700, doi: 10.1104/pp.114.245811.

Lawrence, C. H. (1990). Induction of common scab symptoms in aseptically cultured potato tubers by the vivotoxin, thaxtomin, *Phytopathology* 80(7), 606, doi: 10.1094/Phyto-80-606.

Lewin, A., Wentzel, A. and Valla, S. (2013). Metagenomics of microbial life in extreme temperature environments, *Current Opinion in Biotechnology* 24(3), 516–525, doi: 10.1016/J.COPBIO.2012.10.012.

Lewin, G. R., Carlos, C., Chevrette, M. G., Horn, H. A., McDonald, B. R., Stankey, R. J., Fox, B. G. and Currie, C. R. (2016). Evolution and ecology of *Actinobacteria* and their bioenergy applications, *Annual Review of Microbiology* 70(1), 235–254, doi: 10.1146/annurev-micro-102215-095748.

Li, H. and Li, Z. (2022). The exploration of microbial natural products and metabolic interaction guided by mass spectrometry imaging, *Bioengineering* 9(11), 707, doi: 10.3390/bioengineering9110707.

Li, S., Yang, B., Tan, G. Y., Ouyang, L. M., Qiu, S., Wang, W., Xiang, W. and Zhang, L. (2021). Polyketide pesticides from actinomycetes, *Current Opinion in Biotechnology* 69, 299–307, doi: 10.1016/j.copbio.2021.05.006.

Liu, M. J., Hwang, B. S., Jin, C. Z., Li, W. J., Park, D. J., Seo, S. T. and Kim, C. J. (2019). Screening, isolation and evaluation of a nematicidal compound from actinomycetes against the pine wood nematode, *Bursaphelenchus xylophilus*, *Pest Management Science* 75(6), 1585–1593, doi: 10.1002/ps.5272.

Lugtenberg, B. J. J., Caradus, J. R. and Johnson, L. J. (2016). Fungal endophytes for sustainable crop production, *FEMS Microbiology Ecology* 92(12), fiw194, doi: 10.1093/femsec/fiw194.

Marín, O., González, B. and Poupin, M. J. (2021). From microbial dynamics to functionality in the rhizosphere: a systematic review of the opportunities with synthetic microbial communities, *Frontiers in Plant Science* 12, 650609, doi: 10.3389/fpls.2021.650609.

Martínez-Hidalgo, P., García, J. M. and Pozo, M. J. (2015). Induced systemic resistance against *Botrytis cinerea* by *Micromonospora* strains isolated from root nodules, *Frontiers in Microbiology* 6, 922, doi: 10.3389/fmicb.2015.00922.

Medema, M. H., Blin, K., Cimermancic, P., de Jager, V., Zakrzewski, P., Fischbach, M. A., Weber, T., Takano, E. and Breitling, R. (2011). AntiSMASH: rapid identification, annotation and analysis of secondary metabolite biosynthesis gene clusters in bacterial and fungal genome sequences, *Nucleic Acids Research* 39(suppl_2), W339–W346, doi: 10.1093/nar/gkr466.

Mendes, R., Kruijt, M., de Bruijn, I., Dekkers, E., van der Voort, M., Schneider, J. H. M., Piceno, Y. M., DeSantis, T. Z., Andersen, G. L., Bakker, P. A. and Raaijmakers, J. M. (2011). Deciphering the rhizosphere microbiome for disease-suppressive bacteria, *Science* 332(6033), 1097–1100, doi: 10.1126/science.1203980.

Merzaeva, O. V. and Shirokikh, I. G. (2006). Colonization of plant rhizosphere by actinomycetes of different genera, *Microbiology* 75(2), 226–230, doi: 10.1134/S0026261706020184.

Miller, S. A., Ferreira, J. P. and LeJeune, J. T. (2022). Antimicrobial use and resistance in plant agriculture: a one health perspective, *Agriculture* 12(2), 289, doi: 10.3390/agriculture12020289.

Mohite, O. S., Weber, T., Kim, H. U. and Lee, S. Y. (2019). Genome-scale metabolic reconstruction of actinomycetes for antibiotics production, *Biotechnology Journal* 14(1), e1800377, doi: 10.1002/biot.201800377.

Newitt, J. T., Prudence, S. M. M., Hutchings, M. I. and Worsley, S. F. (2019). Biocontrol of cereal crop diseases using Streptomycetes, *Pathogens* 8(2), 78, doi: 10.3390/pathogens8020078.

Nicot, P. C., Avril, F., Duffaud, M., Leyronas, C., Troulet, C., Villeneuve, F. and Bardin, M. (2019). Differential susceptibility to the mycoparasite *Paraphaeosphaeria minitans* among *Sclerotinia sclerotiorum* isolates, *Tropical Plant Pathology* 44(1), 82–93, doi: 10.1007/s40858-018-0256-7.

Novakova, R., Knirschova, R., Farkasovsky, M., Feckova, L., Rehakova, A., Mingyar, E. and Kormanec, J. (2013). The gene cluster *aur1* for the angucycline antibiotic auricin is located on a large linear plasmid pSA3239 in *Streptomyces aureofaciens* CCM 3239, *FEMS Microbiology Letters* 342(2), 130–137, doi: 10.1111/1574-6968.12095.

Núñez-Montero, K., Lamilla, C., Abanto, M., Maruyama, F., Jorquera, M. A., Santos, A., Martinez-Urtaza, J. and Barrientos, L. (2019). Antarctic *Streptomyces fildesensis* So13.3 strain as a promising source for antimicrobials discovery, *Scientific Reports* 9(1), 7488, doi: 10.1038/s41598-019-43960-7.

O'Sullivan, C. A., Roper, M. M., Myers, C. A. and Thatcher, L. F. (2021). Developing actinobacterial endophytes as biocontrol products for *Fusarium pseudograminearum* in wheat, *Frontiers in Bioengineering and Biotechnology* 9, 691770, doi: 10.3389/fbioe.2021.691770.

Otto-Hanson, L. K., Grabau, Z., Rosen, C., Salomon, C. E. and Kinkel, L. L. (2013). Pathogen variation and urea influence selection and success of *Streptomyces* mixtures in biological control, *Phytopathology* 103(1), 34–42, doi: 10.1094/PHYTO-06-12-0129-R.

Pacios-Michelena, S., Aguilar González, C. N., Alvarez-Perez, O. B., Rodriguez-Herrera, R., Chávez-González, M., Arredondo Valdés, R., Ascacio Valdés, J. A., Govea Salas, M. and Ilyina, A. (2021). Application of Streptomyces antimicrobial compounds for the control of phytopathogens, *Frontiers in Sustainable Food Systems* 5, doi: 10.3389/fsufs.2021.696518.

Prem Anand, K. and Suthindhiran, K. (2022). Methods in actinobacteriology, *Screening and Analysis of Actinobacterial Bioherbicides for Weed Management*, In: Dharumadurai, D. pp. 537–544. Springer, US.

Procópio, R. E., Silva, I. R., Martins, M. K., Azevedo, J. L. and Araújo, J. M. (2012). Antibiotics produced by *Streptomyces*, *The Brazilian Journal of Infectious Diseases* 16(5), 466–471, doi: 10.1016/j.bjid.2012.08.014.

Quinn, G. A., Banat, A. M., Abdelhameed, A. M. and Banat, I. M. (2020). Streptomyces from traditional medicine: sources of new innovations in antibiotic discovery, *Journal of Medical Microbiology* 69(8), 1040–1048, doi: 10.1099/jmm.0.001232.

Ran, Y., Zhang, Y., Wang, X. and Li, G. (2022). Nematicidal metabolites from the Actinomycete *Micromonospora* sp. WH06, *Microorganisms* 10(11), 2274, doi: 10.3390/microorganisms10112274.

Rimbaud, L., Papaïx, J., Rey, J. F., Barrett, L. G. and Thrall, P. H. (2018). Assessing the durability and efficiency of landscape-based strategies to deploy plant resistance to pathogens, *PLoS Computational Biology* 14(4), e1006067, doi: 10.1371/journal.pcbi.1006067.

Roper, M., Thatcher, L., Anderson, J. and Myers, C. (2018). *Methods and Compositions for the Biological Control of Plant Pathogens*, 11 September, US11286458B2.

Sakaguchi, K. (1990). Invertrons, a class of structurally and functionally related genetic elements that includes linear DNA plasmids, transposable elements, and genomes of adeno-type viruses, *Microbiological Reviews* 54(1), 66–74, doi: 10.1128/mr.54.1.66-74.1990.

Sapbamrer, R., Kitro, A., Panumasvivat, J. and Assavanopakun, P. (2023). Important role of the government in reducing pesticide use and risk sustainably in Thailand: current situation and recommendations, *Frontiers in Public Health* 11, 1141142, doi: 10.3389/fpubh.2023.1141142.

Scott, I. M., Vickruck, J., Hann, S., Krolikowski, S., MacKinley, P., Stokes-Rees, J., Hatten, G. and Moffat, C. (2023). Regional differences in susceptibility to spinosyn insecticides registered for Colorado potato beetle management in Canada, *Pesticide Biochemistry and Physiology* 193, 105459, doi: 10.1016/j.pestbp.2023.105459.

Selim, M. S. M., Abdelhamid, S. A. and Mohamed, S. S. (2021). Secondary metabolites and biodiversity of actinomycetes, *Journal, Genetic Engineering and Biotechnology* 19(1), 72, doi: 10.1186/s43141-021-00156-9.

Seshadri, R., Roux, S., Huber, K. J., Wu, D., Yu, S., Udwary, D., Call, L., Nayfach, S., Hahnke, R. L., Pukall, R., White, J. R., Varghese, N. J., Webb, C., Palaniappan, K., Reimer, L. C., Sardà, J., Bertsch, J., Mukherjee, S., Reddy, T. B. K., Hajek, P. P., Huntemann, M., Chen, I. A., Spunde, A., Clum, A., Shapiro, N., Wu, Z. Y., Zhao, Z., Zhou, Y., Evtushenko, L., Thijs, S., Stevens, V., Eloe-Fadrosh, E. A., Mouncey, N. J., Yoshikuni, Y., Whitman, W. B., Klenk, H. P., Woyke, T., Göker, M., Kyrpides, N. C. and Ivanova, N. N. (2022). Expanding the genomic encyclopedia of Actinobacteria with 824 isolate reference genomes, *Cell Genomics* 2(12), 100213, doi: 10.1016/j.xgen.2022.100213.

Sgobba, E., Daguerre, Y. and Giampà, M. (2021). Unravel the local complexity of biological environments by MALDI mass spectrometry imaging, *International Journal of Molecular Sciences* 22(22), 12393, doi: 10.3390/ijms222212393.

Sharma, J., Manjunatha, N., Pokhare, S. S., Patil, P. G., Agarrwal, R., Chakranarayan, M. G., Aralimar, A., Devagire, P. and Marathe, R. A. (2022). Genetic diversity and streptomycin sensitivity in *Xanthomonas axonopodis* pv. punicae causing oily spot disease in pomegranates, *Horticulturae* 8(5), 441, doi: 10.3390/horticulturae8050441.

Sharma, S. B., Sayyed, R. Z., Trivedi, M. H. and Gobi, T. A. (2013). Phosphate solubilizing microbes: sustainable approach for managing phosphorus deficiency in agricultural soils, *SpringerPlus* 2(1), 587, doi: 10.1186/2193-1801-2-587.

Silva, G. da C., Kitano, I. T., Ribeiro, I. A. de F. and Lacava, P. T. (2022). The potential use of Actinomycetes as microbial inoculants and biopesticides in agriculture, *Frontiers in Soil Science* 2, doi: 10.3389/fsoil.2022.833181.

Sparks, T. C., Crouse, G. D., Benko, Z., Demeter, D., Giampietro, N. C., Lambert, W. and Brown, A. V. (2021). The spinosyns, Spinosad, spinetoram, and synthetic spinosyn mimics - discovery, exploration, and evolution of a natural product chemistry and the impact of computational tools, *Pest Management Science* 77(8), 3637–3649, doi: 10.1002/ps.6073.

Stockwell, V. O., Temple, T. N., Johnson, K. B. and Loper, J. E. (2008). Integrated control of fire blight with antagonists and oxytetracycline, *Acta Horticulturae* 793(793), 383–390, doi: 10.17660/ActaHortic.2008.793.57.

Sutherland, E. D., Baker, K. K. and Lockwood, J. L. (1984). Ultrastructure of *Phytophthora megasperma* f. sp. glycinea oospores parasitized by *Actinoplanes missouriensis* and *Humicola fuscoatra*, *Transactions of the British Mycological Society* 82(4), 726–729, doi: 10.1016/S0007-1536(84)80117-5.

Syed Ab Rahman, S. F., Singh, E., Pieterse, C. M. J. and Schenk, P. M. (2018). Emerging microbial biocontrol strategies for plant pathogens, *Plant Science* 267, 102–111, doi: 10.1016/j.plantsci.2017.11.012.

Takano, H. K. and Dayan, F. E. (2020). Glufosinate-ammonium: a review of the current state of knowledge, *Pest Management Science* 76(12), 3911–3925, doi: 10.1002/ps.5965.

Tan, G. Y., Deng, K., Liu, X., Tao, H., Chang, Y., Chen, J., Chen, K., Sheng, Z., Deng, Z. and Liu, T. (2017). Heterologous biosynthesis of Spinosad: an omics-guided large polyketide synthase gene cluster reconstitution in *Streptomyces*, *ACS Synthetic Biology* 6(6), 995–1005, doi: 10.1021/acssynbio.6b00330.

Teixidó, N., Segarra, G., Casals, C., Usall, J. and Torres, R. (2020). Formulations to improve biocontrol products shelf-life and/or ecosystem adaptation. In: De Cal, A., Melgarejo, P. and Magan, N. *How Research Can Stimulate the Development of Commercial Biological Control against Plant Diseases. Progress in Biological Control* (vol. 21), Springer Nature, Switzerland. pp. 257–273.

Teixidó, N., Usall, J. and Torres, R. (2022). Insight into a successful development of biocontrol agents: production, formulation, packaging, and shelf life as key aspects, *Horticulturae* 8(4), 305, doi: 10.3390/horticulturae8040305.

Tran, P. N., Yen, M. R., Chiang, C. Y., Lin, H. C. and Chen, P. Y. (2019). Detecting and prioritizing biosynthetic gene clusters for bioactive compounds in bacteria and fungi, *Applied Microbiology and Biotechnology* 103(8), 3277–3287, doi: 10.1007/s00253-019-09708-z.

Traxler, M. F., Watrous, J. D., Alexandrov, T., Dorrestein, P. C. and Kolter, R. (2013). Interspecies interactions stimulate diversification of the *Streptomyces coelicolor* secreted metabolome, *mBio* 4(4), doi: 10.1128/mBio.00459-13.

Trivedi, P., Leach, J. E., Tringe, S. G., Sa, T. and Singh, B. K. (2020). Plant–microbiome interactions: from community assembly to plant health, *Nature Reviews Microbiology* 18(11), 607–621, doi: 10.1038/s41579-020-0412-1.

Trivedi, P., Mattupalli, C., Eversole, K. and Leach, J. E. (2021). Enabling sustainable agriculture through understanding and enhancement of microbiomes, *New Phytologist* 230(6), 2129–2147, doi: 10.1111/nph.17319.

Viaene, T., Langendries, S., Beirinckx, S., Maes, M. and Goormachtig, S. (2016). *Streptomyces* as a plant's best friend?, *FEMS Microbiology Ecology* 92(8), fiw119, doi: 10.1093/femsec/fiw119.

Walsh, F., Smith, D. P., Owens, S. M., Duffy, B. and Frey, J. E. (2013). Restricted streptomycin use in apple orchards did not adversely alter the soil bacteria communities, *Frontiers in Microbiology* 4, 383, doi: 10.3389/fmicb.2013.00383.

Wang, B., Li, L., Lin, Y., Shen, D., Shao, X., Zhong, C. and Qian, G. (2022). Targeted isolation of biocontrol agents from plants through phytopathogen co-culture and pathogen enrichment, *Phytopathology Research* 4(1), 19, doi: 10.1186/s42483-022-00124-2.

Winding, A., Binnerup, S. J. and Pritchard, H. (2004). Non-target effects of bacterial biological control agents suppressing root pathogenic fungi, *FEMS Microbiology Ecology* 47(2), 129–141, doi: 10.1016/S0168-6496(03)00261-7.

Yashiro, E. and McManus, P. S. (2012). Effect of streptomycin treatment on bacterial community structure in the apple phyllosphere, *PLoS ONE* 7(5), e37131, doi: 10.1371/journal.pone.0037131.

Zheng, Q., Hu, Y., Zhang, S., Noll, L., Böckle, T., Dietrich, M., Herbold, C. W., Eichorst, S. A., Woebken, D., Richter, A. and Wanek, W. (2019). Soil multifunctionality is affected by the soil environment and by microbial community composition and diversity, *Soil Biology and Biochemistry* 136, 107521, doi: 10.1016/j.soilbio.2019.107521.

Chapter 5

Exploring *Clonostachys rosea*: a versatile fungal biocontrol agent for effective plant disease management

Cameron Scott and Zamir K. Punja, Simon Fraser University, Canada

1 Introduction

Clonostachys rosea is a fungal biocontrol agent which has the capacity to manage a wide range of plant pathogens which affect numerous economically important crops such as cereals, vegetables, berries and others. This chapter will focus on the applied use of *C. rosea* in agriculture, including its diverse mechanisms of action and how it has been applied to effectively manage crop diseases. Some of the variables that alter the efficacy of *C. rosea* as a biocontrol agent will also be discussed. Although the research and practical use of *C. rosea* has predominantly focused on its ability to manage plant disease, there is emerging research showing that it may also be effective at controlling plant parasitic nematodes, as well as insects. These pests may damage crops and act as vectors of plant diseases. The study of *C. rosea* has elucidated many genetic components that underline its biocontrol mechanisms and the capabilities of other biocontrol agents. This may in turn lead to more targeted and effective use of *C. rosea* to manage plant diseases. *C. rosea* may also be utilized alongside

http://dx.doi.org/10.19103/AS.2024.0142.15

other disease management approaches in an integrated pest management strategy, which may enhance the efficacy of the entire program. Although topics such as the genetic components of biocontrol are discussed in this chapter it does not offer a comprehensive overview of subjects such as this, rather it is meant to briefly introduce their relevance to the biocontrol abilities of *C. rosea*.

2 Taxonomy, identification and sources

Clonostachys rosea is a filamentous fungus and the anamorph of the teleomorph *Bionectria ochroleuca* (Schroers et al., 1999). *Clonostachys rosea* is a member of the order Bionectriacaeae and phylum Ascomycota. Previously, *C. rosea* may have been referred to as *Gliocladium roseum*, *Gliocladium catenulatum*, or *Clonostachys rosea* f. *catenulata*, although these are currently understood to all be the same species (Jensen et al., 2022). Biocontrol product labels or publications focusing on biocontrol may still refer to *C. rosea* by one of these previously utilized classifications. The taxonomical classification of these fungi as *C. rosea* is based on DNA sequencing data, such as the gene sequences of internal transcribed spacer (ITS), translation elongation factor 1-alpha (TEF-1α) and beta-tubulin (TUB) genes as well as their morphology compared to other *Gliocladium* spp. (Schroers et al., 1999; Schroers, 2001; Moreira et al., 2016).

Clonostachys rosea is distributed nearly worldwide and is most frequently isolated from soil, plant debris and leaf litter, where it exists as a saprophyte (Sun et al., 2020). It has also commonly been isolated from roots, stems and other tissues of living plants, including various crops (Costa et al., 2012). *C. rosea* has also been isolated from nematodes and insects (Verdejo-Lucas et al., 2002; Toledo et al., 2006). Numerous isolates of *C. rosea* have been screened for their potential as biocontrol agents, with some being developed into commercial products (Table 1). In addition to being able to be purchased as a preformulated product, *C. rosea* may also be produced on various scales using

Table 1 Commercial biocontrol products containing *Clonostachys rosea*

Product	Strain	Manufacturer
Prestop/Prestop Mix/Prestop WG[1]	J1446	Danstar Ferment AG[2]/Verdera[2]
Prestop 4B (bee vectored)	J1446	Danstar Ferment AG[2]
Vectorite with CR-7	CR-7	Bee-Vectoring Technology Inc.
Pvent Microbial WP	Not specified	BioSafe Systems
GlioMix	J1446	Verdera[2]
Kamoi	CPQBA 040-11 DRM 07	Agrivalle Brasil Indústria

[1]Rebranded as Lalstop G46 WG.
[2]Danstar Ferment AG and Verdera are subsidiaries of Lallemand Inc.

solid-state or liquid fermentation techniques (Zhang et al., 2013; Sun et al., 2014; de Andrade Carvalho et al., 2018; Mascarin et al., 2022).

3 Mechanisms of action

The ability of *C. rosea* to act as a biocontrol agent (BCA) against a range of potential plant pathogens, including bacteria, fungi and oomycetes, is due to its evolution as a generalist capable of surviving in different niches (Piombo et al., 2023). In the rhizosphere, and soil in general, *C. rosea* can compete with other organisms for resources through direct mechanisms, such as the secretion of antibiotic compounds, enzymes and other proteins (Sun et al., 2020). The secretion of these compounds by *C. rosea* may also allow for it to thrive as a saprophyte in soil or on plant debris (Monkai et al., 2013; Moreira et al., 2016; Atanasova et al., 2018). Other microorganisms may also try to compete for space and resources in this manner. However, *C. rosea* isolates have demonstrated an ability to tolerate and detoxify otherwise toxic compounds (Dubey et al., 2014; Dubey et al., 2016). *Clonostachys rosea* may interact with other fungi as a necrotrophic mycoparasite, capable of feeding on and destroying, a broad range of hosts (Hasan et al., 2022). In the rhizosphere and the phyllosphere, *C. rosea* can colonize plant tissues epiphytically and endophytically, limiting the establishment of potential pathogens (Rojas et al., 2020; Türkölmez et al., 2023). Interactions between *C. rosea* and their plant hosts can trigger a defense response in the plant, known as induced systemic resistance (ISR) (Gong et al., 2017). Differing combinations of these mechanisms may be utilized by the BCA, resulting in its diverse biocontrol capabilities. The growing understanding of the mechanisms of action of *C. rosea*, and biocontrol agents in general, is key to improving their use and efficacy in agriculture.

3.1 Mycoparasitism

Clonostachys rosea was initially investigated as a necrotrophic mycoparasite, in the late 1950s and early 1960s, based on its ability to overgrow fungi on agar media (Barnett and Lily, 1962). In these preliminary experiments, *C. rosea* was observed to directly contact and kill mycelium, conidia and sclerotia of numerous fungi (Makkonen and Pohjakallio, 1960; Barnett and Lily, 1962). A spectrum of susceptibility or resistance to mycoparasitism by *C. rosea* was also observed. However, dual culture plate assays are not necessarily indicative of the ability of *C. rosea* to control a pathogen in an actual field or greenhouse environment (Jensen et al., 2007).

Other early observations of the interactions between *C. rosea* and its hosts were made using light microscopy and staining techniques. These observations have been confirmed and researched further in several more recent studies using light microscopy, confocal fluorescence microscopy and scanning electron microscopy (Figure 1) (Li et al., 2002; Chatterton and Punja, 2009; Mamarabadi et al., 2009; Abdellatif et al., 2022; Geiger et al., 2022; Hasan et al., 2022; Pimentel et al., 2022). A *C. rosea* isolate expressing green fluorescent protein (GFP) grown in the presence of *B. cinerea* was observed to grow in the direction of its host, and once near its host, small papillae-like growths were extended from the ends of its hyphae (Hasan et al., 2022). The BCA then attached to its host, using hook-like structures in conjunction with appressoria, which it used to penetrate and kill spores and mycelium of *B. cinerea*. In contrast to this, work by Li et al. (2002), using scanning electron microscopy, showed no appressoria formation, instead *B. cinerea* was directly penetrated by *C. rosea* hyphal tips and subsequently disintegrated. It is unclear what factors contribute to the presence or absence of appressoria formation when *C. rosea* interacts with a potential host. Although *C. rosea* was originally considered a necrotrophic mycoparasite, emerging research suggests that it may be more accurately characterized as a hemibiotroph, due to its initial non-destructive parasitic relationship with certain fungal hosts (Abdellatif et al., 2022).

The plant pathogens *Pythium aphanidermatum, Pythium ultimum, Fusarium oxysporum* and *Fusarium graminearum* have also been observed to be parasitized by *C. rosea* (Chatterton and Punja, 2009; Mamarabadi et al., 2009; Gimeno et al., 2021a). When *C. rosea* was grown in the presence of sclerotia of *Sclerotinia sclerotiorum* gene encoding signal transduction proteins involved in the recognition and attachment of a mycoparasite to its host were upregulated (Sun et al., 2015). Cell wall degrading enzymes (CWDEs) and toxin-encoding genes were also upregulated in the presence of *S. sclerotiorum*. The production of CWDEs and other metabolites is key to the process of mycoparasitism,

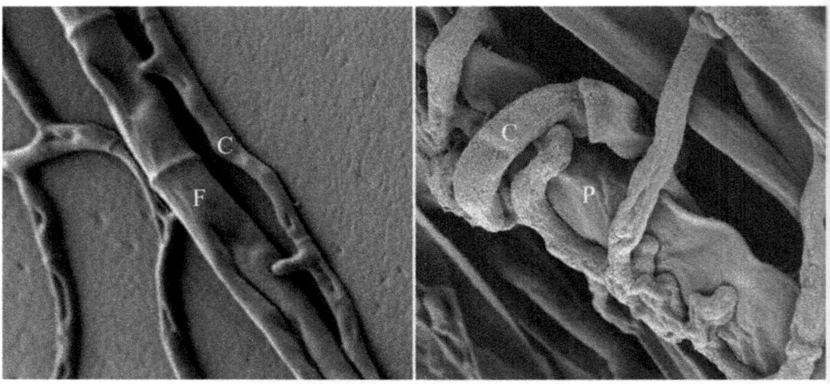

Figure 1 Mycoparasitic activity of *Clonostachys rosea* (C) against the mycelium of (F) *Fusarium oxysporum* or (P) *Pythium* from in vitro dual cultures on water agar. The hyphae of the biocontrol agent can be seen penetrating the mycelium of the pathogens. The figure is adapted from Chatterton and Punja (2009).

as they allow the mycoparasite to disarm its host and degrade its structures (Chatterton and Punja, 2009).

It is unclear to what degree mycoparasitism plays a role in interactions between *C. rosea* and its hosts in a natural environment, and what conditions might optimize its efficacy. When cultured with *B. cinerea* on agar media, with and without additional nutrients, parasitism by *C. rosea* was delayed by 12 h when other sources of nutrients were readily available (Hasan et al., 2022). Research by Chatterton and Punja (2009) showed how the activity of *C. rosea* CWDEs was suppressed by glucose or induced by components of fungal cell walls, such as chitin or fungal cell wall extracts. Similar trends of suppression or induction were seen by Mamarabadi et al. (2009). Yu and Sutton (1997b) observed that mycoparasitism by *C. rosea* played a role in the suppression of *B. cinerea* in raspberry. In that study, the ability of the BCA to antagonize or parasitize the pathogen changed based on the tissues the biocontrol agent and pathogen were applied to (leaves, stems and stamens), possibly dictated by the nutrients available on those tissues. In other BCAs, such as *Trichoderma* spp., the availability of nutrients is known to play a role in the regulation of metabolic pathways involved in the recognition and attachment of the BCA to its fungal host, and in the expression of genes such as those encoding proteinases and other enzymes (Flores et al., 1997; Olmedo-Monfil et al., 2002; Seidl et al., 2009). The effect of nutrient availability on the mycoparasitic behavior and gene expression of *C. rosea* requires more research. An increased understanding of the factors influencing a key mechanism involved in the biocontrol ability of *C. rosea* may lead to more efficacious use of the BCA.

3.2 Cell wall-degrading enzyme production

As previously mentioned, the production of CWDEs plays a key role in the ability of *C. rosea* to parasitize its host. These enzymes include chitinases, glucanases and proteases which catabolize the cell walls of both fungal and oomycete pathogens, such as *Fusarium* and *Pythium* spp. (Chatterton and Punja, 2009; Mamarabadi et al., 2009; Sun et al., 2020; Piombo et al., 2023). For example, the *chiC1* chitinase gene is induced by chitin, a compound that is ubiquitous in nature and a key component of fungal cell walls and invertebrate exoskeletons (Tzelepis et al., 2015). This gene encodes for a group C killer toxin-like chitinase, which is suggested to catabolize host cell walls, allowing for the entry of toxic metabolites (Tzelepis and Karlsson, 2019). The deletion of the *chiC2* gene in *C. rosea* reduced the ability of culture filtrates to inhibit the growth of *Rhizoctonia solani* and *B. cinerea* (Tzelepis et al., 2015). However, the *chiC2* deletion isolate did not have a significantly different relative lesion area, compared to the wild-type strain, when applied to manage *B. cinerea* on *Arabidopsis thaliana* leaves in a laboratory setting (Tzelepis et al. 2015). Similar results were seen by Mamarabadi et al. (2008) when they deleted three different endochitinase genes in *C. rosea* and tested the biocontrol capabilities of these three deletion isolates against *Fusarium culmorum* on barley and *Alternaria radicina* on carrot. Although the biocontrol capabilities of these isolates were not affected by the loss of individual chitinase genes, RT-PCR analysis showed enhanced *F. culmorum* activity, and lowered *C. rosea* activity, when the deletion strains were co-inoculated in liquid culture with *F. culmorum*, compared to the wild-type isolate (Mamarabadi et al., 2008). It is likely that the *C. rosea* isolates in which the chitinase genes were deleted still had adequate biocontrol capabilities because of the presence of various other CWDE genes.

3.3 Antibiotic secretion

Clonostachys rosea is known to produce various antibiotic compounds which can play a role in its mycoparasitic behavior and enhance its ability to compete for space against a diverse range of microbes, as well as nematodes (Thines et al., 1998; Okuda et al., 2000; Dong et al., 2005; Zhai et al., 2016; Han et al., 2020; Saraiva et al., 2020). Consequently, the genome of *C. rosea* has expanded gene copy numbers of gene families associated with the synthesis and transport of secondary metabolites, in some instances, greater than *Trichoderma* spp. or pathogenic *Fusarium* spp. (Karlsson et al., 2015; Iqbal et al., 2018b; Broberg et al., 2021). This emphasizes the likely importance of the synthesis and secretion of secondary metabolites in the biocontrol abilities of *C. rosea*, including in mycoparasitism and competition. These gene families encode for proteins such as membrane transport proteins, polyketide synthase, serine proteases, cytochrome P450 and lyase enzymes.

Peptaibols produced by *C. rosea* showed antifungal activity against *S. sclerotiorum*, inhibiting mycelial growth, reducing sclerotia production and viability, and likely contributing to the lysis of pathogen hyphae (Rodriguez et al., 2011). The formation of appressoria by the pathogen *Magnaporthe grisea*, the causal agent of blast disease on rice, was inhibited by glisoprenins produced by *C. rosea* grown in submerged culture (Thines et al., 1998). However, these glisoprenins only halted the formation of appressoria, they did not inhibit conidia germination or the mycelial growth of the pathogen, nor were other antimicrobial effects observed. Saraiva et al. (2020) demonstrated that other uncharacterized metabolites of *C. rosea* may inhibit the germination of *B. cinerea* conidia as well as the development of its mycelium and sclerotia.

Compounds such as polyketides also play a role in the biocontrol ability of *C. rosea*. Two polyketides produced by *C. rosea*, Clonorosein A and B, had antifungal activity against *B. cinerea* and *F. graminearum*, inhibiting germ tube growth (Fatema et al., 2018). Expression levels of the polyketide synthase gene *pks22* were highly elevated during interactions with both pathogens. Deletion of *pks22* or *pks29* did not have a significant effect on the antagonistic ability of the fungi in dual culture assays with pathogens, including *Alternaria alternata, B. cinerea*, *R. solani* and *F. graminearum*, although the *pks29* deletion strain was no longer able to suppress *B. cinerea* growth in liquid culture (Fatema et al., 2018). When these two deletion strains were applied to barley seeds to control foot rot caused by *F. graminearum*, the *pks22* deletion strain showed no difference compared to the wild-type strain, while treatment with the *pks29* deletion strain resulted in significantly increased disease severity on seedlings (Fatema et al., 2018).

The potential effects of antibiotics secreted by *C. rosea* isolate IK726 were also observed when foliar treatments were made to manage spot blotch (*Bipolaris sorokiniana*) on barley (Jensen et al., 2016). Treatments applied 24 h before, or at the time of, pathogen inoculation resulted in a significant reduction in disease severity, whereas treatments made 24 h after pathogen inoculation had little effect. Microscopy confirmed that the germination of conidia and the formation of appressoria were inhibited by preventative *C. rosea* treatments, which authors attributed to antibiosis potentially in combination with other biocontrol mechanisms such as mycoparasitism and competition (Jensen et al., 2016). The antibiotic capabilities of many of the compounds produced by *C. rosea*, and their roles in its efficacy as a biocontrol agent, are not well understood and require more research (Abdel-Wahab et al., 2019; Kim et al., 2021; Supratman et al., 2021).

3.4 Tolerance to toxic compounds

C. rosea has developed the ability to detoxify toxins as well as actively transport them out of their cells in order to tolerate secondary metabolites from potential prey, antagonistic microbes and host plants during colonization or other sources in their environment including pesticides (Roberti et al., 2006; Dubey et al., 2016; Broberg et al., 2021; Demissie et al., 2020). The secretion of endogenous antibiotics and exogenous toxic compounds from *C. rosea* is mediated by transporters, such as ATP-binding cassettes (ABCs) and major facilitator superfamily (MFS) transporters, which shuttle metabolites across cell membranes (Dubey et al. 2014). Analysis of the genome of *C. rosea* showed high numbers of ABC and MFS genes, including ABC genes in groups with functions including drug and antibiotic resistance, and tolerance to reactive oxygen species (Karlsson et al., 2015).

Treatment of *C. rosea* with a *F. graminearum* filtrate containing products of its secretome, as well as live macroconidia, resulted in the upregulation of four ABC transporters (Demissie et al., 2020). Macroconidia were utilized as the culture filtrate alone would not have allowed for detection of these genes (Demissie et al., 2020). Exposure to other toxic compounds such as bacterial metabolites, the mycotoxin zearalenone (ZEN) and the fungicide boscalid also resulted in the induction of ABC transporter genes (Kosawang et al., 2014a; Karlsson et al., 2015). However, treatment with the *F. graminearum* mycotoxin deoxynivalenol (DON) alone was not sufficient to upregulate these transporters. Instead, genes involved with the production of metabolic enzymes, such as cytochrome c oxidase, and stress response proteins were upregulated by DON treatment (Kosawang et al., 2014a). This suggests that transporter genes may be regulated by multiple factors, including aspects of the pathogen secretome and interactions between *C. rosea* and living *F. graminearum* cells, and that transporter genes are not necessarily solely involved in the ability of *C. rosea* to tolerate toxins (Demissie et al., 2020). For example, *C. rosea* can detoxify ZEN and DON directly, through the enzymes ZHD101 or UGT-glyocosyltransferases, respectively (Kosawang et al., 2014b; Khairullina et al., 2023). Subsequently, the metabolites of the detoxification process may be transported out of the cell through transporters such as the ABC transporters previously mentioned (Dubey et al., 2014). In a greenhouse trial, *C. rosea* treatment resulted in the reduction of DON by 80% in oat kernels (Khairullina et al., 2023). DON and ZEA can contaminate grains and have severe effects on human and animal health (Khairullina et al., 2023). Toxins, such as the mycotoxin fumonisin B1, may also be transported out of the cell without being detoxified first (Chatterjee et al., 2016).

3.5 Endophytic colonization, induced systemic resistance and growth-promoting effects

Through direct interactions with plants such as endophytic colonization, the induction of plant defenses, and to a lesser degree growth promotion, *C. rosea* can contribute to disease management. *Clonostachys rosea* has been demonstrated to endophytically colonize plant tissues such as roots, stems, leaves or flowers in a range of plants including tomato, pepper, eggplant, coffee, onion, geranium, cucumber and others (Figure 2) (Chatterton et al., 2008; Chatterton and Punja, 2012; Muvea et al., 2014; Kapeua-Ndacnou et al., 2023; Türkölmez et al., 2023). The ability of *C. rosea* to colonize plant tissues, especially in advance of a pathogen, may allow it to prevent or minimize pathogen infection in otherwise potentially susceptible tissues, through competition for space and resources.

In the process of colonization, or otherwise interacting with host plant tissues, *C. rosea* can induce a defense response in the plant, called induced systemic resistance (ISR) (Mouekouba, et al., 2014; Gong et al., 2017; Sun et al., 2020). Plants can recognize microbes through specific receptors located on the surfaces of their cells, called pattern recognition receptors (PRRs). These receptors recognize molecules that are not present in the host, and which

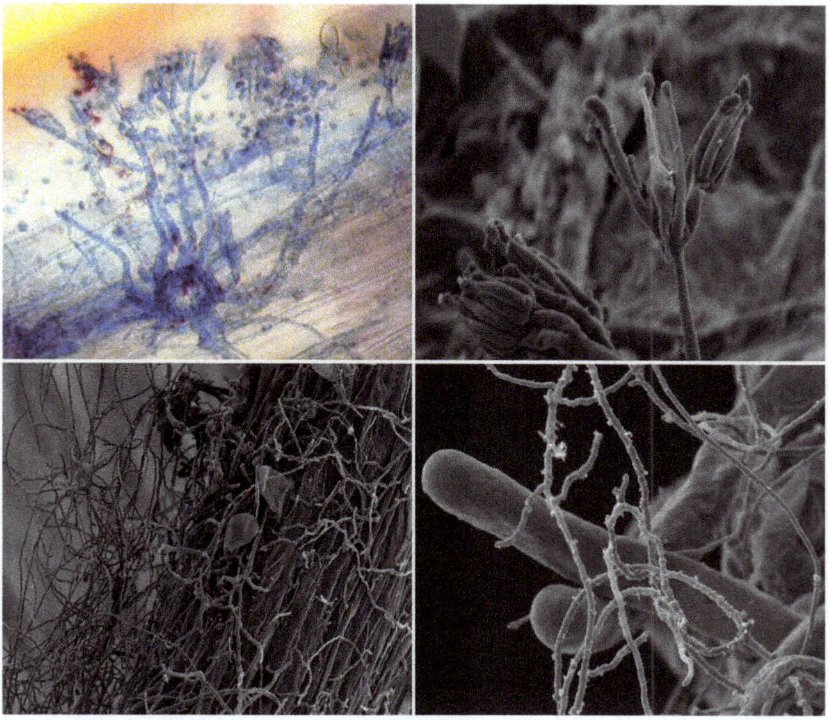

Figure 2 The colonization of cucumber roots by *C. rosea*. Top left: The sporulation of *Clonostachys rosea* on the surface of cucumber roots. Trypan blue staining was used to show conidiophores and conidia development. Top right: Scanning electron microscope image of a *C. rosea* conidiophore. Bottom right: Colonization of a cucumber root by *C. rosea*. Bottom left: Extensive colonization of the surface of a cucumber root by the biocontrol agent. The figure is adapted from Chatterton et al. (2008).

are often highly conserved as they can be key to the basic functions of these microbes. These molecules are referred to as microbe or pathogen associated molecular patterns (MAMPs or PAMPs, respectively) (Newman et al., 2013; Latz et al., 2018; Köhl et al., 2019). Typical examples of MAMPS or PAMPs include components of microbial cell walls or membranes (chitin, β-glucans, peptidoglycan and lipopolysaccharides), flagellin and elongation factor tu (Boller and Felix, 2009; Köhl et al., 2019). However, a broad range of other compounds such as antibiotics, siderophores, enzymes, fatty acids, proteins and volatile organic compounds may also act as ISR elicitors (Boller and Felix, 2009; Pieterse et al., 2014; Mauch-Mani et al., 2017; Köhl et al., 2019). When these patterns are detected by the plant, signaling pathways are activated, namely jasmonic acid (JA) and ethylene (ET) dependent pathways, although other signaling molecules such as reactive oxygen species (ROS), phytohormones (salicylic acid, gibberellins and others) and enzymes can also play a role in

ISR induction (Boller and Felix, 2009; Wang et al., 2019). The activation of these pathways results in altered gene expression and increased resistance to pathogens in both local and distal tissues. Plants exposed to these stimuli, from biocontrols or other sources, may be considered "primed" or in a priming phase. Plants in this phase can subsequently react to challenges by pathogens with more rapid and effective defense responses, compared to those which were not primed (Mauch-Mani et al., 2017).

Disease resistance can be enhanced through a variety of mechanisms such as the increased production of pathogenesis-related proteins and defense enzymes, the accumulation of ROS, increased callose deposition, lignification of cell walls and upregulation of mitogen activated protein kinase (MAPK) pathways, among others (Roberti et al., 2008; Wiesel et al., 2014; Gong et al., 2017;). For example, *B. cinerea* and *C. rosea* treated tomato plants had increased levels of two enzymes, polyphenol oxidase (PPO) and phenylalanine ammonia-lyase (PAL), compared to plants which were only treated with the pathogen (Wang et al., 2019). PPO and PAL are involved in the production of antimicrobial compounds, including phytoalexins, phenolics and quinones (Zhang and Sun, 2021). Mouekouba et al. (2014) reported that tomato leaves treated with *C. rosea* and *B. cinerea* had elevated PAL, PPO and glutathione S-transferase (GST) levels, compared to leaves that received either the pathogen or the biocontrol agent alone. GSTs can suppress oxidative stress and detoxify metabolites that accumulate during infection (Gullner et al., 2018). Work by Gong et al. (2017) and Meng et al. (2022) showed similarly elevated levels of defense enzymes in tomato induced by *C. rosea* treatment to manage *B. cinerea* infection. Induced systemic resistance was also observed on tobacco plants that had their roots treated with *C. rosea* (Lahoz et al., 2004). Elevated levels of enzymes including glucanases and chitinases were detected in the leaves of treated tobacco plants. Tobacco plants that received the *C. rosea* treatment prior, and during, inoculation with *Erysiphe orontii*, a causal agent of powdery mildew, had significantly fewer fungal colonies per leaf and less severe disease symptoms, compared to plants that did not receive *C. rosea* (Lahoz et al., 2004).

Biocontrol agents can also promote plant growth, although it is not clear to what extent this influences their ability to manage disease. Roberti et al. (2008) showed beneficial effects of *C. rosea* treatment on wheat seedling growth, including increased root length and plant dry weight. The growth promoting effects of *C. rosea* have also been demonstrated on wheat, tomato, pepper, cucumber, geranium, rose and pine (Johansen et al., 2005; Sutton et al., 2008; Moraga-Suazo and Sanfuentes, 2017; Han et al., 2022; Türkölmez et al., 2023). *C. rosea* has been shown to stimulate root growth in tomato plants possibly through secretion of the auxin indole-3-acetic acid (IAA) (Han et al., 2022). However, some biocontrol agents, such as isolates of particular *Trichoderma* spp., can synthesize these growth promoting phytohormones, but not enhance the growth of treated plants (Nieto-Jacobo et al., 2017). Other mechanisms such as the production of volatile organic compounds, solubilization of nutrients, and increased tolerance to abiotic stress are also thought to play a role in plant growth promotion by biocontrol agents, but this has not been demonstrated for *C. rosea* (Altomare et al., 1999; Hung et al., 2013; Contreras-Cornejo et al., 2014). Further research is required to determine what other possible mechanisms play a role in the observed plant growth promoting effects of *C. rosea*, and if different isolates of *C. rosea* have differing capacities to promote plant growth.

4 Methods of application for disease management

Clonostachys rosea has been shown to effectively manage a number of plant diseases when applied through various methods (Table 2). These methods include seed treatment and applications to the growing media (through sprays to the bases of plants, drenches or inoculants that are mixed into the growing medium itself) typically to manage seed and soilborne pathogens. The BCA may also be applied to aboveground plant parts through sprays or dispersed by pollinators, such as bumblebees or honeybees, to manage diseases including fusarium head blight and grey mould. Although less researched than other methods, *C. rosea* may also be applied postharvest to manage diseases that affect the quality of fruits, tubers and seeds during storage. The method of biocontrol application should be chosen based on what particular disease is being targeted for management since the biocontrol agent must be present in sufficient concentrations in the appropriate tissues to have an effect on disease development. For example, seed treatment with *C. rosea* is unlikely to be the optimal method of biocontrol delivery to manage *B. cinerea* infections on tomato fruits, as infection occurs considerably later in the growing season, and on tissues, the biocontrol was not directly applied to. Each method of application also has its own associated costs and logistical considerations. For example. treating seeds with a biocontrol agent prior to planting may be more practical than trying to drench the growing area after planting. Other potential factors that can influence biocontrol efficacy must also be considered, including biocontrol dose, abiotic conditions, biocontrol application timing and the target plant and pathogen.

4.1 Factors that affect the efficacy of C. rosea

The efficacy of biocontrol agents can be variable, although in general they are more effective when conditions are less conducive to disease development (Hannusch and Boland, 1996; Boland, 1997). Factors such as inoculum density, temperature, humidity or leaf wetness, light exposure, ultraviolet light (UV),

Table 2 Examples of diseases managed by application of the biocontrol agent *Clonostachys rosea*

Plant	Disease (Pathogen)	Experimental environment	Application method	Reference
Alfalfa				
	Blossom blight (*Botrytis cinerea*)	Field	Spray	Li et al., 2004
Almond				
	Fungal canker (*Neofusicoccum parvum*, *Eutypa lata*)	Field	Spray	Travadon et al., 2023
Apple				
	Core rot (*Fusarium avenaceum*, *B. cinerea*)	Field	Bee vector	Hokkanen et al., 2015
Begonia				
	Leaf spot (*Myrothecium roridum*)	Greenhouse	Spray	Fujinawa et al., 2020
	Fusarium wilt (*Fusarium foetens*)	Greenhouse	Drench	Tian and Zheng, 2013
Cabbage				
	Clubroot (*Plasmodiophora brassicae*)	Field	Drench at sowing	Peng et al., 2011
	Head rot (*Pythium tracheiphilum*)	Field	Spray	Møller et al., 2003
	Head rot (*Sclerotinia sclerotiorum*)	Greenhouse	Spray	Venkatesan et al., 2022
Canola, Oilseed rape				
	Clubroot (*P. brassicae*)	Field	Seed treatment and drench at sowing	Peng et al., 2011
	Clubroot (*P. brassicae*)	Greenhouse	Drench at sowing	Lahlali and Peng, 2014
	Clubroot (*P. brassicae*)	Growth chamber	Drench	Gossen et al., 2016
	Clubroot (*P. brassicae*)	Greenhouse	Seed treatment	Andersen et al., 2018
Cannabis				
	Damping-off (*Fusarium oxysporum*)	Indoor	Drench at propagation	Scott and Punja, 2023
	Root rot (*Pythium myriotylum*)	Indoor	Drench	Scott and Punja, 2023

Carrot				
	Seedling blight (*Alternaria dauci*, *Alternaria radicina*)	Growth chamber	Seed treatment	Jensen et al., 2004
	Seedling blight (*A. dauci*, *A. radicina*)	Field	Seed treatment	Koch et al., 2020
Cereals				
	Seedling blight, foot and root rot (*Fusarium culmorum*)	Greenhouse, field	Seed treatment	Knudsen et al., 1995
	Seedling blight, foot and root rot (*F. culmorum*)	Growth chamber, field	Seed treatment	Jensen et al., 2000
	Seedling blight, foot and root rot (*F. culmorum*)	Field	Seed treatment	Roberti et al., 2000
	Seedling blight, foot and root rot (*F. culmorum*)	Field	Seed treatment	Roberti et al., 2008
	Damping-off (*Fusarium graminearum*)	Growth chamber	Seed treatment, drench before sowing	Yesim, 2020
	Head blight (*F. graminearum*)	Greenhouse, field	Spray	Hue et al., 2009
	Head blight (*F. graminearum*)	Field	Spray	Xue et al., 2009
	Head blight (*F. graminearum*)	Field	Spray	Xue et al., 2014b
	Head blight (*F. graminearum*)	Greenhouse	Spray	Rojas et al., 2014
	Leaf blotch (*Bipolaris sorokiniana*)	Growth chamber	Spray	Jensen et al., 2016
	Rusts (*Puccinia triticina*, *Puccinia coronata* f. sp. *avenacea*, *Puccinia hordei*)	Growth chamber	Spray	Wilson et al., 2020
Cherry				
	Fungal canker (*Calosphaeria pulchella*, *E. lata*)	Field	Spray	Travadon et al., 2023
Cinnamon				
	Root rot (*Phytophthora* spp.)	Greenhouse	Drench prior to sowing and spray	Dang et al., 2023

(*Continued*)

Table 2 (*Continued*)

Plant	Disease (Pathogen)	Experimental environment	Application method	Reference
Cocoa				
	Frosty pod rot (*Moniliophthora roreri*)	Field	Spray	Krauss and Soberanis, 2001
Coffee				
	Coffee leaf rust (*Hemileia vastatrix*)	Greenhouse	Spray and soil amendment	Kapeua-Ndacnou et al., 2023
Corn				
	Seedling blight (*Fusarium subglutinans*)	Growth chamber	Seed treatment	Koch et al., 2020
	Seedling blight (*F. culmorum*, *F. graminearium*, *Globisporangium ultimum*)	Greenhouse	Seed treatment	Pfeiffer et al., 2021
Cucumber				
	Damping-off, root rot (*Pythium aphanidermatum*, *Fusarium oxysporum* f. sp. *radicis-cucumerinum*)	Growth chamber, greenhouse	Drench	Rose et al., 2004
	Root and stem rot (*F. oxysporum* f. sp. *radicis cucumerinum*)	Greenhouse	Drench at sowing	Chatterton et al., 2008
	Grey mould (*B. cinerea*)	Greenhouse	Drench prior to sowing	Silva et al., 2020
Geranium				
	Botrytis blight (*B. cinerea*)	Greenhouse	Spray	Elmhirst et al., 2011
	Botrytis blight (*B. cinerea*)	Greenhouse	Spray	Shrestha and Hausbeck, 2021
Ginseng				
	Damping-off (*Pythium* spp., *Fusarium* spp., *Phytophthora cactorum*, *Rhizoctonia solani*)	Outdoor bed	Seed treatment and drench	Rahman and Punja, 2007
Grapevine				
	Grapevine trunk disease (*Cylindrocarpon* spp., *Botryosphaeriaceae* spp., *Eutypa lata*, others)	Greenhouse	Soil drench	Geiger et al., 2022

Lettuce				
	Root rot (*P. aphanidermatum*)	Greenhouse	Added to hydroponic system	Correa et al., 2010
	Lettuce drop (*S. sclerotiorum*)	Greenhouse	Soil amendment	Rodriguez et al., 2011
Onion				
	Leaf blight (*B. cinerea*)	Field	Spray	James and Sutton, 1996
Pea				
	Pea root rot complex (*Alternaria alternata*, *Fusarium* spp., *Pythium* spp., *R. solani*, *S. sclerotiorum*)	Growth chamber, Field	Seed treatment	Xue 2003a; Xue 2003b
Pepper				
	Grey mould (*B. cinerea*)	Greenhouse	Bee vector	Kapongo et al., 2008
	Crown and root rot (*F. oxysporum*)	Greenhouse	Drench	Cerkauskas, 2017
Potato				
	Dry rot (*F. avenaceum*, *Fusarium coeruleum*)	Growth chamber	Liquid inoculation to wounds	Samils et al., 2016
	Silver scurf (*Helmintosporium solani*)	Growth chamber	Liquid treatment to tubers	Lysøe et al.,2017
	Early blight (*Alternaria grandis*, *Alternaria solani*)	Greenhouse	Spray	Silva et al., 2020
Raspberry				
	Grey mould (*B. cinerea*)	Field	Bee vector, spray	Yu and Sutton, 1997a
Rose				
	Powdery mildew (*Podosphaera pannosa*)	Field	Spray	Elmhirst et al., 2011
	Grey mould (*B. cinerea*)	Growth chamber	Liquid inoculation	Morandi et al., 2000
Soybean				
	Root and stem rot (*S. sclerotiorum*)	Greenhouse	Soil amendment	Rodriguez et al., 2011
	Root and stem rot (*S. sclerotiorum*)	Growth chamber	Spray	Sun et al., 2017

(Continued)

Table 2 (*Continued*)

Plant	Disease (Pathogen)	Experimental environment	Application method	Reference
Strawberry				
	Grey mould (*B. cinerea*)	Field	Bee vector, spray	Peng et al., 1992
	Grey mould (*B. cinerea*)	Field	Spray	Cota et al., 2008
	Grey mould (*B. cinerea*)	Greenhouse	Bee vector	Van delm et al., 2015
	Grey mould (*B. cinerea*)	Field	Spray	Nechet et al., 2017
Tomato				
	Grey mould on stems (*B. cinerea*)	Greenhouse	Spray	Sutton et al., 2002
	Grey mould on stems (*B. cinerea*)	Growth chamber	Spray	Borges et al., 2015
	Root rot (*F. oxysporum* f. sp. *radicis lycopersici*)	Growth chamber	Plant dip	Karlsson et al., 2015
	Grey mould on fruits (*B. cinerea*)	Growth chamber	Spray	Gong et al., 2017
	Early blight (*A. alternata, A. solani*)	Greenhouse, field	Spray	Egel et al., 2019
Zucchini				
	Powdery mildew (*Podosphaera xanthii*)	Greenhouse, field	Spray	Tesfagiorgis et al., 2014

soil type, soil saturation, product formulation, application timing during the day and application timing relative to pathogen inoculation have all been shown to affect the efficacy of *C. rosea* at managing disease (Hannusch and Boland, 1996; Zhang et al., 1996; Yu and Sutton, 1997b; 1998; 1999; Morandi et al., 2006; 2008; Xue et al., 2014b; Borges et al., 2015; Gossen et al., 2016; Gimeno et al., 2021b). Genetic resistance to disease, and the host genotype in general, may also play a role in the efficacy of *C. rosea* as discussed later in this chapter. Consequently, the efficacy of biocontrol products must be assessed in a variety of conditions, not only those that are favourable for its growth and development, and not conducive to disease formation.

When *C. rosea* was applied to the petals of bean plants, to manage *B. cinerea*, suppression ranged from 22% to 100%, depending on temperature and relative humidity (RH) (Hannusch and Boland, 1996). In this trial, temperatures of 20° C and 28° C coupled with high RH (90%-100%) resulted in the highest suppression of grey mould by *C. rosea*. These were also the conditions under which the lowest disease severity was observed on pathogen-inoculated control plants. In contrast, greenhouse trials by Morandi et al. (2006, 2008) showed that *C. rosea* was able to suppress *B. cinerea* growth on rose over a range of environmental conditions, including those that were most conducive to disease development.

Morandi et al. (2008) also demonstrated the effect of application timing during the day and sunlight exposure on the ability of *C. rosea* to manage *B. cinerea* on rose. As the exposure to sunlight increased, the viability of *C. rosea* conidia decreased, even though the trials took place in plastic or glass greenhouses treated to filter UV light. Similarly, Costa et al. (2012; 2013) demonstrated the negative effects of UV-B radiation on the germination of *C. rosea*, its ability to colonize plant tissues, and its subsequent capacity to control *B. cinerea*. To offset the effects of UV-B radiation higher concentrations of biocontrol inoculum may be applied (Costa et al., 2012) or the biocontrol products may be formulated with components that protect conidia from UV radiation (Gimeno et al., 2021b). Contrarily, the ability of *C. rosea* to manage *B. cinerea* on strawberries was not affected by increased UV-B radiation (Nechet et al., 2017). The germination of sufficient quantities of *C. rosea* conidia is key for its biocontrol effects, as it must be able to colonize tissues and compete with potential pathogens, as demonstrated by the reduced efficacy of the biocontrol agent in trials where conidia viability was reduced. For this same reason, the concentration at which the biocontrol agent is applied may influence its ability to manage disease.

The concentration at which *C. rosea* is applied has been shown to have an effect on its ability to manage several diseases. When applied at 1×10^8 cfu mL^{-1} *C. rosea* was effective at managing fusarium head blight on wheat, comparable to fungicide treatments, the disease was significantly reduced by rates as low as 10^6 cfu mL^{-1} (Xue et al., 2014). Rates lower than this could significantly reduce fusarium head blight in field and greenhouse trials but the results were less effective than when higher rates were applied. When *C. rosea* was applied to control *B. cinerea* on raspberry shoots concentrations of biocontrol inoculum between 10^4 and 10^8 conidia mL^{-1} eliminated pathogen sporulation on leaves (Yu and Sutton, 1998). However, sufficient suppression of disease on other tissues such stems, stamens and pistils required higher concentrations of biocontrol inoculum. The inoculum concentration of *C. rosea* has also been demonstrated to be a factor on *B. cinerea* control on tomato plants and black spruce seedlings (Zhang et al., 1996; Sutton et al., 2002; Borges et al., 2015).

4.2 Seed treatment

Seed treatment is a practical method for delivery of a biocontrol agent to a large number of potential plants and is a commonly used strategy for the application of *C. rosea* and other BCAs. This method of application is primarily used when seeking to manage seed-borne pathogens that can cause diseases such as damping-off and seedling blights, as well as soilborne pathogens that affect the roots and crowns of plants. For example, seed treatment of field pea with *C. rosea* strain ACM941 was effective at managing root rot complex of field pea in field trials, with treatments having significant positive effects on emergence and yield (Xue, 2003a; Xue, 2003b). When seeds of field pea, of visibly poor quality harvested from infected plants, were treated with ACM941, the recovery of the pathogens that contribute to root rot complex (*Alternaria alternata, Fusarium* spp., *Asochyta* spp. and *Rhizoctonia solani*) was significantly reduced (Xue, 2003b). In this same experiment, seed germination and emergence were significantly increased, and root rot severity was significantly decreased, by treatment with strain ACM941.

Seed treatment with *C. rosea* has also been shown to effectively manage other pathogens on cereal crops such as wheat and barley. *Clonostachys rosea* significantly reduced seed-borne *Fusarium culmorum*, which causes seedling blight, root rot and head blight of wheat, when applied to field-sown wheat seeds (Knudsen et al., 1995; Jensen et al., 2000; Roberti, 2000). Treatment of wheat and barley seeds with *C. rosea* also significantly reduced disease caused by *Bipolaris sorokiniana*, which causes seedling blight, root rot as well as disease on the foliage and grains (Knudsen et al., 1995; Jensen et al., 2002). Growth parameters including emergence, plant dry weight and yield were positively affected by applications of the BCA to the seeds of these cereal crops.

In addition to the previously mentioned fungal pathogens, *C. rosea* applications have demonstrated efficacy at managing clubroot, caused by the protist *Plasmodiophora brassicae*. Applications of Prestop to canola seeds significantly reduced disease in a greenhouse trial (Peng et al., 2011).

However, drench applications of the biocontrol were shown to be generally more effective in these experiments, potentially due to the higher amount of inoculum being delivered by a drench compared to the amount of biocontrol inoculum that each seed received. A similar result was obtained by Andersen et al. (2018), where applications of *C. rosea* strain IK726 to canola seed reduced disease severity in a greenhouse trial. In canola field trials, Prestop seed treatment reduced clubroot severity, comparable to fungicide treatment, although neither was significantly effective in this particular trial (Peng et al., 2011). However, treatments with Prestop significantly reduced clubroot severity in field-grown Chinese cabbage (Peng et al., 2011). Soil dryness and disease pressure are thought to have played a role in the reduced efficacy of Prestop in these field trials.

4.3 Drenches and applications to growing media

C. rosea may be applied to growing media prior to seeding, transplanting or propagation, as well as to established crops through drip irrigation and drenches. This method of application of *C. rosea* has been shown to control diseases caused by *Fusarium* spp., such as post-emergence wilt on spinach (Cummings et al., 2009), *Fusarium* wilt on begonia (Tian and Zheng, 2013), stem and root rot on greenhouse cucumber (Rose et al., 2003), and crown and root rot on greenhouse pepper (Cerkauskas, 2017). Other diseases such as clubroot on canola, caused by *Plasmodiophora brassicae* (Lahlali and Peng, 2014), pitch canker on pine, caused by *Fusarium cicrinatum* (Moraga-Suazo et al., 2016) and *Pythium aphanidermatum* on greenhouse cucumber (Punja and Yip, 2003) have also been managed through drenches of *C. rosea*. In these studies, *C. rosea* was predominantly applied as a preventative treatment, to maximize its potential to outcompete other microbes, and increase its biocontrol activity. Although the previously mentioned studies were conducted under greenhouse or controlled environment conditions, a limited number of field trials with drenches of *C. rosea* have shown its ability to control plant pathogens in more variable environments and on other crops. (Rahman and Punja 2007; Peng et al., 2011; Lahlali and Peng, 2014).

4.4 Spray application

In addition to controlling seed and soilborne plant pathogens, *C. rosea* also can control diseases affecting foliar and flower tissues through spray applications. When applied to field grown strawberry plants twice a week *C. rosea* was more effective at managing grey mould (*Botrytis cinerea*), and had a greater positive effect on yield, compared to a weekly application of the fungicides procymidone or captan (Cota, et al., 2008). In trials to manage powdery mildew (*Podosphaera pannosa*) on container-grown roses, biweekly sprays of *C. rosea* significantly reduced infected leaf area to levels comparable to those observed on myclobutanil treated plants (Elmhirst et al., 2011). Sprays of *C. rosea* also significantly reduced powdery mildew (*Podosphaera xanthii*) infection on zucchini in greenhouse and field trials (Tesfagiorgis and Laing 2010; Tesfagiorgis et al., 2014). Other diseases have also been controlled through spray applications of *C. rosea*., including grey mould on raspberry (Yu and Sutton, 1997a), grey mould on tomato stems (Sutton et al., 2002; Borges et al., 2015), early blight (*Alternaria* spp.) on tomato (Egel et al., 2019), fusarium head blight (*Fusarium graminearum*) and rust pathogens (*Puccinia* spp.) on wheat (Xue et al., 2009; Xue et al., 2014b; Wilson et al, 2020), and leaf and head rot (*Sclerotinia sclerotiorum*) of Chinese cabbage (Møller et al., 2003).

4.5 Postharvest application

A limited number of studies have demonstrated that *C. rosea* may have promising capabilities for managing postharvest plant diseases when applied after harvest and prior to storage. When applied to manage silver scurf on potato, caused by *Helminthosporium solani*, *C. rosea* greatly reduced disease severity. Potato tubers that received the BCA developed symptoms on less than 5% of their surfaces, while the pathogen-only treatment group had symptoms on over 90% of their surfaces (Lysøe et al., 2017). When applied to tomato fruits to control tomato grey mould disease *C. rosea* significantly reduced visible lesions (Gong et al., 2017). Treatments with *C. rosea* on potato tubers and tomato fruits were also observed to induce resistance in these tissues. Additionally, cell free culture filtrates of *C. rosea* may be used to manage disease, as they contain antifungal compounds. When *C. rosea* filtrates were inoculated in wounds on tomato fruits they suppressed anthracnose-induced decay, caused by *Colletotrichum coccodes* (Hassine et al. 2022). Applications of *C. rosea* isolate IK726 to English oak acorns prior to storage resulted in increased germination rates, with or without disinfection by hot water treatment, and reduced infection by *Ciboria batschiana*, a major pathogen of stored acorns (Knudsen et al., 2004). Postharvest diseases affect a wide array of crops in addition to those mentioned here, such as citrus fruits, stone fruits, apple, strawberry and others (Lastochkina et al., 2019). Applications of *C. rosea* or its byproducts could play an increased role in postharvest disease management in these crops, but further research is required.

4.6 Entomovectoring

The use of honey bees, bumble bees or other pollinators as vectors for biocontrol agents is an innovative and effective approach for the delivery of inoculum to crops at key stages in their development, while also enhancing pollination. Initially, to investigate the possibility of using honey bees as a vector of *C. rosea*, Peng et al. (1992) created a rudimentary inoculum dispenser that bees would travel through before leaving the hive. As they travelled through the dispenser the bees became covered in *C. rosea* spores mixed with a carrier, such as talc. When tested in a field trial to manage *B. cinerea* on strawberries, the honey bees successfully transferred the inoculum to flowers, which resulted in subsequent suppression of disease (Peng et al., 1992). From this initial work, entomovectoring has been researched with other biocontrol agents, plant pathogens and bee vectors, employing a range of dispenser designs (Mommaerts and Smagghe, 2011). Commercial dispensers have also been developed for use with bumble bees and honey bees (Table 1).

In addition to the initial research performed by Peng et al. (1992), *C. rosea* has been applied through bee vectors to successfully manage botrytis blossom blight on rabbiteye blueberry (Smith et al., 2012), grey mould on raspberry (Yu and Sutton, 1997a), greenhouse-grown tomato and pepper (Kapongo et al., 2008) and field-grown strawberry (Mommaerts et al., 2010; Karise et al., 2016). Preliminary research has also shown that this may be a possible method for the management of apple core rot, caused by *Fusarium avenaceum* (Hokkanen, et al., 2015).

4.7 Combined application with other biocontrol agents

Biocontrol agents may be applied through many of the previously described methods either alone, or in combination with other biocontrol agents (Collinge et al., 2022). This combined application is referred to as applying biocontrol agents in 'consortia' or using mixed biocontrol inocula. These mixes may consist of different isolates of the same species of BCA, or of microbes from different kingdoms entirely. The aim of this approach is to provide a mix of inocula that may have additive or synergistic effects to manage multiple pests or pathogens or to manage a specific problem more effectively.

Metarhizium spp. and *C. rosea* were applied together to wheat seeds to manage *F. culmorum* and root feeding insects (Keyser et al., 2016). In this study, treatment with *C. rosea* and *Metarhizium* spp. was shown to significantly reduce disease symptoms as well as result in significant infection of root feeding insects, respectively, when applied alone or in combination. Although *C. rosea* has been observed to have mycoparasitic activity against *M. anisopliea in vitro*, it is unclear as to how this translates to the potential for *C. rosea* to inhibit its efficacy (Krauss et al., 2004). In the study by Keyser et al. (2016) cotreated seeds resulted in lower virulence to the insect pest *Tenebrio molitor* compared to the *Metarhizium*-only treatments. However, treatments with *M. brunneum* and *C. rosea* still resulted in up to 80% insect mortality, suggesting that this may be an effective and practical approach to managing fungal and insect pests with a combined treatment.

The potential for biocontrol agents to act as antagonists, or otherwise inhibit one another, is a key consideration when selecting which microbes to apply in consortia. This may depend on factors such as which species, or isolates of species, are being used as well as the concentrations and ratio of the inocula (Hoopen et al., 2010; Xu et al., 2010; Xu et al., 2011b; Krauss et al., 2013). For example, when *Trichoderma* spp. and *C. rosea* were applied together at a ratio of 5:1, to cocoa pods in a field setting, *C. rosea* was the dominant species, suppressing the growth of *Trichoderma* spp. (Krauss et al., 2013). When the concentration of *Trichoderma* spp. was increased to 10 times that of *C. rosea*, only temporary suppression of *C. rosea* was observed; two weeks after the application of this mixture *C. rosea* was again the dominant species. In this study, *C. rosea* maintained its populations on cocoa pods under field conditions for the duration of the trial and was also the strongest colonizer of surface-sterilized pods. Based on these findings, Krauss et al. (2013) suggested that incompatibilities between microbes may be partially overcome by adjusting their inoculum ratios. However, it is not clear if mixes of these biocontrol agents would have any increased efficacy at managing cocoa pod diseases compared to applications solely of the more dominant *C. rosea*. The *Trichoderma* spp. studied by Krauss et al. (2013) were poor colonizers of cocoa pods, especially compared to the biocontrol agent they were being applied with. The authors go on to suggest that further work should focus on *Trichoderma* spp. which act through another mechanism, namely epiphytic colonization of plant tissues.

The mechanisms through which different biocontrol agents act should be considered when selecting microbes for in consortia applications, as this may improve the efficacy of the mix, be a source of antagonistic interactions, or most commonly, provide no benefit compared to using only the more efficacious BCAs (Robinson-Boyer et al., 2009; Xu et al., 2011a; Xu et al., 2011b; Pertot et al., 2017). For example, a combination of *C. rosea* and *Talormyces apiculatus* resulted in greater control of basal stem rot of oil palm compared to the application of either agent alone (Goh et al., 2020). It was suggested by the authors that this may be due to the fact that these BCAs act through different mechanisms; *C. rosea* acted as an aggressive antagonist while *T. apiculatus* promoted plant growth, acted as a biotrophic parasite and improved the nutrient status of the plants. Conversely, when *C. rosea* was paired with either *Serratia rubidaea* or *Pseudomonas chlororaphis* to manage tomato foot and root rot, caused by *F. oxysporum* f. sp. *Radices-lycopersici* (Forl), the reduction in disease was comparable to when *C. rosea* was applied alone (Kamou et al., 2016). It is possible that the mix of BCAs was not more effective at managing disease compared to *C. rosea* alone because the BCAs competed for the same resources. Bacteria such as these may use hyphae of Forl as a source of nutrients and *C. rosea* has been shown to parasitize Forl hyphae (Karlsson et al., 2015; Kamou et al., 2016). Some inhibition between *C. rosea* and *P.*

chlororaphis was also observed in a series of *in vitro* assays (Kamou et al., 2016). However, a combination of *C. rosea* and *Pseudomonas chlororaphis* had an additive effect when used to manage tomato foot and root rot caused by Forl, resulting in significantly greater control of this disease than either BCA alone, as demonstrated by Karlsson et al. (2015). This may be due to the ability of *C. rosea* to tolerate the *P. chlororaphis*-produced antibiotic, phenazine-1-carboxamide (PCN), which in the same tests completely inhibited Forl growth (Karlsson et al., 2015). It is unclear if the difference in the biocontrol efficacy of *C. rosea* combined with *P. chlororaphis* at managing Forl shown in these two studies is due to the difference in *P. chlororaphis* isolate being tested, or some other aspect of the pathosystem such as the growing environment.

Applications of *C. rosea* in consortia with other microbes have also been studied. When *C. rosea* IK 726 was applied with the mycorrhizal fungus *Glomus intraradices* to tomato plants, growth promotion comparable to or better than treatment with *C. rosea* only, was observed (Ravnskov et al., 2006). This occurred despite inhibition between the two fungi.

5 Use in integrated pest management

Biocontrol products may also be used alongside other disease management approaches, as part of an integrated pest management (IPM) strategy. Examples of other disease management strategies include cultural controls which can change aspects of the growing environment (temperature, humidity, nutrient levels, sanitation and crop debris management, etc.) to make it less conducive to disease, chemical control (pesticides) and the use of disease-resistant varieties. When these different strategies are utilized together in the correct manner they can have an additive effect, enhancing the overall efficacy of the disease management program. *C. rosea* has been shown to benefit from being combined with other disease management strategies, however further research may help researchers and growers understand the optimal manner in which to combine these approaches.

5.1 Application with pesticides

Clonotschays rosea isolates exhibit a spectrum of resistance to pesticides. Fungal isolates showed lower sensitivity to fungicides such as carboxin, metalaxyl, sedaxane, thiram, fluopyram and Switch® (cyprodinil and fludioxonil), whereas they were more sensitive to fungicides including prochloraz, Pristine® (boscalid and pyraclostrobin) and Maestro® (Captan) (Roberti et al., 2006; Reeh and Cutler, 2013; Song et al., 2023; Pimentel et al., 2022). Low sensitivity to herbicides was also observed (Roberti et al., 2006). Although the growth of *C. rosea* was negatively affected by direct exposure to higher doses of the soil fumigant dazomet, pretreatment of soil with the fumigant followed by treatment with the BCA resulted in significantly greater control of *F. oxysporum* f. sp. *cucumerinum*, the causal agent of fusarium wilt on cucumbers, than either treatment applied alone (Tian et al., 2014). Similarly, when used alongside fluxapyroxad or fluopyram to control grey mould on greenhouse grown tomatoes, the combination of *C. rosea* and a fungicide resulted in greater disease control than either treatment alone (Song et al., 2023). This combination also reduced the levels of both products that needed to be used. Significant control of *R. solani* on hydroponically grown tobacco was achieved by using a fungicide-tolerant isolate of *C. rosea* and iprodine, allowing for a lower dose of the fungicide to be used (Tarantino et al., 2007).

This research shows that *C. rosea* may be successfully used with full or reduced doses of fungicides or alternated with fungicides to control plant diseases. Consequently, the addition of biocontrol products may help reduce the use of fungicides. This approach can be beneficial by reducing potential off-target ecological effects, and by reducing the potential for the development of fungicide resistance in the pathogen. However, further work is required to establish what other pathosystems may benefit from this approach. Future work may also involve the editing of *C. rosea* genomes to increase fungicide tolerance rather than screening preexisting isolates (Muñoz et al., 2019). These new isolates may potentially be used as part of a pest management program that focuses on the reduced use of fungicides through the addition of a biocontrol agent.

5.2 Application to crop debris

The application of *C. rosea* to crop residues may also be a facet of integrated pest management. Crop debris or residues, such as stalks, stems, leaves, petals and grains may be colonized by pathogens during, or after, the growing season and act as a source of inoculum (spores, mycelium, etc.) for future crops (Luongo et al., 2005; Osborne and Stein, 2007). Treatment of crop residues with a BCA can result in lower pathogen populations and subsequently reduce future inoculum levels, as the BCA may colonize plant tissues more quickly than the pathogen or directly suppress its growth through the previously discussed mechanisms (Morandi et al., 2001; Palazzini et al., 2013; Xue et al., 2014a). The application of *C. rosea* has been shown to reduce the colonization of wheat, maize and soybean debris by *Fusarium* spp. (Luongo et al., 2005; Palazzini et al., 2013; Xue et al., 2014a) and to reduce airborne inoculum levels (Gimeno et al., 2020). In work by Xue et al. (2014a), *C. rosea* isolate ACM941 was sprayed on wheat residues (peduncles, stems and spikelets) that were naturally infected with *F. graminearum* in the autumn, to manage fusarium head blight on wheat. This significantly reduced the amount of pathogen inoculum (perithecia) present in the next growing season to levels comparable to what was observed on tebuconazole-treated residues. Treatment of maize crop residues with *C. rosea* also significantly reduced fusarium head blight disease severity and mycotoxin (ZEN and DON) contamination of grains in a subsequently grown wheat crop (Gimeno et al., 2020). *Clonostachys rosea* can also be applied to plant debris to reduce sporulation by *B. cinerea* on rose and strawberry crops (Sutton and Peng, 1993; Morandi et al., 2001, 2003, 2006, 2008; Nobre et al., 2005). However, further research is required to determine to what extent this may reduce disease severity.

5.3 Use with silicon

The addition of silicon directly to plants or to their growing media, through sprays or drenches, has been shown to benefit plants by alleviating abiotic and biotic stresses, including reducing symptoms of various diseases (Debona et al., 2017). Sprays of *C. rosea* significantly reduced powdery mildew on zucchini in greenhouse and field trials, and when applied in tandem with potassium silicate drenches, disease control was improved compared to either treatment alone (Tesfagiorgis et al., 2014). In these trials, *C. rosea* was shown to be effective at temperatures that were not favourable for disease management using potassium silicate application alone. When *C. rosea* was applied as a drench in combination with silicon silicon-containing nutrient solution, to manage *B. cinerea* on greenhouse-grown cucumber, disease took significantly longer to develop and disease development overall was significantly reduced (Silva et al., 2020). However, in these experiments, the combination of silicon and *C. rosea* did not produce significantly different results than applications of *C. rosea* alone, in terms of disease management. The combination of biocontrol agents and silicon has also been shown to work synergistically to manage post-harvest diseases (Tian et al., 2005; Qin and Tian, 2005; Farahani et al., 2012). However, research about the potential for applications of biocontrol agents alongside silicon to have enhanced effects at managing disease is otherwise very limited.

5.4 Resistant varieties

The use of cultivars that are resistant to disease is an inexpensive, efficient and fundamental strategy in integrated disease management (Nelson et al., 2018). The use of resistant varieties may also improve the efficacy of other disease management approaches, including the application of biocontrol agents. When the *C. rosea* isolate CLO-1 was applied to control fusarium head blight on three different varieties of wheat, cultivar resistance played a role in the efficacy of the biocontrol agent (Xue et al., 2014b). Applications of *C. rosea* were most effective on the moderately resistant varieties and least effective on the two more susceptible varieties that were tested. *C. rosea* was also effective at managing *B. cinerea* on geranium cultivars that had high and low disease susceptibility, whereas some tested biocontrol agents were only effective on cultivars with lower susceptibility (Shrestha and Hausbeck, 2021). *C. rosea* also effectively reduced pitch canker development, caused by *Fusarium circinatum*, on a disease-resistant pine variety (Moraga-Suazo et al., 2016). The same effects were not observed on susceptible pine varieties, highlighting the potential effects of host genotype on ISR and biocontrol efficacy in general.

6 Interactions with nematodes and other plant pests

Biocontrol agents may also be useful for managing plant diseases caused by other pathogenic organisms, such as plant parasitic nematodes (PPNs). Nematodes including root-knot nematodes (*Meloidogyne* spp.), cyst nematodes (*Heterodera* and *Globodera* spp.), lesion nematodes (*Pratylenchus* spp.) and stem and bulb nematodes (*Diteylenchus dipsaci*), among others, cause significant damage to crops and result in economic losses worldwide (Pires et al., 2022). The application of *C. rosea* strain IK726 to naturally nematode-infested soil decreased the populations of PPNs, while having no obviously deleterious effects on nonparasitic nematode populations (Iqbal et al., 2018a). Consequently, carrot and wheat plants grown in the biocontrol treated soils had improved growth, compared to plants grown in untreated soil. The addition of *C. rosea* inoculum to the growing media of greenhouse grown tomato plants was also effective at controlling root knot nematodes (Cristóbal-Alejo et al., 2021). Research on the efficacy of *C. rosea* at managing PPNs is otherwise extremely limited, and no field trials have been done to date. However, this is a potentially promising area of research since numerous other biocontrol agents, such as *Bacillus* spp., *Trichoderma* spp. and others, have shown efficacy in managing a range of PPN species (Pocurull et al., 2020; Huang et al., 2021; Pires et al., 2022).

These biocontrol agents affect PPNs through mechanisms such as endoparasitism, toxin and enzyme production, and through indirect mechanisms such as the induction of plant defenses (Ghahremani et al., 2020; Pires et al., 2022). The mechanisms through which *C. rosea* may affect PPNs requires further study, although initial research has shown that the production of enzymes and nematocidal compounds can have a role (Dong et al., 2005; Li et al., 2006; Zou et al., 2010; Iqbal et al., 2018b, 2019, 2020). Using green fluorescent protein as a marker, *C. rosea* conidia could be seen germinating on and penetrating nematodes before killing them (Zhang et al., 2008).

Clonostachys rosea also has the potential to act as an entomopathogenic fungus. It has shown promising effects on a wide range of insects that can cause substantial damage to crops and that may act as vectors of plant pathogens. These include the Asian citrus psyllid (Yang et al., 2021), khapra beetle, red flour beetle and cowpea weevil, which cause damage to stored grains and other products (Mohammed et al., 2021). Also negatively affected are whitefly (Anwar et al., 2018), tomato leafminer (Mahmoud et al., 2021), leafhoppers and mango hoppers (Toledo et al., 2006; Tamta et al., 2022), in addition to green peach aphids and bean aphids (Mohammed et al., 2022). Additional studies are needed to determine the range of insects affected by *C. rosea*, the conditions that may optimize its effects and the practical limitations of this method of insect pest management on a large scale.

When infecting Asian citrus psyllid (*Diaphorina citri*) adults, conidia of *C. rosea* were observed to germinate, penetrate and subsequently kill the insect. The appressorium and penetration peg, which are key structures by which entomopathogenic fungi attack and gain entry to their hosts, are aided by the production of enzymes, such as proteases and chitinases, which degrade the host's cuticle (Shin et al., 2020). Additionally, *C. rosea* was shown to suppress the immune response of *D. citri*, through the downregulation of genes encoding cuticle proteins and the upregulation of genes associated with apoptosis (Yang et al., 2021). However, other possible mechanisms that are involved in the entomopathogenic activity of *C. rosea* require further research.

7 *Clonostachys rosea* as a plant pathogen

Although *C. rosea* is an effective biocontrol agent against numerous pathogens and pests, there are also reported instances of isolates acting as plant pathogens. *C. rosea* has been reported to cause root rot on soybean, orchids and garlic as well as on the traditional Chinese medicinal plants *Astagalus membranaceus* and *Angelica sinensis* (Bienapfl et al., 2012; Lee et al., 2020; Diaz et al., 2022; Ma et al., 2022; Qi et al., 2022). On faba bean and gladiolus plants, *C. rosea* was reported to cause crown rot and wilt, and basal stem rot and corm rot, respectively (González-Pérez et al., 2008; Afshari and Hemmati, 2017). Additionally, it has been reported to cause dry rot on potatoes, leaf spot on tea (*Camellia sinensis*) and fruit rot on avocado (Theron and Holz, 1991; Coyotl-Pérez et al., 2022; Linghu et al., 2023).

Using nuclear ribosomal internal transcribed spacer (ITS) and translation elongation factor 1-alpha (TEF-1α) sequencing, as well as analysis of the morphological characteristics of their isolates, authors were able to identify the pathogens as *C. rosea* (Afshari and Hemmati, 2017; Coyotl-Pérez et al., 2022). When the ITS and TEF-1α sequences of pathogenic *C. rosea* reported by Coyotl-Pérez et al. (2022) were compared to other *C. rosea* isolates, they showed the highest homology with sequences reported to cause diseases by Ma et al. (2022) and Qi et al. (2022). Similarly, when ITS sequences of pathogenic *C. rosea* isolates were compared to sequences of *C. rosea* isolates from around the world, pathogenic isolates were shown to cluster together (Jensen et al., 2022). Although the inclusion of TEF-1α sequences in the previous work helped to confirm the identity of *Clonostachys* spp., further phylogenetic analyses using genetic markers such as beta-tubulin (TUB), ATP citrate lyase (acl1) and RNA polymerase II large subunit (rpb1), are at times necessary for higher resolution identification and differentiation at the species level (Moreira et al., 2016; Wang et al., 2023). The genetic differences between pathogenic and beneficial isolates of *C. rosea*, or what may make an otherwise beneficial isolate have a pathogenic effect on a plant host, is currently not well understood.

8 Application of *C. rosea* for disease management on *Cannabis sativa*

Cannabis sativa is a diverse and valuable plant which is grown worldwide for its nutritious seeds, strong fibre or potent secondary metabolites, such as delta-9-tetrahydrocannabinol (THC) and cannabidiol (CBD). Cannabis grown for its secondary metabolites, or drug type cannabis, can be grown in a variety of production systems, including outdoors, in greenhouses with or without environmental controls, and completely indoors. Cannabis is affected by numerous diseases, the most common of which are powdery mildew, damping-off and root and crown rot caused by *Fusarium* spp. and *Pythium* spp. and bud rot caused mainly by *B. cinerea* (Punja, 2021). *C. rosea* is registered for use as a fungicide in Canada, in the form of Prestop, to suppress crown and root rot caused by *Pythium* spp. and *B. cinerea* on cannabis grown indoors. However, there is little or no data regarding the efficacy of *C. rosea* in controlling these diseases on cannabis, under any production system. Initial research by Scott and Punja (2023) showed that when applied preventatively during the propagation stage to cannabis cuttings, *C. rosea* significantly reduced the severity of damping-off, caused by *F. oxysporum* (Figure 3). In the same study, the BCA also had a significant effect against *P. myriotylum* when applied preventatively, reducing disease severity and increasing fresh root weight compared to the pathogen-only treatment group (Figure 4). Endophytic colonization of cannabis stems by *C. rosea* in biocontrol-treated cuttings was demonstrated (Scott and Punja, 2023) (Figure 5); however the mechanisms through which *C. rosea* reduced these diseases on cannabis remain to be studied.

As discussed previously, *C. rosea* is effective at managing *B. cinerea* on a range of crops, both under field and greenhouse conditions. It is unclear if this BCA will be effective in reducing diseases on cannabis grown under a range of production conditions. Cannabis must undergo microbial testing for total yeast and mould levels based on the number of colony-forming units (cfu) per gram of dried flower tissue. The cfu/g limit varies greatly by state and country (Punja, 2021). It is not clear what the possible implications of biocontrol treatment may have for microbial testing on cannabis. The proper timing of applications of BCAs such as *C. rosea* will be key for the successful

Figure 3 Biocontrol of Fusarium damping off on cannabis cuttings by application of (A) *Clonostachys rosea* (Lalstop) to the rockwool blocks 48 hr prior to application of the pathogen compared to (B) the pathogen-only control. Photo was taken 14 days after pathogen inoculation. The graph shows the disease severity ratings at 18 days post inoculation with the pathogen. Different letters above each bar denote significant differences in the disease severity ratings as determined through a Kruskal-Wallis with *post hoc* Conover-Iman test and Bonferonni correction ($P < 0.05$). The figure is adapted from Scott and Punja (2023).

and practical management of *B. cinerea* on cannabis. The biocontrol will likely require time to colonize flower tissues before disease pressure becomes too severe, while simultaneously not contributing too greatly to microbial testing done postharvest. Under *in vitro* conditions, application of Prestop to detached cannabis inflorescences 48 h preceding inoculation with *B. cinerea* reduced bud rot development (Punja and Ni, 2021). The biocontrol agent sporulated extensively on these tissues.

Clonostachys rosea may potentially play other roles in cannabis production. As previously discussed, applications of *C. rosea* may be effective at managing PPNs, such as root-knot nematode which is a reported pest of *C. sativa* (Bernard et al., 2022). Similarly, *C. rosea* has been shown to have entomopathogenic effects on leaf hopper species which are a pest of cannabis and may vector pathogens such as beet curly top virus (Giladi et al., 2020; Melgarejo et al., 2022). Applications of *C. rosea* on other crops have effectively controlled a range of plant pathogens, some of which are also reported to infect cannabis, including *Fusarium graminearum, Sclerotinia sclerotiorum*, *Rhizoctonia solani*

Figure 4 The effects of Lalstop (*Clonostachys rosea*) at managing *Pythium myriotylum* on cannabis. The plant on the left is the pathogen only treatment, while the plant on the right received Lalstop 7 days prior to pathogen inoculation. The photo was taken 18 days post inoculation with the pathogen. The figure is adapted from Scott and Punja (2023).

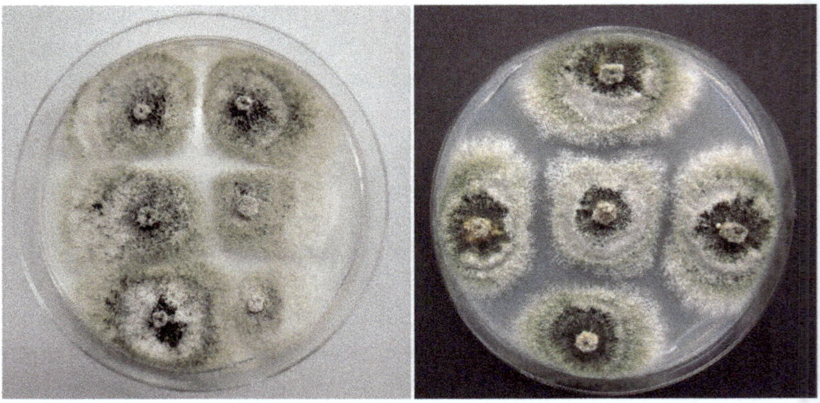

Figure 5 Recovery of *Clonostachys rosea* from cannabis cuttings with (left) and without (right) surface sterilization. Stem sections were plated onto potato dextrose agar one week after treatment with the biocontrol agent. The figure is adapted from Scott and Punja (2023).

and *Bipolaris* spp. Further studies are needed to assess the extent to which *C. rosea* can reduce these diseases affecting cannabis crops.

9 Conclusion and future trends

C. rosea is an impactful biocontrol agent, showing strong effects against a range of fungal, bacterial and oomycete pathogens in various agricultural environments. Application methods including seed treatments, drenches, sprays and entomovectoring can be effectively utilized to deliver *C. rosea,* depending on the disease being targeted for management. However, the efficacy of *C. rosea* may be greatly reduced, or can otherwise be unreliable, depending on many factors. Future work can aim to address the variability of *C. rosea* through numerous potential avenues, including an increased understanding of how various conditions affect *C. rosea* efficacy, as well as the mechanisms through which *C. rosea* acts and their underlying genetic components. Enhanced understanding of these areas may allow for more effective application of *C. rosea* to manage disease.

The abiotic stresses that affect biocontrol efficacy may also be mitigated using improved product formulations. These formulations may contain protective compounds, nutrients to enhance biocontrol growth and polymers which may encapsulate the microbes (Kim et al., 2008; Wu et al., 2015; Gimeno et al., 2021b). Encapsulation may improve the resistance of biocontrol agents to stresses such as low humidity and extreme temperatures and may increase product shelf life (Marian and Shimizu, 2019). When embedded in sodium alginate, the stability of *C. rosea* was improved, and no reduction in its biocontrol efficacy was observed (Liu et al., 2023). The potential of spray drying technologies to create dry flowable *C. rosea* product formulations, with many desirable characteristics such as good shelf life, fine particle size, resistance to wash off by rain and mixability, requires further research. Work by Hong-qu et al. (2018) demonstrated the potential of these types of formulations, with a dry flowable *C. rosea* formulation containing both conidia and fermentation broth having a stable shelf life and being able to manage *S. sclerotiorum* more effectively on cucumber compared to conidia powder or fermentation broth alone.

Careful screening and selection of *C. rosea* isolates to apply in consortia may also lead to improved biocontrol efficacy and reliability. Biocontrol agents that can excel in a range of environmental conditions and which may suppress disease through differing mechanisms may be combined to manage disease in variable conditions. Screening of *C. rosea* isolates for compatibility with other microbes, tolerances to abiotic stresses, direct or indirect biocontrol capabilities and other traits of interest may also be enhanced using available whole genome sequences of *C. rosea,* alongside genetic markers (Raymaekers et al., 2020). Targeted changes to the genome of *C. rosea*, through more conventional approaches such as protoplast fusion, or modern gene editing technologies such as CRISPR-Cas9, could improve its efficacy by changing the expression of currently existing genes, or by inserting new genes, potentially for characteristics such as fungicide resistance (Wang et al., 2022). The efficacy of *C. rosea* may also be improved by editing plant genomes to alter characteristics such as their innate disease resistance or their sensitivity to ISR induced by biocontrol agents (Leibman-Markus et al., 2021). Research into these areas, and others, can enhance the reliability and efficacy of *C. rosea* to manage plant pathogens, and other pests, as well as potentially lead to the development of new *C. rosea*-based biocontrol products, thus increasing the effective uses of *C. rosea* in agriculture and our overall understanding of biocontrol agents.

10 Where to look for further information

- The review by Jensen et al. (2022) contains information about some of the other aspects of *C. rosea* as a biocontrol agent, including information about the genetic components of *C. rosea* as a biocontrol agent and the process of biocontrol commercialization.
- Further information about the development, formulation and commercialization of *C. rosea* and other biocontrol products, including safety and potential regulatory steps required for product development, can be found in Jensen et al. (2007); Droby et al. (2016); Hung-qu et al. (2013); Marian and Shimizu (2019); Zaki et al. (2020); Bettiol et al. (2021); Collinge et al. (2022); Texeidó et al. (2022) Ptaszek et al. (2023).
- A comprehensive overview of the use of insect vectors to control plant diseases and enhance pollination is provided by Smagghe et al. (2020).

11 References

Abdellatif, L., et al. (2022). Mode of action of potential biocontrol agents against *Fusarium* species and *Cochliobolus sativus*. *Mycologia* 114(3): 476–486.

Abdel-Wahab, N.M., et al. (2019). Cyclic heptapeptides from the soil-derived fungus *Clonostachys rosea*. *Bioorgan. Med. Chem.* 27: 3954–3959.

Afshari, N. and Hemmati, R. (2017). First report of the occurrence and pathogenicity of *Clonostachys rosea* on faba bean. *Australas. Plant Pathol.* 46(3): 231–234.

Altomare, C., et al. (1999). Solubilization of phosphates and micronutrients by the plant-growth promoting and biocontrol fungus *Trichoderma harzianum rifai*. *Appl. Eviron. Microbiol.* 65: 2926–2933.

Andersen, C., et al. (2018). Seed coating with the fungal biocontrol agent *Clonostachys rosea* controls clubroot in oilseed rape. Proceedings of the IOBC/WPRS Working Group "Integrated Control in Oilseed Crops." Zagreb, Croatia, 18–20 September 2018. *IOBC WPRS Bulletin*. 136: 157–163.

Anwar, W., et al. (2018). Entomopathogenic fungus *Clonostachys rosea* as a biocontrol agent against whitefly (*Bemisia tabaci*). *Biocont. Sci. Tech.* 28(8): 750–760.

Atanasova, L., et al. (2018). Evolution and functional characterization of pectate lyase PEL12, a member of a highly expanded *Clonostachys rosea* polysaccharide lyase 1 family. *BMC Microbiol.* 18(1): 178.

Barnett, H. L. and Lilly, V. G. (1962). A destructive mycoparasite, Gliocladium roseum. *Mycologia* 54(1): 72–77.

Bernard, E.C., et al. (2022). Review of nematode interactions with hemp (*Cannabis sativa*). *J. Nematol.* 54(1): 20220002.

Bettiol, W., et al. (2021). Advances in screening approaches for the development of microbial bioprotectants for control of plant diseases. In: Köhl, J., Ravensberg, W. (Eds) *Microbial Bioprotectants for Plant Disease Management*. Burleigh Dodds Science Publishing, Cambridge, UK.

Bienapfl, J.C., et al. (2012). First report of *Clonostachys rosea* causing root rot of soybean in the United States. *Plant Dis.* 96(11): 1700–1700.

Boland, C.J. (1997). Stability analysis for evaluating the influence of environment on chemical and biological control of white mold (*Sclerotinia sclerotiorum*) of bean. *Biol. Cont.* 9: 7–14.

Boller, T. and Felix, G. (2009). A renaissance of elicitors: perception of microbe-associated molecular patterns and danger signals by pattern-recognition receptors. *Annu. Rev. Plant Biol.* 60: 379–406.

Borges, Á.V., et al. (2015). Biocontrol of gray mold in tomato plants by *Clonostachys rosea*. *Trop. Plant Pathol.* 40(2): 71–76.

Broberg, M., et al. (2021). Comparative genomics highlights the importance of drug efflux transporters during evolution of mycoparasitism in *Clonostachys* subgenus *Bionectria* (fungi, Ascomycota, Hypocreales). *Evol. Appl.* 14(2): 476–497.

Cerkauskas, R.F. (2017). Etiology and management of Fusarium crown and root rot (*Fusarium oxysporum*) on greenhouse pepper in Ontario, Canada. *Can. J. Plant Path.* 39(2): 121–132.

Chatterjee, S., et al. (2016). Interactions among filamentous fungi *Aspergillus niger*, *Fusarium verticillioides* and *Clonostachys rosea*: fungal biomass, diversity of secreted metabolites and fumonisin production. *BMC Microbiol.* 16: 83.

Chatterton, S., et al. (2008). Colonization of cucumber plants by the biocontrol fungus *Clonostachys rosea* f. *catenulata*. *Biol. Cont.* 46: 267–278.

Chatterton, S. and Punja, Z.K. (2009). Chitinase and beta-1,3-glucanase enzyme production by the mycoparasite *Clonostachys rosea* f. catenulata against fungal plant pathogens. *Can. J. Microbiol.* 55(4): 356–367. https://doi.org/10.1139/w08-156.

Chatterton, S. and Punja, Z.K. (2012). Colonization of geranium foliage by *Clonostachys rosea* f. *catenulata*, a biological control agent of botrytis grey mould. *Botany* 90: 1–10.

Collinge, D.B., et al. (2022). Biological control of plant diseases – what has been achieved and what is the direction? *Plant Pathol.* 71: 1024–1047.

Contreras-Cornejo, H.A., et al. (2014). *Trichoderma* spp. improve growth of arabidopsis seedlings under salt stress through enhanced root development, osmolite production, and Na+ elimination through root exudates. *Mol. Plant-Microbe Interact.* 27(6): 503–514.

Correa, E.B., et al. (2010). Biological control of *Pythium aphanidermatum* root rot and growth promotion of hydroponic lettuce by *Clonostachys rosea*. *Trop. Plant Pathol.* 35: 248–252.

Costa, L.B., et al. (2012). Impact of UV-B radiation on *Clonostachys rosea* germination and growth. *World J Microbiol Biotechnol*. 28(7): 2497–504.

Cota, L.V., et al. (2008). Biological control of strawberry gray mold by *Clonostachys rosea* under field conditions. *Biol. Cont*. 46: 515–522.

Coyotl-Pérez, W.A., et al. (2022). First report of *Clonostachys rosea* associated with avocado fruit rot in Puebla, Mexico. *Mex. J. Phytopathol*. 40(2): 298–307.

Cristóbal-Alejo, J., et al. (2021). *Clonostachys rosea* selected by nematicidal screening and its efficacy against *Meloidogyne incognita* in a greenhouse. *Biocont. Sci. Tech*. 31(12): 1283–1297.

Cummings, J.A., et al. (2009). Greenhouse evaluation of seed and drench treatments for organic management of soilborne pathogens of spinach. *Plant Dis*. 93: 1281–1292.

Dang, Q.N., et al. (2023). Fungal and bacterial endophytes antagonistic to *Phytophthora* species causing root rot in *Cinnamomum cassia*. *Mycological Prog*. 22: 28.

de Andrade Carvalho, A. L., et al. (2018). Optimizing the mass production of *Clonostachys rosea* by liquid-state fermentation. *Biol. Cont*. 118: 16–25.

Debona, D., et al. (2017). Silicon's role in abiotic and biotic plant stresses. *Ann. Rev. Phytopathol*. 55: 85–107.

Demissie, Z.A., et al. (2020). Transcriptomic and exometabolomic profiling reveals antagonistic and defensive modes of *Clonostachys rosea* action against *Fusarium graminearum*. *Mol. Plant. Microbe Interact*. 33(6): 842–858.

Diaz, R., et al. (2022). First report of *Clonostachys rosea* causing root rot on garlic in Mexico. *Plant Dis*. 106: 3000.

Dong, J. Y., et al. (2005). Nematicidal epipolysulfanyldioxopiperazines from *Gliocladium roseum*. *J. Nat. Prod*. 68(10): 1510–1513.

Droby, S., et al. (2016). The science, development, and commercialization of postharvest biocontrol products. *Postharv. Biol. Tech*. 122: 22–29.

Dubey, M.K., et al. (2014). An ATP-binding cassette pleiotropic drug transporter protein is required for xenobiotic tolerance and antagonism in the fungal biocontrol agent *Clonostachys rosea*. *Mol. Plant. Microbe Interact*. 27(7): 725–732.

Dubey, M.K., et al. (2016). The ABC transporter ABCG29 is involved in H_2O_2 tolerance and biocontrol traits in the fungus *Clonostachys rosea*. *Mol. Genet. Genomics* 291(2): 677–686.

Egel, D.S., et al. (2019). Efficacy of organic disease control products on common foliar diseases of tomato in field and greenhouse trials. *Crop Prot*. 122: 90–97.

Elmhirst, J.F., et al. (2011). Evaluation of biological control agents for control of botrytis blight of geranium and powdery mildew of rose. *Can. J. Plant Path*. 33(4): 499–505.

Farahani, L., et al. (2012). Biocontrol of blue mold of apple by *Candida membranifaciens* in combination with silicon. *Arch. Phytopathol. Plant Prot*. 45(3): 310–317.

Fatema, U., et al. (2018). Functional analysis of polyketide synthase genes in the biocontrol fungus *Clonostachys rosea*. *Sci. Rep*. 8(1): 15009.

Flores, A., et al. (1997). Improved biocontrol activity of *Trichoderma harzianum* by over-expression of the proteinase-encoding gene *prb1*. *Curr. Genet*. 31: 30–37.

Fujinawa, M.F., et al. (2020). Biological control of gray mold and myrothecium leaf spot in begonias. *Crop Prot*. 133: 105138.

Geiger, A., et al. (2022). Mycoparsitism capability and growth inhibition activity of *Clonostachys rosea* isolates against fungal pathogens of grapevine trunk dsieases suggest potential for biocontrol. *PLoS One* 17(9). https://doi.org/10.1371/journal.pone.0273985

Ghahremani, Z., et al. (2020). *Bacillus firmus* strain I-1582, a nematode antagonist by itself and through the plant. *Front. Plant Sci*. 11: 796.

Giladi, Y., et al. (2020). First report of beet curly top virus infecting *Cannabis sativa* in Western Colorado. *Plant Disease* 104: 999.

Gimeno, A., et al. (2020). From laboratory to the field: biological control of *Fusarium graminearum* on infected maize crop residues. *J. Appl. Microbiol*. 129: 680–694.

Gimeno, A., et al. (2021a). A versatile microfluidic platform measures hyphal interactions between *Fusarium graminearum* and *Clonostachys rosea* in real-time. *Commun. Bio*. 4: 1–10.

Gimeno, A., et al. (2021b). UV protection and shelf life of the biological control agent *Clonostachys rosea* against *Fusarium graminearum*. *Biol. Cont*. 158: 104600.

Goh, K.T., et al. (2020). Biocontrol and plant-growth-promoting traits of *Talaromyces apiculatus* and *Clonostachys rosea* consortium against ganoderma basal stem rot disease of oil palm. *Microorganisms* 8: 1138.

Gong, C., et al. (2017). Analysis of *Clonostachys rosea*-induced resistance to grey mould disease and identification of the key proteins induced in tomato fruit. *Postharvest Biol. Technol*. 123: 83–93.

González-Pérez, E., et al. (2008). First report of *Acremonium strictum* and *Gliocladium roseum* causing basal stem and corm rot of *Gladiolus grandiflorus* in Mexico. *J. Plant Pathol*. 90(3): 586–586.

Gossen, B.D., et al. (2016). Effect of soil type, organic matter content, bulk density and saturation on clubroot severity of biofungicide efficacy. *Plant Pathol*. 65: 1238–1245.

Gullner, G., et al. (2018). Glutathione s-transferase enzymes in plant-pathogen interactions. *Front. Plant Sci*. 9: 1836.

Han, P., et al. (2020). Metabolites from *Clonostachys* fungi and their biological activities. *J. Fungi* 6(4): 229.

Han, Z., et al. (2022). *Clonostachys rosea* promotes root growth in tomato by secreting auxin produced through the tryptamine pathway. *Fungi 8*: 1166.

Hannusch, D.J. and Boland, G.J. (1996). Interactions of air temperature, relative humidity and biological control agents on grey mold of bean. *Euro. J. Plant Pathol*. 102: 133–142.

Hasan, R., et al. (2022). Monitoring mycoparasitism of *Clonostachys rosea* against *Botrytis cinerea* using GFP. *J. Fungi* 8: 567.

Hassine, M., et al. (2022). Soil-borne and compost-borne *Penicillium* sp. and *Gliocladium* spp. as potential microbial biocontrol agents for the suppression of anthracnose-induced decay on tomato fruits. *Egypt J. Biol. Pest Cont*. 32: 20.

Hong-qu, W., et al. (2018). Preparation of dry flowable formulations of *Clonostachys rosea* by spray drying and application for *Sclerotinia sclerotiorum* control. *J. Integr. Agric*. 17(3): 613–620.

Hokkanen, H.M.T., et al. (2015). Managing bees for delivering biological control agents and improved pollination in berry and fruit cultivation. *Sustain. Agri. Res*. 4(3): 89–102.

Hoopen, G.M., et al. (2010). Compatibility between *Clonostachys* isolates with a view to mixed inocula for biocontrol. *Mycologia* 102(5): 1204–1215.

Huang, M., et al. (2021) *Bacillus firmus* I-1582 promotes plant growth and impairs infection and development of the cyst nematode *Heterodera schachtii* over two generations. *Sci. Rep*. 11: 1–15.

Hue, A.G., et al. (2009). Biological control of fusarium head blight of wheat with *Clonostachys rosea* strain ACM941. *Can. J. Plant Path*. 31(2): 169–179.

Hung, R., et al. (2013). *Arabidopsis thaliana* as a model system for testing the effect of *Trichoderma* volatile organic compounds. *Fungal Ecol*. 6(1): 19–26.

Iqbal, M., et al. (2018a). Evaluation of *Clonostachys rosea* for control of plant-parasitic nematodes in soil and in roots of carrot and wheat. *Phytopath*. 108(1): 52–59.

Iqbal, M., et al. (2018b). Comparative evolutionary histories of fungal proteases reveal gene gains in the mycoparasitic and nematode-parasitic fungus *Clonostachys rosea*. *BMC Evol. Biol*. 18(1): 171.

Iqbal, M., et al. (2019). Deletion of the nonribosomal peptide synthetase gene *nps1* in the fungus *Clonostachys rosea* attenuates antagonism and biocontrol of plant pathogenic *Fusarium* and nematodes. *Phytopath*. 109(10): 1698–1709.

Iqbal, M., et al. (2020). Natural variation of root lesion nematode antagonism in the biocontrol fungus *Clonostachys rosea* and identification of biocontrol factors through genome-wide association mapping. *Evol. Appl*. 13(9): 2264–2283.

James, T.D.W., Sutton, J.C. (1996). Biological control of botrytis blight of onion by *Gliocladium roseum* applied as sprays and with fabric applicators. *Euro. J. Plant Path*. 102: 265–275.

Jensen, B., et al. (2000). Biological seed treatment of cereals with fresh and long-term stored formulations of *Clonostachys rosea*: biocontrol efficacy against *Fusarium culmorum*. *Eur. J. Plant Pathol*. 106(3): 233–242.

Jensen, B., et al. (2002). Survival of conidia of *Clonostachys rosea* on stored barley seeds and their biocontrol efficacy against seed-borne *Bipolaris sorokiniana*. *Sci. Technol*. 12(4): 427–441.

Jensen, B., et al. (2016). *Clonostachys rosea* reduces spot blotch in barley by inhibiting prepenetration growth and sporulation of *Bipolaris sorokiniana* without inducing resistance. *Pest. Manag. Sci*. 72: 2231–2239.

Jensen, D.F., et al. (2022). *Clonostachys rosea* to control plant diseases. In: Köhl, J., Ravensberg, W. (Eds) *Microbial Bioprotectants for Plant Disease Management*. Burleigh Dodds Science Publishing, Cambridge, UK.

Jensen, D.F., et al. (2007). Development of a biocontrol agent for plant disease control with special emphasis on the near commercial fungal antagonist *Clonostachys rosea* strain'IK726'. *Australas. Plant Pathol*. 36(2): 95–101.

Johansen, A., et al. (2005). Non-target effects of the microbial control agents *Pseudomonas fluorescens* DR54 and *Clonostachys rosea* IK726 in soils cropped with barley followed by sugar beet: a greenhouse assessment. *Soil Biol. Biochem*. 37(12): 2225–2239.

Kamou, N.N., et al. (2016). Investigating the compatibility of the biocontrol agent *Clonostachys rosea* IK726 with prodigiosin-producing *Serratia rubidaea* S55 and phenazine-producing *Pseudomonas chlororaphis* ToZa7. *Arch. Microbiol*. 198: 369–377.

Kapeua-Ndacnou, M., et al. (2023). Assessing the biocontrol potential of *Clonostachys* species isolated as endophytes from *Coffea* ppecies and as mycoparasites of *Hemileia* rusts of coffee in Africa. *J. Fungi* 9(2): 248.

Kapongo, J.P., et al. (2008). Co-vectoring of *Beauveria bassiana* and *Clonostachys rosea* by bumble bees (*Bombus impatiens*) for control of insect pests and suppression of grey mould in greenhouse tomato and sweet pepper. *Biol. Cont*. 46: 508–514.

Karise, R., et al. (2016). Reliability of the entomovector technology using presto-mix and *Bombus terrestris* L. as a fungal disease biocontrol method in open field. *Sci. Rep.* 6: 31650.

Karlsson, M., et al. (2015). Insights on the evolution of mycoparasitism from the genome of *Clonostachys rosea*. *Genome Biol. Evol.* 7(2): 465–480.

Keyser, C.A., et al. (2016). Dual effects of *Metarhizium* spp. and *Clonostachys rosea* against an insect and a seed-borne pathogen in wheat. *Pest Manag. Sci.* 72(3): 517–526.

Khairullina, A., et al. (2023). Biocontrol effect of *Clonostachys rosea* on *Fusarium graminearum* infection and mycotoxin detoxification in oat (*Avena sativa*). *Plants* 12(3): 500.

Kim, C.-K., et al. (2021). Roseabol A, a new peptaibol from the fungus *Clonostachys rosea*. *Molecules* 26: 3594.

Kim, Y.S., et al. (2008) Enhancement of biocontrol activity of antagonistic *Chryseobacterium* strain KJ1R5 by adding carbon sources against *Phytophthora capsici*. *Plant Pathol J.* 24: 164–170.

Knudsen, I.M.B., et al. (1995). Biocontrol of seedling diseases of barley and wheat caused by *Fusarium culmorum* and *Bipolaris sorokiniana*: effects of selected fungal antagonists on growth and yield components. *Plant Pathol.* 44: 467–477.

Knudsen, I.M.B., et al. (2004). Effects of hot water treatment, biocontrol agents, disinfectants and fungicide on storability of English oak acorns and control of the pathogen, *Ciboria batschiana*. *For. Pathol.* 34: 47–64.

Koch, E., et al. (2020). Artificial inoculation methods for testing microorganisms as control agents of seed- and soil-borne *Fusarium*-seedling blight of maize. *J. Plant Dis. Prot.* 127: 883–893.

Köhl, J., et al. (2019). Mode of action of microbial biological control agents against plant diseases: relevance beyond efficacy. *Front. Plant Sci.* 10: 845.

Kosawang, C., et al. (2014a). Transcriptomic profiling to identify genes involved in *Fusarium* mycotoxin deoxynivalenol and zearalenone tolerance in the mycoparasitic fungus *Clonostachys rosea*. *BMC Genomics* 15: 55.

Kosawang, C., et al. (2014b). Zearalenone detoxification by zearalenone hydrolase is important for the antagonistic ability of *Clonostachys rosea* against mycotoxigenic *Fusarium graminearum*. *Fungal Biol.* 118(4): 364–373.

Krauss, U., et al. (2004). Interaction between the entomopathogens *Beauveria bassiana*, *Metarhizium anisopliae* and *Paecilomyces fumosoroseus* and the mycoparasites *Clonostachys* spp., *Trichoderma harzianum* and *Lecanicillium lecanii*. *Biocont. Sci. Tech.* 14(4): 331–346.

Krauss, U., et al. (2013). Mycoparasitism by *Clonostachys byssicola* and *Clonostachys rosea* on *Trichoderma* spp. from cocoa (*Theobroma cacao*) and implication for the design of mixed biocontrol agents. *Biol. Cont.* 67: 317–327.

Krauss, U. and Soberanis, W. (2001). Rehabilitation of diseased cacao fields in Peru through shade regulation and timing of biocontrol measures. *Agroforest. Sys.* 53: 179–184.

Lahlali, R. and Peng, G. (2014). Suppression of clubroot by *Clonostachys rosea* via antibiosis and induced host resistance. *Plant Pathol.* 63(2): 447–455.

Lahoz, E., et al. (2004). Induction of systemic resistance to *Erysiphe orontii* cast in tobacco by application on roots of an isolate of *Gliocladium roseum* Bainier. *J. Phytopathol.* 152: 465–470.

Lastochkina, O., et al. (2019). *Bacillus* Spp.: efficient biotic strategy to control postharvest diseases of fruits and vegetables. *Plants*. 8(4): 97.

Latz, M.A.C., et al. (2018). Endophytic fungi as biocontrol agents: elucidating mechanisms in disease suppression. *Plant Ecol. Divers*. 11: 555–567.

Lee, S.A., et al. (2020). First report of *Clonostachys rosea* causing root rot of *Gastrodia elata* in Korea. *Plant Dis*. 104(11): 3069.

Leibman-Markus, M., et al. (2021). Gene editing of the decoy receptor *LeEIX1* increases host receptivity to *Trichoderma* bio-control. *Front. Fungal Biol*. 2: 678840.

Li, G.Q., et al. (2002). Ultrastructural study of mycoparasitism of *Gliocladium roseum* on *Botrytis cinerea*. *Bot. Bull. Acad. Sin*. 43: 211–218.

Li, G.Q., et al. (2004). Biological control of blossom blight of alfalfa caused by *Botrytis cinerea* under environmentally controlled and field conditions. *Plant Dis*. 88: 1246–1251.

Li, J., et al. (2006). Purification and characterization of an extracellular serine protease from *Clonostachys rosea* and its potential as a pathogenic factor. *Process Biochem*. 41(4): 925–929.

Linghu, M., et al. (2023). First report of leaf spot cause by *Clonostachys rosea* on tea (*Camellia sinensis*) in China. *Plant Dis*. 107(8): 2537.

Liu, J., et al. (2023). Evaluation of immobilized microspheres of *Clonostachys rosea* on *Botrytis cinerea* and tomato seedlings. *Biomat*. 301: 122217.

Luongo, L., et al. (2005). Potential of fungal antagonists for biocontrol of *Fusarium* spp. in wheat and maize through competition in crop debris. *Biocont. Sci. Tech*. 15(3): 229–242.

Lysøe, E., et al. (2017). A three-way transcriptomic interaction study of a biocontrol agent (*Clonostachys rosea*), a fungal pathogen (*Helminthosporium solani*), and a potato host (*Solanum tuberosum*). *Mol. Plant. Microbe Interact*. 30(8): 646–655.

Ma, H., et al. (2022). Root rot of *Angelica sinensis* caused by *Clonostachys rosea* and *Fusarium acuminatum* in China. *Plant Dis*. 106: 2264.

Mahmoud, F.M., et al. (2021). Entomopathogenic efficacy of the endophytic fungi: *Clonostachys* sp. and *Beauveria bassiana* on *Tuta absoluta* (Meyrick) (Lepidoptera: Gelechiidae) larvae under laboratory and greenhouse conditions. *Egypt. J. Biol. Pest Cont*. 31: 43.

Makkonen, R. and Pohjakallio, O. (1960). On the parasites attacking the sclerotia of some fungi pathogenic to higher plants and on the resistance of these sclerotia to their parasites. *Acta Agri. Scand*. 10(2–3): 105–126.

Mamarabadi, M., et al. (2008). Three endochitinase-encoding genes identified in the biocontrol fungus *Clonostachys rosea* are differentially expressed. *Curr. Genet*. 54(2): 57–70.

Mamarabadi, M., et al. (2009). An N-acetyl-beta-D-glucosaminidase gene, *cr-nag1*, from the biocontrol agent *Clonostachys rosea* is up-regulated in antagonistic interactions with *Fusarium culmorum*. *Mycol. Res*. 113(1): 33–43.

Marian, M. and Shimizu, M. (2019). Improving performance of microbial biocontrol agents against plant diseases. *J. General Plant Pathol*. 85: 329–336.

Mascarin, G.M., et al. (2022). *Clonostachys rosea*: production by submerged culture and bioactivity against *Sclerotinia sclerotiorum* and *Bemisia tabaci*. *Front. Microbiol*. 13: 851000.

Mauch-Mani, B., et al. (2017). Defense priming: an adaptive part of induced resistance. *Annu. Rev. Plant Biol*. 68: 485–512.

Melgarejo, T.A., et al. (2022). Curly top disease of hemp (*Cannabis sativa*) in California is caused by mild-type strains of beet curly top virus often in mixed infection. *Plant Dis.* 106(12): 3022–3026.

Meng, F., et al. (2022). Insights into the molecular basis of biocontrol of *Botrytis cinerea* by *Clonostachys rosea* in tomato. *Scientia Horticulturae* 291: 110547.

Mohammed, A.A., et al. (2021). Efficacy of *Clonostachys rosea*, as a promising entomopathogenic fungus, against coleopteran stored product insect pests under laboratory conditions. *Egypt. J. Biol. Pest Cont.* 31: 55.

Mohammed, A.A., et al. (2022). Molecular identification of two entomopathogenic fungus *Clonostachys rosea* strains and their efficacy against two aphid species in Iraq. *J. Gene. Engin. Biotech.* 20: 67.

Møller, K., et al. (2003). Biocontrol of *Pythium tracheiphilum* in Chinese cabbage by *Clonostachys rosea* under field conditions. *Biocont. Sci. Technol.* 13: 171–182.

Mommaerts, V., et al. (2010). Development of a new dispenser for microbiological control agents and evaluation of dissemination by bumblebees in greenhouse strawberries. Pest Manag. Sci. 66: 1199–1207.

Mommaerts, V. and Smagghe, G. (2011). Entomovectoring in plant protection. *Arthropod-Plant Interact.* 5: 81–95.

Monkai, J., et al. (2013). Fungi on decaying leaves of *Magnolialiliifera* and *Cinnamomum iners* show litter fungi to be hyperdiverse. *Mycosphere.* 4: 292–301.

Moraga-Suazo, P. and Sanfuentes, E. (2017). Growth promotion of *Pinus radiata* seedlings by soil inoculation and seed pretreatment with the biological control agent *Clonostachys rosea*. *Gayana Bot.* 74(1): 140–146.

Moraga-Suazo, P., et al. (2016). Induced systemic resistancetriggered by *Clonostachys rosea* against *Fusarium circinatum* in *Pinus radiata*. *Forest Res.* 5: 174.

Morandi, M.A.B., et al. (2000). Effects of host and microbial factors on development of *Clonostachys rosea* and control of *Botrytis cinerea* in rose. *Eur. J. Plant Path.* 106(5): 439–448.

Morandi, M.A.B., et al. (2001). Development of *Clonostachys rosea* and interactions with *Botrytis cinerea* in rose leaves and residues. *Phytoparasitica* 29(2): 103–113.

Morandi, M.A.B., et al. (2003). Suppression of *Botrytis cinerea* sporulation by *Clonostachys rosea* on rose debris: a valuable component in *Botrytis* blight management in commercial greenhouses. *Bio. Cont.* 26: 311–317.

Morandi, M.A.B., et al. (2006). Relationships of microclimatic variables to colonization of rose debris by *Botrytis cinerea* and the biocontrol agent *Clonostachys rosea*. *Biocont. Sci. Tech.* 16: 619–630.

Morandi, M.A.B., et al. (2008). Influence of application time on the establishment, survival, and ability of *Clonostachys rosea* to suppress *Botrytis cinerea* sporulation on rose debris. *Crop Prot.* 27: 77–83.

Moreira, G.M., et al. (2016). Multilocus phylogeny of *Clonostachys* subgenus *Bionectria* from Brazil and description of *Clonostachys chloroleuca* sp nov. *Mycolog. Prog.* 15(10–11): 1031–1039.

Mouekouba, L.D.O., et al. (2014). Analysis of *Clonostachys rosea*-induced resistance to tomato gray mold disease in tomato leaves. *PLoS One* 9(7): e102690.

Muñoz, I. V., et al. (2019). CRISPR-Cas for fungal genome editing: a new tool for the management of plant diseases. *Front. Plant Sci.* 10: 135.

Muvea, A.M., et al. (2014). Colonization of onions by endophytic fungi and their impacts on the biology of *Thrips tabaci*. *PLoS One* 9(9): e108242

Nechet, K.L., et al. (2017). Effect of increased UV-B radiation on biological control of the gray mold by *Clonostachys rosea* and on the expression of strawberry defense-related enzymes. *Australas. Plant Path.* 46(2): 107–113.

Nelson, R., et al. (2018). Navigating complexity to breed disease-resistant crops. *Nat. Rev. Genet.* 19: 21–33.

Newman, M.-A., et al. (2013). MAMP (microbe-associated molecular pattern) triggered immunity in plants. *Front. Plant Sci.* 4: 139.

Nieto-Jacobo, M.F., et al. (2017) Environmental growth conditions of *Trichoderma* spp. affects indole acetic acid derivatives, volatile organic compounds, and plant growth promotion. *Front. Plant Sci.* 8: 102.

Nobre, S.A.M., et al. (2005). Selection of *Clonostachys rosea* isolates from Brazilian ecosystems effective in controlling *Botrytis cinerea*. *Biol. Cont.* 34(2): 132–143.

Okuda, T., et al. (2000). Production of TMC-151, TMC-154 and TMC-171, a new class of antibiotics, is specific to *'Gliocladium roseum'* group. *Mycosci.* 41: 239–253.

Olmedo-Monfil, V., et al. (2002). Multiple environmental signals determine the transcriptional activation of the mycoparasitism related gene *prb1* in *Trichoderma atroviride*. *Mol. Genet. Genomics* 267: 703–712.

Osborne, L.E. and Stein, J.M. (2007). Epidemiology of Fusarium head blight on small-grain cereals. *Int. J. Food Microbiol.* 119: 103–108.

Palazzini, J.M., et al. (2013). Biocontrol and population dynamics of *Fusarium* spp. on wheat stubble in Argentina. *Plant Pathol.* 62: 659–866.

Pimentel, M.F., et al. (2022). Reduction of pythium damping-off in soybean by biocontrol seed treatment. *Plant Dis.* 106: 2403–2414.

Pires, D., et al. (2022). The fight against plant-parasitic nematodes: current status of bacterial and fungal biocontrol agents. *Pathogens* 11(10): 1178.

Peng, G., et al. (1992). Effectiveness of honey-bees for applying the biocontrol agent *Gliocladium roseum* to strawberry flowers to suppress *Botrytis cinerea*. *Can. J. Plant Pathol.* 14(2): 117–129.

Peng, G., et al. (2011). Potential biological control of clubroot on canola and crucifer vegetable crops. *Plant Path.* 60(3): 566–574.

Pertot, I., et al. (2017). Combining biocontrol agents with different mechanisms of action in a strategy to control *Botrytis cinerea* on grapevine. *Crop Prot.* 97: 85–93.

Pfeiffer, T., et al. (2021). Selection of bacteria and fungi for control of soilborne seedling diseases of maize. *J. Plant Dis. Prot.* 128: 1227–1241.

Piombo, E., et al. (2023). Insights into the ecological lifestyle of *Clonostachys* fungi through analysis of their predicted secretomes. *Front. Microbiol.* 14: 1112673.

Pieterse, C.M.J., et al. (2014). Induced systemic resistance by beneficial microbes. *Annu. Rev. Phytopathol.* 52: 347–375.

Pocurull, M., et al. (2020). Commercial formulates of trichoderma induce systemic plant resistance to *Meloidogyne incognita* in tomato and the effect is additive to that of the Mi-1.2 resistance gene. *Front. Microbiol.* 10: 3042.

Ptaszek, M., et al. (2023). Microbial-based products to control soil-borne pathogens: methods to improve efficacy and to assess impacts on microbiome. *Microorganisms* 11: 224.

Punja, Z.K. (2021). Emerging diseases of *Cannabis sativa* and sustainable management. *Pest Manag. Sci.* 77: 3857–3870.

Punja, Z.K. and Ni, L. (2021). The bud rot pathogens infecting cannabis (*Cannabis sativa* L., marijuana) inflorescences: symptomology, species identification, pathogenicity and biological control. *Can. J. Plant Pathol.* 43(6). https://doi.org/10.1080/07060661.2021.1936650

Punja, Z.K. and Yip R. (2003). Biological control of damping-off and root rot caused by *Pythium aphanidermatum* on greenhouse cucumbers. *Can. J. Plant Pathol.* 25(4): 411–417.

Qi, H., et al. (2022). First report disease of *Clonostachys rosea* causing root rot on *Astragalus membranaceus* in China. *Plant Dis.* 106: 1752.

Qin, G.Z. and Tian, S.P. (2005). Enhancement of biocontrol activity of *Cryptococcus laurentii* by silicon and the possible mechanisms involved. *Phytopathol.* 95: 69–75.

Rahman, M. and Punja, Z.K. (2007). Biological control of damping-off on American ginseng (*Panax quinquefolius*) by *Clonostachys rosea* f. *catenulata* (= *Gliocladium catenulatum*). *Can. J. Plant Pathol.* 33: 499–505.

Ravnskov, S., et al. (2006). Soil inoculation with the biocontrol agent *Clonostachys rosea* and the mycorrhizal fungus *Glomus intraradices* results in mutual inhibition, plant growth promotion and alteration of soil microbial communities. *Soil Biol. Biochem.* 38(12): 3453–3462.

Raymaekers, K., et al. (2020). Screening for novel biocontrol agents applicable in plant disease management – a review. *Biol. Cont.* 144. https://doi.org/10.1016/j.biocontrol.2020.104240

Reeh, K.W. and Cutler, G.C. (2013). Laboratory efficacy and fungicide compatibility of *Clonostachys rosea* against *Botrytis* blight on lowbush blueberry. *Can. J. Plant Sci.* 93: 639–642.

Roberti, R., et al. (2006). Sensitivity of *Clonostachys rosea* and *Trichoderma* spp. as potential biocontrol agents to pesticides. *J. Phytopath.* 154(2): 100–109.

Roberti, R., et al. (2000). Evaluation of biological seed treatment of wheat for the control of seed-borne *Fusarium culmorum*. *J. Plant. Dis. Protect.* 107: 484–493.

Roberti, R., et al. (2008). Induction of PR proteins and resistance by the biocontrol agent *Clonostachys rosea* in wheat plants infected with *Fusarium culmorum*. *Plant Sci.* 175(3): 339–347.

Robinson-Boyer, L., et al. (2009). Management of strawberry grey mould using mixtures of biocontrol agents with different mechanisms of action. *Biocont. Sci. Tech.* 19: 1051–1065.

Rodriguez, M.A., et al. (2011). *Clonostachys rosea* BAFC3874 as a *Sclerotinia sclerotiorum* antagonist: mechanisms involved and potential as a biocontrol agent. *J. Appl. Microbiol.* 110(5): 1177–1186.

Rose, S., et al. (2003). Efficacy of biological and chemical treatments for control of *Fusarium* root and stem rot on greenhouse cucumber. *Plant Dis.* 87: 1462–1470.

Rose, S., et al. (2004). Biological control of Fusarium and Pythium root rots on greenhouse cucumbers grown in rockwool. *Acta Horticult.* 635: 73–78.

Rojas, E.C., et al. (2020). Selection of fungal endophytes with biocontrol potential against Fusarium head blight in wheat. *Biol. Cont.* 144: 104222.

Samils, N., et al. (2016). Biological control against postharvest diseases on potato tubers. *IOBC WPRS Bull.* 115: 241–242.

Saraiva, R.M., et al. (2020). Compounds produced by *Clonostachys rosea* deleterious to *Botrytis cinerea*. *Braz. J. Agr.* 95: 34–47.

Schroers, H.J. (2001). A monograph of *Bionectria* (Ascomycota, Hypocreales, Bionectriaceae) and its *Clonostachys* anamorphs. *Stud. Mycol.* 46: 1–214.

Schroers, H.J., et al. (1999). Classification of the mycoparasite *Gliocladium roseum* in *Clonostachys* as *C. rosea*, its relationship to *Bionectria ochroleuca*, and notes on other *Gliocladium*-like fungi. *Mycologia* 91: 365–385.

Scott, C. and Punja, Z.K. (2023). Biological control of *Fusarium oxysporum* causing damping-off and *Pythium myriotylum* causing root and crown rot on cannabis (*Cannabis sativa* L.) plants. *Can. J. Plant Path.* 45(3): 238–252. https://doi.org/10.1080/07060661.2023.2172082

Seidl, V., et al. (2009). Transcriptomic response of the mycoparasitic fungus *Trichoderma atroviride* to the presence of a fungal prey. *BMC Genomics*. 10: 567.

Shin, T.Y., et al. (2020). Pathogenesis-related genes of entomopathogenic fungi. *Arch. Insect Biochem. Phys.* 105(4): 21747.

Shrestha, S. and Hausbeck, M.K. (2021). Evaluation of geranium cultivars and biorational products to control botrytis blight in the greenhouse. *Plant Health Progress* 22: 474–482.

Silva, H.A.O., et al. (2020). Biocontrol of potato early blight and suppression of *Alternaria grandis* sporulation by *Clonostachys* spp. *Plant Path*. 70: 1677–1685.

Silva, P.R.A., et al. (2020). Silicon, *Clonostachys rosea*, and their interaction for gray mold management in cucumber. *J. Plant Pathol.* 102: 1257–1262.

Smagghe, G., et al. (Eds.). (2020). *Entomovectoring for Precision Biocontrol and Enhanced Pollination of Crops*. Springer, Cham.

Smith, B.J., et al. (2012) Efficacy of bumble bee disseminated biological control agents for control of botrytis blossom blight of rabbiteye blueberry. *Int. J. Fruit Sci.* 12: 156–168.

Song, J., et al. (2023). Synergistic effects of *Clonostachys rosea* isolates and succinate dehydrogenase inhibitors fungicides against gray mold on tomato. *Microorganisms* 11(1): 20.

Sun, M.H., et al. (2014). Effects of culture conditions on spore types of *Clonostachys rosea* 67–1 in submerged fermentation. *Lett. Appl. Microbiol.* 58(4): 318–324.

Sun, Z.-B., et al. (2015). Identification of mycoparasitism-related genes in *Clonotachys rosea* 67-1 active against *Sclerotinia sclerotiorum*. *Sci. Rep*. 5: 18169.

Sun, Z.-B., et al. (2017). Transformation of the endochitinase gene *Chi67-1* in *Clonostachys rosea* 67–1 increases its biocontrol activity against *Sclerotinia sclerotiorum*. *AMB Express* 7(1): 1.

Sun, Z.-B., et al. (2020). Biology and applications of *Clonostachys rosea*. *J. Appl. Microbiol.* 129: 486–495.

Supratman, U., et al. (2021) New metabolites produced by endophyte *Clonostachys rosea* B5-2. *Nat Prod Res*. 35(9): 1525–1531.

Sutton, J.C., et al. (2002). Ability of *Clonostachys rosea* to establish and suppress sporulation potential of *Botrytis cinerea* in deleafed stems of hydroponic greenhouse tomatoes. *Sci. Technol*. 12(4): 413–425.

Sutton, J.C., et al. (2008). Evaluation of the fungal endophyte *Clonostachys rosea* as an inoculant to enhance growth, fitness and productivity of crops plants. *Acta Horticulturae* 782: 279–286.

Sutton, J.C. and Peng, G. (1993). Biocontrol of *Botrytis cinerea* in strawberry leaves. *Phytopathol*. 83: 615–621.

Tamta, A.K., et al. (2022). First record of *Clonostachys rosea* (Ascomycota: Hypocreales) entomopathogenic fungus in the mango hopper *Amritodus atkinsoni* (Hemiptera: Cicadellidae). *Pathogens*. 11: 1447.

Tarantino, P., et al. (2007). Control of *Rhizoctonia solani* in a tobacco-float system using low rates of iprodione- and iprodione-resistant strains of *Gliocladium roseum*. *Crop Prot*. 26: 1298–1302.

Theron, D.J. and Holz, G. (1991). Dry rot of potatoes caused by *Gliocladium roseum*. *Plant Pathol*. 40(2): 302–305.

Thines, E., et al. (1998). Glisoprenins C, D and E, new inhibitors of appressorium formation in *Magnaporthe grisea*, from cultures of *Gliocladium roseum*. 1. Production and biological activities. *J. Antibiot*. 51(2): 117–122.

Tian, S., et al. (2005). Synergistic effects of combining biocontrol agents with silicon against postharvest diseases of jujube fruit. *J. Food Pro*. 68(3): 544–550.

Tian, T., et al. (2014). Synergistic effect of dazomet soil fumigation and *Clonostachys rosea* against cucumber fusarium wilt. *Phytopath*. 104(12): 1314–1321.

Tian, X. and Zheng, Y. (2013). Evaluation of biological control agents for *Fusarium* wilt in Hiemalis begonia. *Can J Plant Pathol*. 35(3): 363–370.

Toledo, A.V., et al. (2006). First record of *Clonostachys rosea* (Ascomycota: Hypocreales) as an entomopathogenic fungus of *Oncometopia tucumana* and *Sonesimia grossa* (Hemiptera: Cicadellidae) in Argentina. *J. Invert. Path*. 92: 7–10.

Teixidó, N., et al. (2022). Insights into a successful development of biocontrol agents: production, formulation, packaging, and shelf life as key aspects. *Horticulturae* 8(4): 305.

Tesfagiorgis, H.B. and Laing, M.D. (2010). Isolation and in vitro screening of potential biocontrol agents against powdery mildew of zucchini caused by *Podosphaera xanthii*. *S. Afr. J. Plant Soil* 27(2): 179–186.

Tesfagiorgis, H.B., et al. (2014). Evaluation of biocontrol agents and potassium silicate for the management of powdery mildew of zucchini. *Biol. Cont*. 73: 8–15.

Travadon, R., et al. (2023). Field evaluation of biological wound treatments for the management of almond, cherry, and grapevine fungal canker diseases. *Biol. Cont*. 185. https://doi.org/10.1016/j.biocontrol.2023.105292

Türkölmez, S., et al. (2023). *Clonostachys rosea* strain ST1140: an endophytic plant-growth-promoting fungus, and its potential use in seedbeds with wheat-grain-substrate. *Current Microbiol*. 80: 36.

Tzelepis, G., et al. (2015). Identifying glycoside hydrolase family 18 genes in the mycoparasitic fungal species *Clonostachys rosea*. *Microbiology (Reading)* 161(7): 1407–1419.

Tzelepis, G. and Karlsson, M. (2019). Killer toxin-like chitinase in filamentous fungi: structure, regulation and potential roles in fungal biology. *Fung. Biol. Rev*. 33: 123–132.

Van Delm, T., et al. (2015). Control of *Botrytis cinerea* in strawberries with *Gliocladium catenulatum* vectored by bumblebees. *J. Berry Res*. 5(1): 23–28.

Venkatesan, R.M., et al. (2023). First report of *Clonostachys rosea* as a mycoparasite on *Sclerotinia sclerotiorum* causing head rot of cabbage in India. *Plants* 12: 199.

Verdejo-Lucas, S., et al. (2002). Species of root-knot nematodes and fungal egg parasites recovered from vegetables in Almeria and Barcelona, Spain. *J. Nematol*. 34(4): 405–408.

Wang, Q., et al. (2019). The involvement of jasmonic acid, ethylene and salicylic acid in the signalling pathway of *Clonostachys rosea*-induced resistance to gray mold disease in tomato. *Analyt. Theoret. Plant Path*. 109: 1102–1114.

Wang, Y., et al. (2022). Use of CRISPR-Cas tools to engineer *Trichoderma* species. *Microbial Biotech*. 15: 2521–2532.

Wang, Y., et al. (2023). Phylogeny and systematics of the genus *Clonostachys*. *Front. Microbiol.* 14: 1117753.

Wiesel, L., et al. (2014). Molecular effects of resistance elicitors from biological origin and their potential for crop protection. *Front. Plant Sci.* 5: 655.

Wilson, A., et al. (2020). Investigating hyperparasites as potential biological control agents of rust pathogens on cereal crops. *Australas. Plant Path.* 49(3): 231–238.

Wu, K., et al. (2015) Pectin enhances bio-control efficacy by inducing colonization and secretion of secondary metabolites by *Bacillus amyloliquefaciens* SQY 162 in the rhizosphere of tobacco. *PLoS One*. https://doi.org/10.3410/f.725509196.793507217

Xu, X.M., et al. (2010). Using combinations of biocontrol agents to control *Botrytis cinerea* on strawberry leaves under fluctuating temperatures. *Biocont. Sci. Tech.* 20(4): 359–373.

Xu, X.M., et al. (2011a). Combined use of biocontrol agents to manage plant diseases in theory and practice. *Phytopath.* 101(9): 1024–1031.

Xu, X.M., et al. (2011b). A numerical study of combined use of two biocontrol agents with different biocontrol mechanisms in controlling foliar pathogens. *Phytopath.* 101(9): 1032–1044.

Xue, A.G. (2003a). Efficacy of *Clonostachys rosea* strain ACM941 and fungicide seed treatments for controlling the root rot complex of field pea. *Can. J. Plant Sci.* 83: 519–524.

Xue, A.G. (2003b). Biological control of pathogens causing root rot complex of field pea using *Clonostachys rosea* strain ACM941. *Phytopath.* 93: 329–335.

Xue, A.G., et al. (2009). Biological control of fusarium head blight of wheat with *Clonostachys rosea* strain ACM941. *Can. J. Plant Path.* 31(2): 169–179.

Xue, A.G., et al. (2014a). Efficacy of CLO-1 biofungicide in suppressing perithecial production by *Gibberella zeae* on crop residues. *Can. J. Plant Pathol.* 36(2): 161–169.

Xue, A.G., et al. (2014b). Concentration and cultivar effects on efficacy of CLO-1 biofungicide in controlling fusarium head blight of wheat. *Biol. Cont.* 73: 2–7.

Yang, Z., et al. (2021). Effects of the entomopathogenic fungus *Clonostachys rosea* on mortality rates and gene expression profiles in *Diaphorina citri* adults. *J. Invtert. Path.* 179. https://doi.org/10.1016/j.jip.2021.107539

Yu, H. and Sutton, J.C. (1997a). Effectiveness of bumblebees and honeybees for delivering inoculum of *Gliocladium roseum* to raspberry flowers to control *Botrytis cinerea*. *Biol. Control* 10(2): 113–122.

Yu, H. and Sutton, J.C. (1997b). Morphological development and interactions of *Gliocladium roseum* and *Botrytis cinerea* in raspberry. *Can. J. Plant Pathol.* 19(3): 237–246.

Yu, H. and Sutton, J.C. (1998). Effect of inoculum density, wetness duration, and temperature on control of *Botrytis cinerea* by *Gliocladium roseum* in raspberry. *Can. J. Plant Pathol.* 20: 243–252.

Yu, H. and Sutton, J.C. (1999). Density dynamics of *Gliocladium roseum* in relation to biological control of *Botrytis cinerea* in red raspberry. *Can. J. Plant Pathol.* 21: 1, 23–32.

Zaki, O., et al. (2020). Limiting factors of mycopesticide development. *Biol. Control* 144: 104220.

Zhai, M.M., et al. (2016). Isolation of secondary metabolites from the soil-derived fungus *Clonostachys rosea* YRS-06, a biological control agent, and evaluation of antibacterial activity. *J. Agric. Food Chem.* 64(11): 2298–2306.

Zhang, J. and Sun, X. (2021). Recent advances in polyphenol oxidase-mediated plant stress responses. *Phytochem*. 181: 112588.

Zhang, L., et al. (2008). Investigation on the infection mechanism of the fungus *Clonostachys rosea* against nematodes using the green fluorescent protein. *Appl. Microbiol. Biotechnol*. 78: 983–990.

Zhang, P.G., et al. (1996). Inoculum concentration and time of application of *Gliocladium roseum* in relation to biocontrol of *Botrytis cinerea* in black spruce seedlings. *Can. J. For. Res*. 26: 360–367.

Zhang, Y.Y., et al. (2013). A new two-phase kinetic model of sporulation of *Clonostachys rosea* in a new solid-state fermentation reactor. *Process Biochem*. 48(8): 1119–1125.

Zou, C.-G., et al. (2010). Regulation of subtilisin-like protease *prC* expression by nematode cuticle in the nematophagous fungus *Clonostachys rosea*. *Enviro. Microbiol*. 12(12): 3243–3252.

Chapter 6

The use of *Pythium oligandrum* as a biocontrol agent

Kun Yang, Guangdong Academy of Agricultural Sciences/Guangdong Provincial Key Laboratory of Crop Genetic Improvement, China; Siqi Niu and Qing Yu, Nanjing Agricultural University, China; Xiaobo Li, Guangdong Academy of Agricultural Sciences/Guangdong Provincial Key Laboratory of Crop Genetic Improvement, China; Maofeng Jing, Nanjing Agricultural University, China

1 Introduction

Oomycetes are microorganisms found in diverse ecosystems, including tropical rainforests, oceans, deserts, and even extreme environments, for example in Antarctica (Belonoznikova et al., 2022). They have adapted to survive under a wide range of conditions and play important roles in these ecosystems. Most oomycetes are devastating pathogens of plants and animals. Interestingly, the nonpathogenic oomycete *Pythium oligandrum* is an effective mycoparasite that colonizes the root ecosystems of diverse plants to induce defense responses against pathogens. As a sterol auxotroph, *P. oligandrum* relies on external sources of sterols from other host pathogens or plants, similar to other oomycetes. It also produces antimicrobial compounds, competes for nutrients and space with pathogens, and promotes plant growth via producing the auxin

http://dx.doi.org/10.19103/AS.2024.0142.16

precursor tryptamine (Bělonožníková et al., 2022). Since the initial discovery of *P. oligandrum* in the root system of peas in 1930, it has served as a biological control agent (BCA) for plant disease control and has been widely and successfully used in agricultural production and ecosystem management. The application of *P. oligandrum* is an environmentally friendly crop disease control strategy, is naturally present in the rhizosphere of many plants, and has shown the potential to change the microbial diversity in soil through its interactions with other microorganisms (Bělonožníková et al., 2022). Meanwhile, it is able to spread into the plant root tissues without inducing any symptoms and also beneficial for the plant to defend various stresses (Le Floch et al., 2003; Bělonožníková et al., 2022). Polyversum® is currently the only biopesticide product registered for use in disease control based on *P. oligandrum* oospores. It has been registered as a biopesticide in the US and China and as a fungicide in plant protection products by EU authorities. According to different crop types, Polyversum® can be used in various manners, such as foliar application, root dipping/drenching, seed treatment, and watering. Polyversum® has been successfully applied in preventive and emergency applications in the control of fungal diseases in a wide spectrum of crops, such as lawns, fruit trees, and bushes and ornamental plants.

2 The main biocontrol mechanisms of *Pythium oligandrum*

As an effective mycoparasite, *P. oligandrum* suppresses and potentially kills pathogens as its major biocontrol function. Exhibiting remarkable mycoparasitic properties, *P. oligandrum* has been shown to entwine hyphae around those of various oomycetes and fungi (Table 1). *P. oligandrum* actively grows towards/along fungal hyphae. During the interaction with its hosts, the hyphae of *P. oligandrum* coil around and penetrate the hyphal structures of hosts and form papilla-like structures at the sites of potential penetration into host's hyphae; *P. oligandrum* degrades cell wall polysaccharides of both fungi and oomycete using chitinases, cellulases, endo-β-1,3-glucanases, or various exoglycosidases at penetrated sites to facilitate the entry and provide carbon sources required for active growth and development, resulting in the complete destruction of host hyphae and death of hosts (Daly et al., 2021; You et al., 2019). In addition, volatile organic compounds (VOCs) emitted by *P. oligandrum* also significantly contribute to parasitic interactions (Sheikh et al., 2023b).

P. oligandrum either effectively suppresses the mycelial growth of the pathogen or directly kills the pathogen hyphae through mycoparasitism *in vitro* (Sayed et al., 2021; You et al., 2019). Additionally, *P. oligandrum* exhibits parasitic behavior toward several oomycete pathogens, including *Phytophthora infestans*, *Phytophthora megasperma*, *Pythium ultimum*, *Pythium debaryanum*,

Table 1 Summary of the major plant pathogens parasited by *Pythium oligandrum*

Pathogen type	Pathogen species	Isolates	Reference
Oomycete	*Pythium myriotylum*	GAQ1 (CGMCC no. 17470)	(Sheikh, Zhou et al., 2023a)
Oomycete	*Pythium aphanidermatum*	MS15, MS19, MS31	(Sayed, Abdelmohsen et al., 2021)
Oomycete	*Pythium myriotylum*	GAQ1 (CGMCC no. 17470)	(Daly, Chen et al., 2021)
Oomycete	*Pythium aphanidermatum*, *Pythium myriotylum*	Strain D11	(You, Barraud et al., 2019)
Oomycete	*Phytophthora infestans*	Strain CBS 530.74, CBS 200.184	(Horner, Grenville-Briggs et al., 2012)
Oomycete	*Pythium ultimum*	*P. oligandrum*	(Ali-Shtayeh and Saleh, 1999)
Oomycete	*Pythium ultimum*, *Phytophthora megasperma*	*P. oligandrum* isolate 1010	(Benhamou, Rey et al., 1999)
Oomycete	*Pythium ultimum*	*P. oligandrum* was isolated from soil samples collected from an agricultural field near the Nile shore at El-Minia city, Egypt, in November, 1995.	(Abdelzaher, Elnaghy et al., 1997)
Oomycete	*Pythium ultimum*, *Pythium debaryanum*, *Aphanomyces laevis*	*P. oligandrum*	(Veselý, 1978)
Fungus	*Fusarium graminearum*, *Fusarium verticillioides*	Strain M1/ATCC 38472	(Pellan, Durand et al., 2020)
Fungus	*Microsporum canis*	Strain ATCC 38472	(Gabrielova, Mencl et al., 2018)
Fungus	*Microsporum canis*, *Microsporum gypseum*, *Trichophyton mentagrophytes*	*P. oligandrum*	(Načeradská, Fridrichová et al., 2016)
Fungus	*Fusarium*	*P. oligandrum*	(Benhamou, le Floch et al., 2012)
Fungus	*Rhizoctonia solani* AG-3	NITE P-619	(Ikeda, Shimizu et al., 2012)
Fungus	*Botrytis cinerea*, *Sclerotinia minor*	*P. oligandrum* isolate 1133	(Rey, Le Floch et al., 2005)
Fungus	*Fusarium oxysporum*, *Rhizoctonia solani*	*P. oligandrum* isolate 1010	(Benhamou, Rey et al., 1999)
Fungus	*Verticillium dahliae*	*P. oligandrum*	(Al-Rawahi and Hancock, 1998)
Fungus	*Fusarium oxysporum*	*P. oligandrum* isolate 1010	(Benhamou, Rey et al., 1997)

and *Aphanomyces laevis* (Abdelzaher et al., 1997; Ali-Shtayeh and Saleh, 1999; Benhamou et al., 1999; Horner et al., 2012; Veselý, 1978). *P. oligandrum* exerts mycoparasitic effects against various fungal pathogens. *P. oligandrum* manifests parasitism predominantly against *Fusarium*, including species such as *F. graminearum*, *F. verticillioides*, and *F. oxysporum* (Benhamou et al., 2012; Benhamou et al., 1997; Benhamou et al., 1999; Pellan et al., 2020), followed by *Microsporum* genus, inclusive of *M. canis* and *M. gypseum* (Gabrielova et al., 2018; Načeradská et al., 2016). Moreover, it exhibits parasitic behaviors against various fungi including *Trichophyton mentagrophytes*, *Rhizoctonia solani*, *Botrytis cinerea*, *Sclerotinia minor*, and *Verticillium dahliae* (Al-Rawahi and Hancock, 1998; Ikeda et al., 2012; Načeradská et al., 2016; Rey et al., 2005) (Table 1). Together, *P. oligandrum* is able to protect plants from several filamentous pathogens via mycoparasitism.

P. oligandrum also protects plants from pathogen invasion by inducing resistance. Through their symbiotic relationship with *P. oligandrum*, plants can better resist pathogen attacks and improve their overall resistance levels. During its interactions with plants, *P. oligandrum* can elicit resistance against various pathogens. For example, microbe-associated molecular patterns (MAMPs) from *P. oligandrum*, such as elicitins, are recognized by highly conserved plant pattern recognition receptors (PRRs) to activate pattern-triggered immunity (PTI) which primarily addresses a wide range of potential pathogens (Ngou et al., 2022). Plants are under constant threat from pathogens. In the process of plant evolution, they have formed a multifaceted innate immune system to protect themselves. Plants utilize cell surface PRRs to sense pathogen- or microbe-associated molecular patterns (PAMPs or MAMPs) and initiate effective PTI responses accordingly. Plants can recognize and counteract invaders through their highly conserved PRRs, including leucine-rich repeat receptor-like kinases (LRR-RLKs) and receptor-like proteins (LRR-RLPs). PTI is characterized by many hallmarks, including reactive oxygen species (ROS) burst, mitogen-activated protein kinase (MAPK) activation, and defense-related gene upregulation to defend pathogen invasion. Our recent study also showed that *P. oligandrum* triggers ferroptosis, a cellular death process driven by iron-dependent phospholipid peroxidation, as a defense strategy to protect plants from pathogen invasion (Cheng et al., 2022; Stockwell, 2022). Considerable research efforts have been directed toward investigating the above-mentioned strategies, which serve as the primary means by which *P. oligandrum* achieves the goal of inducing plant resistance.

Plants must both grow and defend in order to reproduce and survive in the natural world; however, plant resource allocation patterns often reveal tradeoffs that favor growth over defense or vice versa. Interestingly, several beneficial microorganisms are able to achieve plant defense–growth balance, which promote nutrient acquisition and immune responses. Growth enhancement by beneficial microorganisms has been extensively documented and is frequently linked to the production of microbial phytohormones and secondary metabolites. *P. oligandrum* also improves plant fitness by supplying the root system with tryptamine (TNH_2), a precursor of the phytohormone auxin (Le Floch et al., 2003; Bělonožníková et al., 2022; Gerbore et al., 2014). Auxins in plants are synthesized in four tryptophan-dependent routes, including TNH_2, indole-3-acetamide, indole-3-pyruvic acid, and indole-3-acetaldoxime pathways. *P. oligandrum* metabolises Trp and indole-3-acetaldehyde (IAAld) into TNH_2 through the TNH_2 pathway for plant growth (Le Floch et al., 2003; Bělonožníková et al., 2022). Recently, numerous studies have documented the ability of *P. oligandrum* to stimulate plant growth. *P. oligandrum* VOCs have been shown to enhance ginger growth, specifically in terms of root volume, quantity of root tips, cumulative root length, and mean root diameter (Sheikh et al., 2023a). *P. oligandrum* strains MS15, MS19, and MS31 have been found to significantly increase the shoot and root lengths of soybeans (Sayed et al., 2021). Furthermore, *P. oligandrum* not only improves growth parameters such as plant height, fresh weight, and dry weight in tomatoes and eggplants (Habib et al., 2021) but also enhances attributes of the red clover such as tap-root diameter, root branching, standing height, and forage yield (Pisarčik et al., 2020) (Table 2).

3 *Pythium oligandrum* in plant disease control

Multiple studies have yielded extensive evidence demonstrating the efficacy of using *P. oligandrum* in mitigating plant diseases induced by oomycetes, fungi, and bacteria (Table 3). The use of *P. oligandrum* as a means to combat oomycete diseases is effective in mitigating the detrimental effects inflicted

Table 2 Summary of plant growth promoted by *Pythium oligandrum*

Plant species	Growth-promoting characteristics	Isolates	Reference
Ginger (*Zingiber officinale* Roscoe)	Root volume, number of root tips, total root length, and average root diameter	GAQ1 (CGMCC no. 17470)	(Sheikh, Zhou et al., 2023b)
Soybean (*Glycine max* L.)	Shoot length, root length	MS15, MS19, MS31	(Sayed, Abdelmohsen et al., 2021)
Tomato (*Solanum lycopersicum* L.), Eggplant (*Solanum melongena* L.)	Plant height, fresh weight, dry weight	Polyversum	(Habib, Saab et al., 2021)
Red clover (*Trifolium pratense* L.)	Tap-root diameter, root branching, stand height, forage yield	Polyversum 1, Polyversum 3	(Pisarčik, Hakl et al., 2020)

by these pathogens. As early as 1997, Abdelzaher et al. reported the isolation of *P. oligandrum* from Egyptian soil, demonstrating its remarkable efficacy in controlling damping-off disease caused by *Pythium ultimum* in wheat (Abdelzaher et al., 1997). In 1999, Ali-Shtayeh and Saleh reported that *P. oligandrum* was also effective in controlling diseases caused by *P. ultimum* in cucumber (Ali-Shtayeh and Saleh, 1999). *P. oligandrum* has also been shown to effectively control the diseases caused by *Pythium aphanidermatum*. For example, the study conducted by Sayed et al. demonstrated that three strains of *P. oligandrum*, namely MS15, MS19, and MS31, were effective in controlling diseases caused by *P. aphanidermatum* in soybeans (Sayed et al., 2021). The *P. oligandrum* strain D11 was also effective in controlling soybean damping-off caused by *P. aphanidermatum* and *Pythium myriotylum* (You et al., 2019). *P. oligandrum* GAQ1 strain significantly inhibited *Pythium* soft rot in ginger caused by *P. myriotylum* (Daly et al., 2021; Sheikh et al., 2023a). Cultural filtrates of *P. oligandrum* (JU0328 and JU0329) significantly inhibit mycelial growth and the production of zoospores and oospores of *P. aphanidermatum* and *Pythium diclinum* (Moustafa, 2019). Polyversum (*P. oligandrum*) has also been used to control *P. aphanidermatum* on tomato and eggplant (Habib et al., 2021). Furthermore, three strains of *P. oligandrum*, namely LMSA 1.01.631, CBS 109981, and CBS 530.74, have been identified as effective agents for the control of diseases caused by *Pythium dissotocum* in tomatoes (Vallance et al., 2009). Together, *P. oligandrum* is currently being used in the control of various pathogenic oomycetes, including *P. ultimum, P. aphanidermatum, P. myriotylum, P. diclinum*, and *P. dissotocum*, in several plants including wheat, soybean, tomato, eggplant, and cucumber (Table 3).

Among the studies conducted on *P. oligandrum*, its efficacy in controlling *Fusarium* disease. *P. oligandrum* has been utilized in wheat, barley, and tomatoes to combat various *Fusarium* species, including *F. culmorum*, *F. oxysporum*, and *F. graminearum* (Benhamou et al., 1997; Cwalina-Ambroziak and Nowak, 2012; Ng et al., 2021; Postulkova et al., 2018). Yacoub et al. found that *P. oligandrum* strains Po1, Po2, Po3, and Po37 could be used to control both the pathogenic species *Phaeomoniella chlamydospora* and *Neofusicoccum parvum* on grapevines (Yacoub et al., 2016, 2020a, 2020b). Gerbore et al. discovered that *P. oligandrum* culture filtrate or oligandrincan could be used to effectively control *Erysiphe necator* on grapevines (Gerbore et al., 2014). *P. oligandrum* applied to peppers can effectively control *V. dahliae* and *V. wilt* (Al-Rawahi and Hancock, 1998; Rekanovic et al., 2007). The treatment of sugar beets with *P. oligandrum* can effectively control seedling and taproot diseases caused by *Aphanomyces cochlioides* (Takenaka and Ishikawa, 2013). In addition, *P. oligandrum* can also prevent and control a variety of other fungal pathogens, including *Sclerotinia sclerotiorum*, *R. solani*, *B. cinerea*, *Sphaerotheca macularis*, and *Mycosphaerella*

Table 3 Summary of *Pythium oligandrum* as biological control agent

Pathogen type	Pathogen species	Plant species	Isolates	Reference
Oomycete	*Pythium myriotylum*	Ginger (*Zingiber officinale* Roscoe)	GAQ1 (CGMCC no. 17470)	(Sheikh, Zhou et al., 2023b)
Oomycete	*Pythium myriotylum*	Ginger (*Zingiber officinale* Roscoe)	GAQ1 (CGMCC no. 17470)	(Daly, Chen et al., 2021)
Oomycete	*Pythium aphanidermatum*	Soybean (*Glycine max* L.)	MS15, MS19, MS31	(Sayed, Abdelmohsen et al., 2021)
Oomycete	*Pythium aphanidermatum*	Tomato (*Solanum lycopersicum* L.), Eggplant (*Solanum melongena* L.)	Polyversum	(Habib, Saab et al., 2021)
Oomycete	*Pythium aphanidermatum*, *Pythium myriotylum*	Soybean (*Glycine max* L.)	Strain D11	(You, Barraud et al., 2019)
Oomycete	*Pythium aphanidermatum*, *Pythium diclinum*	Tomato (*Solanum lycopersicum* L.)	JU0328, JU0329	(Moustafa and Abdelzaher, 2018, Moustafa, 2019)
Oomycete	*Pythium dissotocum*	Tomato (*Solanum lycopersicum* L.)	LMSA 1.01.631, CBS 109981, CBS 530.74	(Vallance, Le Floch et al., 2009)
Oomycete	*Pythium ultimum*	Cucumber (*Cucumis sativus* L.)	*P. oligandrum*	(Ali-Shtayeh and Saleh, 1999)
Oomycete	*Pythium ultimum*	Wheat (*Triticum aestivum L.*)	*P. oligandrum* was isolated from soil samples collected from an agricultural field near the Nile shore at El-Minia city, Egypt, in November, 1995.	(Abdelzaher, Elnaghy et al., 1997)
Oomycete	*Phytophthora sojae*	Soybean (*Glycine max* L.)	CBS 530.74	(Cheng et al., 2021)
Oomycete	*Phytophthora infestans*	Tomato (*Solanum lycopersicum* L.)	CBS 530.74 CBS 200.184	(Horner et al., 2012)
Fungus	*Fusarium culmorum*	Wheat (*Triticum aestivum* L.)	*P. oligandrum*	(Ng, Pernica et al., 2021)
Fungus	*Neofusicoccum parvum*	Grapevine (*Vitis vinifera* L.)	Po37	(Yacoub, Haidar et al., 2020a)
Fungus	*Phaeomoniella chlamydospora*	Grapevine (*Vitis vinifera* L.)	Po2	(Yacoub, Magnin et al., 2020b)

(Continued)

Table 3 (*Continued*)

Pathogen type	Pathogen species	Plant species	Isolates	Reference
Fungus	*Sclerotinia sclerotiorum*	Carrot roots (*Daucus carota* L.)	Polyversum WP	(Baturo-Cieśniewska, Łukanowski et al., 2018)
Fungus	*Fusarium oxysporum*, *Fusarium culmorum*, *Fusarium graminearum*	Barley (*Hordeum vulgare* L.)	Polyversum	(Postulkova, Rezanina et al., 2018)
Fungus	*Phaeomoniella chlamydospora*	Grapevine (*Vitis vinifera* L.)	Po1, Po2, and Po3	(Yacoub, Gerbore et al., 2016)
Fungus	*Erysiphe necator*	Grapevine (*Vitis vinifera* L.)	*P. oligandrum* culture filtrate or oligandrin	(Gerbore, Vallance et al., 2014)
Fungus	*Aphanomyces cochlioides*	Beet (*Beta vulgaris* L.)	MMR1, MMR2, MMR3, MMR7, MMR8, MMR10, MMR11, MMR13, MMR14, MMR15, MMR17, MMR18, MMR19, MMR20, MMR21, NBRC32559, OPU425[1)]	(Takenaka and Ishikawa, 2013)
Fungus	*Fusarium* species	Tomato (*Solanum lycopersicum* L.)	Polyversum	(Cwalina-Ambroziak and Nowak, 2012)
Fungus	*Rhizoctonia solani* AG-3	Potato (*Solanum tuberosum* L.)	MMR2	(Ikeda, Shimizu et al., 2012)
Fungus	*Botrytis cinerea*, *Sphaerotheca macularis*, *Mycosphaerella fragariae*	Strawberry (*Fragaria × ananassa*)	Polyversum WP	(Meszka and Bielenin, 2010)
Fungus	*Verticillium wilt*	Pepper (*Capsicum annuum* L.)	Polyversum	(Rekanovic, Milijasevic et al., 2007)
Fungus	*Botrytis cinerea*	Tomato (*Solanum lycopersicum* L.)	*P. oligandrum* strain No1133 (Denmark)	(Le Floch, Rey et al., 2003)
Fungus	*Fusarium oxysporum* f. sp. *radicislycopersici*	Tomato (*Solanum lycopersicum* L.)	*P. oligandrum* isolate 1010	(Benhamou, Rey et al., 1997)

Fungus	*Verticillium dahliae*	Pepper (*Capsicum annuum*)	*P. oligandrum*	(Al-Rawahi and Hancock, 1998)
Bacteria	*Ralstonia solanacearum*	Tomato (*Solanum lycopersicum* L.), sugar beet (*Beta vulgaris* L.), potato (*Solanum tuberosum* L.)	From fieldsoils in Hokkaido, Japan	(Takenaka, 2015)
Bacteria	*Ralstonia solanacearum*	Tomato (*Solanum lycopersicum* L.)	MMR2	(Masunaka, Nakaho et al., 2009)
Bacteria	*Ralstonia solanacearum*	Tomato (*Solanum lycopersicum* L.)	*P. oligandrum*	(Takenaka, Sekiguchi et al., 2008)
Bacteria	*Ralstonia solanacearum*	Tomato (*Solanum lycopersicum* L.)	MMR2	(Hase, Shimizu et al., 2006)
Root diseases	Root diseases	Red clover (*Trifolium pratense* L.)	Polyversum 1, Polyversum 3	(Pisarčik, Hakl et al., 2020)
Root diseases	Root diseases	Lucerne (*Medicago sativa* L.)	Polyversum	(Pisarčik, Hakl et al., 2019)

fragariae (Baturo-Cieśniewska et al., 2018; Ikeda et al., 2012; Le Floch et al., 2003; Meszka and Bielenin, 2010) (Table 3).

Furthermore, *P. oligandrum* can stimulate plant defense against bacterial pathogens. Treatment of tomato plants with *P. oligandrum* demonstrates a significant potential to impede the onset of the bacterial pathogen, *Ralstonia solanacearum*, the causative agent of wilt disease (Hase et al., 2006; Masunaka et al., 2009; Takenaka, 2015; Takenaka et al., 2008). *P. oligandrum* has also been used to control and prevent root diseases. Pisarčik et al. found that a product of *P. oligandrum*, Polyversum® (Biopreparaty spol. s r. o., Czech Republic), can effectively control root diseases in lucerne (Pisarčik et al., 2019). Polyversum® is also effective in controlling root diseases in red clover (Pisarčik et al., 2020). *P. oligandrum* can also be used to prevent and treat some unknown pathogens in rapeseed and tomato (Bělonožníková et al., 2022; Le Floch et al., 2003). However, the mechanism of *P. oligandrum* in promoting host resistance against bacteria is not clear yet, we speculate that *P. oligandrum* promotes plants to defend against bacterial diseases via secreting antimicrobial components, inducing plant immune responses, competing spaces and nutrients with bacteria, etc.

Together, the use of *P. oligandrum* to induce plant resistance can prevent and treat various diseases, including those caused by oomycetes, fungi, and bacteria, as well as pathogens that are yet to be identified (Table 3). The above research indicates that the main pathways for the prevention and control of plant diseases are through the strain itself and the culture filtrate of the strain. And we can clearly see that *P. oligandrum* has great potential and a broad-spectrum in the prevention and control of plant diseases. The use of *P. oligandrum* to treat plants can significantly enhance their resistance to various pathogens. However, the biological control mechanism of *P. oligandrum* need further investigation.

4 Identification of plant immunity-inducing proteins from *Pythium oligandrum*

Advancements in plant science and biotechnology have gradually revealed the mechanisms underlying plant immunity (Yang et al., 2022). The innate immune system of plants consists of two distinct phases. PTI, induced by microbe/pathogen/damage-associated molecular patterns, is the primary line of defense in plants. Plants recognize and overcome invaders through highly conserved PRRs (Ngou et al., 2022). The PTI response primarily involves a wide range of pathogens. MTI/PTI is characterized by multiple hallmarks, including callose deposition, ROS bursts, defense-related gene expression, and activation of MAPKs (Ngou et al., 2022). Effector-triggered immunity (ETI) is considered the second line of defense in plants (Ngou et al., 2022). At this stage, plants are capable of recognizing and combating the effector molecules of specific pathogens. This response is typically stronger and more effective as it can lead to a hypersensitive cell death response (HR) and prevent further spread of pathogenic microorganisms (Jones and Dangl, 2006). ETI also collaborates with PTI to defend plants against pathogens (Ngou et al., 2022). Increasing public concern about the environmental and health issues associated with synthetic chemicals has caused a shift toward more sustainable and eco-friendly disease management practices. Therefore, several BCAs have been developed and used. Plant immunity is widely used for the prevention and control of crop diseases and insect pests. MAMPs are potential biological factors that can activate plant immunity and are collectively known as plant immunity inducers that confer broad-spectrum resistance against pathogens. Improving plant resistance using plant immunity inducers is an environment-friendly method for pest management.

To obtain putative immunity-inducing proteins in *P. oligandrum*, as shown in Figure 1, first, candidate target genes from the *P. oligandrum* genome were searched and identified using a bioinformatics pipeline. Second, the genes were agroinfiltrated into *Nicotiana benthamiana* leaves to determine whether they could induce HR, ROS, or other immune responses *in planta*. Third, the screened HR/ROS-inducing genes were expressed in *E. coli* and purified for functional assays, including direct infiltration into *N. benthamiana*, to determine their threshold concentrations for necrosis induction and disease suppression. Fourth, the selected bioactive immunity-inducing proteins were used to infiltrate and spray leaves, fruits, or whole plants, and the disease severity was evaluated after pathogen inoculation. Finally, immunity-inducing proteins were identified and isolated from *P. oligandrum* by comparison with the control. This strategy has been successfully used against *P. oligandrum* and *Pythium periplocum* (Yang et al., 2021, 2022, 2023).

Candidate genes were identified from *P. oligandrum* genomes using a bioinformatics pipeline. They were then agroinfiltrated in *N. benthamiana* leaves to screen HR/ROS-inducing genes. Bioactive proteins screened out were expressed in *E. coli* and purified for functional assays, including direct infiltrations in *N. benthamiana* for the determination of their threshold concentrations for necrosis induction and disease suppression evaluation. Finally, the bioactive plant immunity-inducing proteins selected were used to infiltrate and spray plants.

The identification of immunity-inducing proteins is of great significance and has significant implications for the entire field of biological control research. The traditional application method of *P. oligandrum* is mainly to use the strain itself or its spores directly on plants, and there are also ways to use *P. oligandrum* culture filtrate. Through advanced molecular biology experiments, identify immunity-inducing proteins, obtain a large amount of this protein using in vitro protein expression technology, and then directly apply the protein to plants.

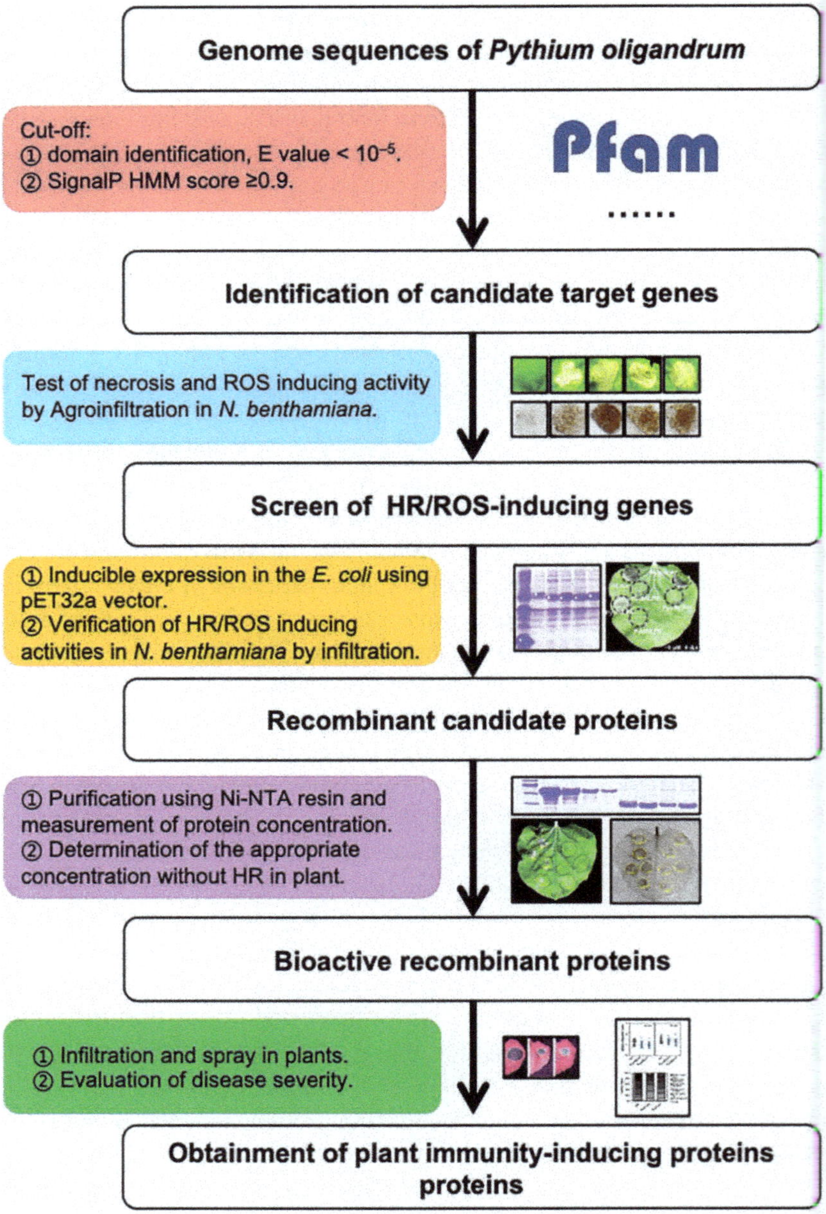

Figure 1 Strategy for identifying immunity-inducing proteins from *Pythium oligandrum*.

The direct use of immunity-inducing proteins has two advantages. First, it only retains the effective components that induce plant resistance in *P. oligandrum* and amplifies its induction signal by expressing a large amount of this protein, thereby improving the induction effect. The second is the direct use of proteins in vitro, which avoids the controversy caused by the application of functional genes in transgenic plants.

5 Main types of plant immunity-inducing proteins in *Pythium oligandrum*

Currently, two types of MAMP/plant immunity-inducing proteins, elicitins and necrosis- and ethylene-inducing peptide 1 (Nep1)-like proteins (NLPs), have been identified and applied in *P. oligandrum* to control several plant diseases (Table 4).

5.1 *Elicitin-type immunity-inducing proteins*

Elicitins are MAMPs specifically produced by oomycetes that activate plant defenses (Derevnina et al., 2016). All elicitins share a highly conserved 98-amino-acid (aa) domain with six cysteine residues forming three disulfide bridges. Elicitin-like proteins also have six structurally conserved cysteine residues; however, the lengths of their elicitin domains and the cysteine spacing patterns vary within each phylogenetic clade member (Derevnina et al., 2016). *P. oligandrum* genome encodes only elicitin-like proteins, including three cell wall protein elicitors (CWPs): POD-1, POD-2, and POS-1, and two extracellular proteins: oligandrin (Oli)-D1 and Oli-D2 (Masunaka et al., 2009). Oligandrins have been shown to induce disease resistance in various plants. Oligandrin was then purified from the culture filtrate of *P. oligandrum* (Picard et al., 2000). Upon the application of oligandrin to tomato plants, this protein induced plant defense mechanisms, thereby playing a pivotal role in confining the invasion of stem cells by the deleterious oomycete *P. parasitica* (Picard et al., 2000). Oligandrin is effective in instigating plant defense against fungal pathogens. The application of oligandrin triggered the expression of various defense-associated transcripts and, via diverse mechanisms, amplified resistance against powdery mildew in a susceptible tomato genotype (Satková et al., 2017). Oligandrin also has the capacity to manage gray mold and may have a significant influence on the stimulation of resistance against *B. cinerea* (Lou et al., 2011; Mohamed et al., 2007; Wang et al., 2011). Furthermore, oligandrin initiates systemic resistance to *Fusarium* crown and root rot in tomato plants (Benhamou et al., 2001). Ouyang et al. found that Oli-D1 and Oli-D2 could induce immune responses in tobacco and tomato (Ouyang et al., 2015) (Table 4). POD-1 and POD-2 could induce defense-related gene expression in both sugar beets and tomatoes and enhance the resistance of sugar beets against the fungus *A. cochlioides* (Takenaka et al., 2006, 2011). Moreover, our team recently identified a novel elicitin-like protein, PoEli8, from *P. oligandrum*, which can confer resistance to *P. capsici* in Solanaceae plants, as well as to

Table 4 Summary the plant immunity-inducing proteins from *Pythium oligandrum*

	CWP/MAMPs Type	Pathogen type	Pathogen species	Plant species	Isolates	Reference
MAMPs (Elicitin like, and NLP)	Oligandrin	Fungus	*Oidium neolycopersici*	Tomato (*Solanum lycopersicum* L.)	*P. oligandrum*	(Satková, Starý et al., 2017)
	Oligandrin	Fungus	*Botrytis cinerea*	Tomato (*Solanum lycopersicum* L.)	Strain RCU1	(Lou, Wang et al., 2011)
	Oligandrin	Fungus	*Botrytis cinerea*	Tomato (*Solanum lycopersicum* L.)	Strain RCU1	(Wang, Lou et al., 2011)
	Oligandrin	Fungus	*Botrytis cinerea*	Grapevine (*Vitis vinifera* L.)	Strain Py 7	(Mohamed, Lherminier et al., 2007)
	Oligandrin	Fungus	*Fusarium oxysporum*	Tomato (*Solanum lycopersicum* L.)	*P. oligandrum*	(Benhamou, Bélanger et al., 2001)
	Oligandrin	Oomycete	*Phytophthora parasitica*	Tomato (*Solanum lycopersicum* L.)	Strain 1010	(Picard, Ponchet et al., 2000)
	Oligandrin (Oli-D1, Oli-D2)	Fungus	*Botrytis cinerea*	Tomato (*Solanum lycopersicum* L.)	CBS 530.74	(Ouyang, Li et al., 2015)
	Oli-D2	Oomycetes	*Phytophthora capsici* and *Phytophthora nicotianae*	Tobacco (*Nicotiana benthamiana*), Tomato (*Solanum lycopersicum* L.), Pepper (*Capsicum annuum*)	CBS 530.74	(Yang, Wang et al., 2022)
	POD-1 and POD-2	No	No	Tomato (*Solanum lycopersicum* L.)	MMR2	(Takenaka, Yamaguchi et al., 2011)
		Fungus	*Aphanomyces cochlioides*	Sugar beet (*Beta vulgaris* L.)	MMR2	(Takenaka, Nakamura et al., 2006)
		Oomycetes	*Phytophthora capsici* and *Phytophthora nicotianae*	Tobacco (*Nicotiana benthamiana*), Tomato (*Solanum lycopersicum* L.), Pepper (*Capsicum annuum*)	CBS 530.74	(Yang, Wang et al., 2022)
	PoEli8	Oomycetes	*Phytophthora capsici* and*Phytophthora nicotianae*	Tobacco (*Nicotiana benthamiana*), Tomato (*Solanum lycopersicum* L.), Pepper (*Capsicum annuum*)	CBS 530.74	(Yang, Wang et al., 2022)
	PyolNLP5, PyolNLP7	Oomycetes	*Phytophthora capsici* and *Phytophthora nicotianae*	Tobacco (*Nicotiana benthamiana*), Tomato (*Solanum lycopersicum* L.), Pepper (*Capsicum annuum*)	CBS 530.74	(Yang, Dong et al., 2021)
Effector (RxLR)	PyolRXLR32	Oomycete	*Phytophthora capsici*	Tobacco (*Nicotiana benthamiana*)	CBS 530.74	(Ai, Yang et al., 2020)

Oli-D2 and POD-1 (Yang et al., 2022). Further research revealed that the cell surface RLP REli in *N. benthamiana* and its co-receptor RLKs BAK1 and SOBIR1 can perceive the immune response of elictin-like proteins of *P. oligandrum*, such as Oli-D2, POD-1, and PoEli8 (Figure 2) (Yang et al., 2022).

P. oligandrum elicitin PoEli8 exhibits plant immunity-inducing activity in plants, which is recognized by the plant cell surface RLP REli. Cell death and ROS accumulation triggered by the PoEli8 protein were dependent on REli and its co-receptors BAK1 and SOBIR1. Spraying PoEli8 proteins in the plants is able to protect Solanaceae plants against *Phytophthora* diseases.

5.2 (Nep1)-like-protein-type immunity-inducing proteins

NLPs represent another category of MAMPs that are widely distributed among eukaryotic and prokaryotic microorganisms, including bacteria, fungi, and oomycetes. Phylogenetic analysis of their amino acid sequences distinguishes between Type I, Type II, and Type III NLPs, which have one, two, and three pairs of conserved cysteine residues, respectively (Gijzen and Nurnberger, 2006; Seidl and Van den Ackerveken, 2019). All three types of NLPs are found in bacteria and fungi, whereas oomycetes produce only Type I or Type II NLPs. In oomycetes, only Type II NLPs are found in *P. oligandrum* and its sister species *P. periplocum* (Seidl and Van den Ackerveken, 2019; Yang et al., 2021). Multiple NLPs are known to behave as virulence factors and MAMPs to activate immune responses in dicotyledons (Seidl and Van den Ackerveken, 2019). NLPs feature

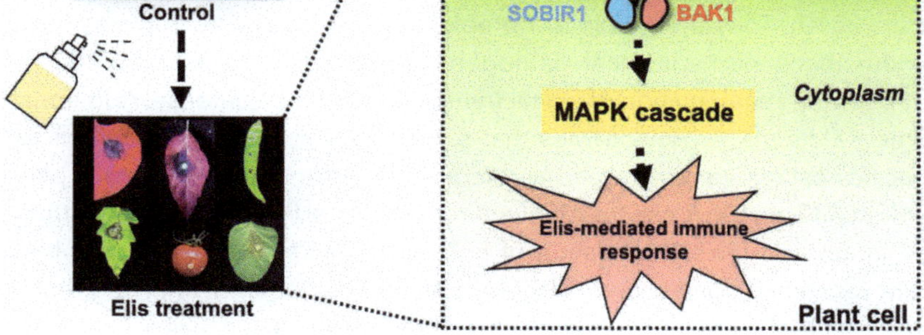

Figure 2 PoElis confers disease resistance against *Phytophthora capsici* in Solanaceae plants.

an N-terminal signal peptide that directs proteins to the secretory pathway for delivery into the extracellular environment, and a C-terminal conserved NPP1 domain (PF05630) with a conserved heptapeptide motif (GHRHDWE) in its central region. The cytotoxic NLPs bind glycosyl inositol phosphoryl ceramide (GIPC) sphingolipids as cytolytic, to induce necrosis in eudicots but not in monocots (Seidl and Van den Ackerveken, 2019). Plants such as *Arabidopsis thaliana* have evolved to recognize the conserved NLP peptides (nlp20 and nlp24) of NLPs as MAMPs to trigger immunity responses including MAPK activation and ROS burst. Type II NLPs, PyolNLP5 and PyolNLP7, derived from *P. oligandrum*, significantly reduced disease severity and suppressed the *in planta* growth of *P. capsici* in solanaceous plants, including tobacco (*N. benthamiana*), tomato (*S. lycopersicum*), and pepper (*C. annuum*) (Yang et al., 2021). Further investigation revealed that the plant disease resistance conferred by PyolNLP5 was independent of its necrosis-inducing capabilities and the generation of ROS (Yang et al., 2022) (Table 4). All pathogenic *Pythium* NLPs examined contain the nlp24 peptide of 24–27 amino acids in length, which include a conserved "GHRHDF/LE" motif (Seidl and Van den Ackerveken, 2019). A unique "GHxFAYYFxKDQ" motif occurs in the N-terminal regions of NLPs from *P. oligandrum* and *P. periplocum*, whereas pathogenic *Pythium* NLPs contain an "AIMYSWYFPKDSP" motif typically found in other type 1 and type 2 NLPs. The unique "GHxFAYYFxKDQ" motif is required for the function of *P. oligandrum* NLPs (Yang et al., 2021).

5.3 RxLR effectors

To manipulate host defenses, pathogenic oomycetes secrete and translocate RxLR effectors into host cells. RxLR effectors are a type of effectors secreted by oomycetes that are translocated into plant cells during interaction and have been reported to widely exist in oomycetes. The conserved N-terminal RxLR-dEER motif is required to mediate delivery into host cells, probably via specific infection structures called haustoria or directly from hyphae (Kale et al., 2010). They are found to be able to promote successful infection via various mechanisms in plant–pathogens interaction, such as suppression of PTI pathway, manipulation of signal transduction of host defense. Additionally, some RXLRs play avirulent roles that exhibit gene-for-gene interactions with specific host resistance proteins and trigger hypersensitive cell death in plants. PyolRXLR32, an RXLR effector derived from *P. oligandrum*, was found to trigger cell death and the accumulation of ROS in *N. benthamiana* and to enhance plant resistance against *P. capsici* (Ai et al., 2020) (Table 4), implying that the RxLR effector PyolRXLR32 confers plant defense-inducing activity.

6 The use of recombinant immunity-inducing proteins in *Phytophthora* disease control

Several immunity-inducing proteins from *P. oligandrum* have been successfully used to control plant diseases. Recombinant immunity-inducing proteins are the main form of application. To obtain recombinant proteins using PoEli8 as an example (Yang et al., 2022, 2023), the *PoEli8* gene without signal peptides was cloned, inserted into a pET-32a vector (His-tag), and then transformed into *Escherichia coli* strain BL21. Expression of 6xHis-tagged protein was induced by adding 0.1 mM isopropyl-β-d-thiogalactopyranoside (IPTG) and incubating at 16°C for 12 h. Soluble His-tagged proteins were purified by affinity chromatography using Ni-NTA Agarose according to the manufacturer's instructions. Protein standard bovine serum albumin (BSA) solution with a range of 0.5 μg/μL to 2.0 μg/μL was used to make a standard curve for protein sample evaluation. Partially purified proteins (5 μL) and a standard BSA solution in PBS were separated by SDS-PAGE (Yang et al., 2021). Gels were stained with Coomassie solution (1 g of Coomassie brilliant blue R250, 250 mL of isopropanol, 100 mL of acetic acid, and 650 mL of deionized water) for 2 h with gentle shaking and then destained in a solution (10% (v/v) acetic acid and 5% (v/v) ethanol) with gentle shaking (Yang et al., 2021). Transparent SDS-PAGE gels showing clear bands were photographed under natural light. The protein concentration in the SDS-PAGE gel bands was quantified using ImageJ software (http://rsb.info.nih.gov/ij/) (Yang et al., 2021).

To use recombinant proteins, their stability under different temperatures and pH conditions should first be evaluated. Cell death triggered by recombinant PoEli8 under different conditions in *N. benthamiana* usually acts as the main index of immune-inducing activity. Before testing this activity, the recombinant protein was incubated at 4–75°C for between 3 h and 2 months or was dissolved in the PBS buffers under pH 2, 4.5, 7, and 9.5. The thermal stability results indicated that the activity of the recombinant PoEli8 protein could be maintained for 1 month at 4°C, 1 week at 25°C, 1 week at 37°C, and only 3 h at 50°C. The pH stability results showed that PoEli8 was not sensitive to pH (Yang et al., 2022). Therefore, if PoEli8 recombinant protein can be used as a candidate for the development of induced resistance protein products, the pH of the buffer should be maintained between 2 and 9.5 and low temperature during storage and transportation should be ensured. Furthermore, recombinant proteins can be used to spray leaves, fruits, whole plants, or dipped roots. For example, tomatoes, pepper fruits, and leaves pre-sprayed with PoEli8 and PoNLP5/7 proteins showed substantially enhanced plant resistance to *P. capsici*. Spraying leaves and fruits and dipping roots with PpEli2 proteins from *P. periplocum* enhanced broad-spectrum resistance (BSR) in tomato, pepper, tobacco, and soybean against both fungal and oomycete pathogens (Figure 2) (Yang et al., 2023). In summary, we found that the recombinant immunity-inducing proteins mentioned above have broad-spectrum properties but require low-temperature cold-chain transportation during storage and transportation. Further research on recombinant immunity-inducing proteins needs to explore more effective and convenient usage methods.

7 Conclusion

P. oligandrum acts as a BCA that is non-toxic and environmentally friendly to humans and mammals and has been found to control a wide range of pathogens and provide additional plant growth benefits. It has been applied to crops such as potatoes, cotton, and tomatoes and has been shown to decrease disease incidence. At the same time, the use of immunity-inducing proteins not only avoids controversy over genetically modified applications but also has the potential to become a new product that can replace chemical pesticides. However, their biocontrol efficiency may be lower than or comparable to that of chemical pesticides, depending on the complex environmental conditions in the field. Compared to the application of other biocontrol agents such as *Bacillus* spp. and *Trichoderma* spp., there are still many gaps that need to be filled and improved. The specific issues and prospects for future research include the following:

1 Currently, the registration and application scope of *P. oligandrum* are relatively limited. Field experiments and applications to other crops still need to be studied, and their application range expanded.
2 The current dosage form of Polyversum® is a wettable powder, and further development of different dosage forms is required. For example, the aqueous or oil solution is easier than powder in field application.
3 The production of oospores, persistence in the environment, culture conditions, and stability and activity of oospores in storage environments must be studied and optimized. Researchers can screen the key factors such as metabolic products in oospores formation stage and apply the related factors to promote the production.
4 The integration and application of *P. oligandrum* with other BCAs or control measures (e.g. chemical pesticides) require further exploration and validation. To date, there is still a lack of research in the comparisons or applications with chemical pesticides and other BCAs.
5 The application of plant immunity-inducing proteins and chemical compounds based on *P. oligandrum* is currently lacking and may become an important direction for research and manufacture in the future.

8 Where to look for further information

Centres of expertise include:

- Centre National de la Recherche Scientifique (CNRS), Universite de Pau et des Pays de l'Adour, Université de Toulouse, Université Toulouse III in France
- State Key Laboratory of Agricultural and Forestry Biosecurity, Nanjing Agricultural University, Jiangsu Academy of Agricultural Sciences in China
- Swedish University of Agricultural Sciences (SLU) in Sweden
- Czech University of Life Sciences Prague in Czechia
- International Centre for Aquaculture Research and Development (ICARD), University of Aberdeen in UK

Key conferences include:

- Oomycete Molecular Genetics Network Annual Meeting

9 References

Abdelzaher, H. M. A., et al. (1997). "Isolation of Pythium oligandrum from Egyptian soil and its mycoparasitic effect on Pythium ultimum var. ultimum the damping-off organism of wheat." *Mycopathologia* 139: 97–106.

Ai, G., et al. (2020). "Pythium prediction and characterization of RXLR effectors in species." *Molecular plant-microbe interactions: MPMI* 33(8): 1046–1058.

Al-Rawahi, A. K. and Hancock, J. G. (1998). "Parasitism and biological control of verticillium dahliae by pythium oligandrum." *Plant disease* 82: 1100–1106.

Ali-Shtayeh, M. S. and Saleh, A. S. F. (1999). "Isolation of Pythium acanthicum, P. oligandrum, and P. periplocum from soil and evaluation of their mycoparasitic activity and biocontrol efficacy against selected phytopathogenic Pythium species." *Mycopathologia* 145: 143–153.

Baturo-Cieśniewska, A., et al. (2018). "Development of Sclerotinia sclerotiorum (Lib.) de Bary on stored carrot treated with Pythium oligandrum Drechsler determined by qPCR assay." *Acta Scientiarum Polonorum Hortorum Cultus* 17(5): 111–121.

Belonoznikova, K., et al. (2022). "Pythium oligandrum in plant protection and growth promotion: Secretion of hydrolytic enzymes, elicitors and tryptamine as auxin precursor." *Microbiological Research* 258: 126976.

Benhamou, N., et al. (2001). "Oligandrin, the elicitin-like protein produced by the mycoparasite Pythium oligandrum, induces systemic resistance to Fusarium crown and root rot in tomato plants." *Plant Physiology and Biochemistry* 39: 681–698.

Benhamou, N., et al. (2012). "Pythium oligandrum: an example of opportunistic success." *Microbiology* 158(11): 2679–2694.

Benhamou, N., et al. (1997). "Treatment with the mycoparasite pythium oligandrum triggers induction of defense-related reactions in tomato roots when challenged with Fusarium oxysporum f. sp. radicis-lycopersici." *Phytopathology* 87(1): 108–122.

Benhamou, N., et al. (1999). "Ultrastructural and cytochemical aspects of the interaction between the mycoparasite *Pythium oligandrum* and soilborne plant pathogens." *Phytopathology* 89: 506–517.

Cheng, Y., et al. (2022). "Ferroptosis induced by the biocontrol agent Pythium oligandrum enhances soybean resistance to Phytophthora sojae." *Environmental Microbiology* 24(12): 6267–6278.

Cwalina-Ambroziak, B. and Nowak, M. (2012). "The effects of biological and chemical controls on fungal communities colonising tomato (Lycopersicon esculentum Mill.) plants and soil." *Folia Horticulturae* 24(1): 13–20.

Daly, P., et al. (2021). "Dual-transcriptomic, microscopic, and biocontrol analyses of the interaction between the bioeffector pythium oligandrum and the pythium soft-rot of ginger pathogen pythium myriotylum." *Frontiers in Microbiology* 12. https://doi.org/10.3389/fmicb.2021.765872

Derevnina, L., et al. (2016). "Nine things to know about elicitins." *New Phytologist* 212(4): 888–895.

Gabrielova, A., et al. (2018). "The oomycete pythium oligandrum can suppress and kill the causative agents of dermatophytoses." *Mycopathologia* 183(5): 751–764.

Gerbore, J., et al. (2014). "Characterization of Pythium oligandrum populations that colonize the rhizosphere of vines from the Bordeaux region." *FEMS Microbiology Ecology* 90(1): 153–167.

Gijzen, M. and Nurnberger, T. (2016). "Nep1-like proteins from plant pathogens: Recruitment and diversification of the NPP1 domain across taxa." *Phytochemistry* 67(16): 1800–1807.

Habib, W., et al. (2021). "Evaluation of different control measures against damping off in greenhouse solanaceous crops." *Research Journal of Plant Pathology* 4. https://doi.org/10.15740/has/ijpp/11.1/1-9

Hase, S., et al. (2006). "Induction of transient ethylene and reduction in severity of tomato bacterial wilt by Pythium oligandrum." *Plant Pathology* 55(4): 537–543.

Horner, N. R., et al. (2012). "The oomycete Pythium oligandrum expresses putative effectors during mycoparasitism of Phytophthora infestans and is amenable to transformation." *Fungal Biology* 116(1): 24–41.

Ikeda, S., et al. (2012). "Biocontrol of black scurf on potato by seed tuber treatment with Pythium oligandrum." *Biological Control* 60(3): 297–304.

Jones, J. D. and Dangl, J. L. (2006). "The plant immune system." *Nature* 444(7117): 323–9.

Kale, S. D., et al. (2010). "External lipid PI3P mediates entry of eukaryotic pathogen effectors into plant and animal host cells." *Cell* 142(2): 284–95.

Le Floch, G., et al. (2003). "Enhancement of development and induction of resistance in tomato plants by the antagonist,Pythium oligandrum." *Agronomie* 23(5–6): 455–460.

Lou, B.-G., et al. (2011). "Enhancement of defense responses by oligandrin against Botrytis cinerea in tomatoes." *African Journal of Biotechnology* 10: 11442–11449.

Masunaka, A., et al. (2009). "Visualization of Ralstonia solanacearum cells during biocontrol of bacterial wilt disease in tomato with Pythium oligandrum." *Journal of General Plant Pathology* 75(4): 281–287.

Meszka, B. and Bielenin, A. (2010). "A new biological product against strawberry grey mould." *Phytopathologia* 58: 13–19.

Mohamed, N., et al. (2007). "Defense responses in grapevine leaves against botrytis cinerea induced by application of a pythium oligandrum Strain or Its Elicitin, oligandrin, to roots." *Phytopathology* 97: 611–620.

Moustafa, S. M. N. (2019). "Influence of Pythium oligandrum on Mycelia and production of Zoo-and Oo-spores of two phytopathogenic Pythium spp." *Journal of Pure and Applied Microbiology* 13(1): 233–240.

Moustafa, S. M. N. and Abdelzaher, H. M. A. (2018). "Increasing of Tomato Yield grown in hydroponic system using pythium oligandrum Isolated from Khoaa, Aljouf, Saudi Arabia." *Egyptian Journal of Microbiology* 53: 1–8.

Načeradská, M., et al. (2016). "Antifungal effects of the biological agentPythium oligandrumobserved in vitro." *Journal of Feline Medicine and Surgery* 19(8): 817–823.

Ng, C. A., et al. (2021). "Biocontrol effect of Pythium oligandrum on artificial Fusarium culmorum infection during malting of wheat." *Journal of Cereal Science* 100: 103258.

Ngou, B. P. M., Ding, P., and Jones, J. D. (2022). "Thirty years of resistance: Zig-zag through the plant immune system." *Plant Cell* 00: 1–32.

Ouyang, Z., et al. (2015). "Elicitin-like proteins Oli-D1 and Oli-D2 fromPythium oligandrumtrigger hypersensitive response inNicotiana benthamianaand induce resistance againstBotrytis cinereain tomato." *Molecular Plant Pathology* 16(3): 238–250.

Pellan, L., et al. (2020). "Commercial biocontrol agents reveal contrasting comportments against two mycotoxigenic fungi in cereals: fusarium graminearum and fusarium verticillioides." *Toxins* 12(3). https://doi.org/10.3390/toxins12030152

Picard, K., et al. (2000). "Oligandrin: a proteinaceous molecule produced by the mycoparasite pythium oligandrum induces resistance to phytophthora parasitica infection in tomato plants." *Plant Physiology* 124: 379–395.

Pisarčik, M., et al. (2020). "Effect of Pythium oligandrum and poly-beta-hydroxy butyric acid application on root growth, forage yield and root diseases of red clover under field conditions." *Crop Protection* 127. https://doi.org/10.1016/j.cropro.2019.104968

Pisarčik, M., et al. (2019). "Biological control in lucerne crops can negatively affect the development of root morphology, forage yield and quality." *Plant, Soil and Environment* 65(10): 477–482.

Postulkova, M., et al. (2018). "Suppression of fungal contamination byPythium oligandrumduring malting of barley." *Journal of the Institute of Brewing* 124(4): 336–340.

Rekanovic, E., et al. (2007). "Possibilities of biological and chemical control of verticillium wilt in pepper." *Phytoparasitica* 35(5): 436–441.

Rey, P., et al. (2005). "Interactions between the mycoparasite Pythium oligandrum and two types of sclerotia of plant-pathogenic fungi." *Mycological Research* 109(7): 779–788.

Satková, P., et al. (2017). "Diverse responses of wild and cultivated tomato to BABA, oligandrin and Oidium neolycopersici infection." *Annals of Botany* 119: 829–840.

Sayed, S. R. M., et al. (2021). "Myco-suppression analysis of soybean (Glycine max) damping-off caused by Pythium aphanidermatum." *Plants* 10(4). https://doi.org/10.3390/plants10040788

Sheikh, T. M. M., et al. (2023a). "Volatile organic compounds emitted by the biocontrol agent pythium oligandrum contribute to ginger plant growth and disease resistance." *Microbiology Spectrum* 11(4). https://doi.org/10.1128/spectrum.01510-23

Sheikh, T. M. M., et al. (2023b). "Volatile organic compounds from pythium oligandrum play a role in its parasitism on plant-pathogenic pythium myriotylum." *Applied and Environmental Microbiology* 89(2): e0203622.

Stockwell, B. R. (2022). "Ferroptosis turns 10: Emerging mechanisms, physiological functions, and therapeutic applications." *Cell* 185(14): 2401–2421.

Takenaka, S. (2015). "Studies on biological control mechanisms of Pythium oligandrum." *Journal of General Plant Pathology* 81(6): 466–469.

Takenaka, S. and Ishikawa, S. (2013). "Biocontrol of sugar beet seedling and taproot diseases caused by *Aphanomyces cochlioides* by *Pythium oligandrumt* reatments before transplanting." *Japan Agricultural Research Quarterly* 47: 75–83.

Takenaka, S., et al. (2006). "Novel elicitin-like proteins isolated from the cell wall of the biocontrol agent Pythium oligandrum induce defence-related genes in sugar beet." *Molecular Plant Pathology* 7(5): 325–339.

Takenaka, S., et al. (2008). "Colonization of Pythium oligandrum in the tomato rhizosphere for biological control of bacterial wilt disease analyzed by real-time PCR and confocal laser-scanning microscopy." *Phytopathology* 98: 187–195.

Takenaka, S., et al. (2011). "Implications of oligomeric forms of POD-1 and POD-2 proteins isolated from cell walls of the biocontrol agent Pythium oligandrum in relation to their ability to induce defense reactions in tomato." *Journal of Plant Physiology* 168(16): 1972–1979.

Vallance, J., et al. (2009). "Influence of Pythium oligandrum biocontrol on fungal and oomycete population dynamics in the rhizosphere." *Applied and Environmental Microbiology* 75(14): 4790–4800.

Veselý, D. (1978). "Parasitic Relationships between Pythium oligandrum Drechsler and some other species of the oomycetes class." *Zentralblatt für Bakteriologie, Parasitenkunde, Infektionskrankheiten und Hygiene. Zweite Naturwissenschaftliche Abteilung: Mikrobiologie der Landwirtschaft, der Technologie und des Umweltschutzes* 133(4): 341–349.

Wang, A.-Y., et al. (2011). "Defense responses in tomato fruit induced by oligandrin against Botrytis cinerea." *African Journal of Biotechnology* 10: 4596–4601.

Yacoub, A., et al. (2016). "Ability of Pythium oligandrum strains to protect Vitis vinifera L., by inducing plant resistance against Phaeomoniella chlamydospora, a pathogen involved in Esca, a grapevine trunk disease." *Biological Control* 92: 7–16.

Yacoub, A., et al. (2020a). "Pythium oligandrum induces grapevine defence mechanisms against the trunk pathogen Neofusicoccum parvum." *Phytopathologia Mediterranea* 59: 565–580.

Yacoub, A., et al. (2020b). "The biocontrol root-oomycete, pythium oligandrum, triggers grapevine resistance and shifts in the transcriptome of the trunk pathogenic fungus, phaeomoniella chlamydospora." *International Journal of Molecular Sciences* 21(18): 6876.

Yang, K., et al. (2021). "Type 2 Nep1-like proteins from the biocontrol oomycete pythium oligandrum suppress phytophthora capsici infection in solanaceous plants." *Journal of Fungi* 7(7): 496.

Yang, K., et al. (2022). "Novel eiicitin from pythium oligandrum confers disease resistance against phytophthora capsici in solanaceae plants." *Journal of Agricultural and Food Chemistry* 70(51): 16135–16145.

Yang, K., et al. (2023). "The Pythium periplocum elicitin PpEli2 confers broad-spectrum disease resistance by triggering a novel receptor-dependent immune pathway in plants." *Horticulture Research* 10(2): uhac255.

You, X., et al. (2019). "Suppressive effects of Pythium oligandrum on Soybean Damping off caused by P. aphanidermatum and P. myriotylum." *Annual Report of The Kansai Plant Protection Society* 61(0): 9–13.

Chapter 7

Effective biocontrol of grape diseases: assessing and improving efficacy

V. Altieri, P. Battilani, M. Camardo Leggieri, G. Fedele, T. Ji, V. Rossi and I. Salotti, Università Cattolica del Sacro Cuore, Italy

1 Introduction

Grapevine is one of the most important crops globally, accounting for more than 7.3 million hectares of cultivated area (Roca, 2022). Grapevine is affected

http://dx.doi.org/10.19103/AS.2024.0142.18

by several diseases which vary by region (Bois et al., 2017), with mildews being the major phytosanitary threat in most areas. These diseases negatively impact both grapevine yield and wine quality (Gouadec et al., 2007), resulting in significant economic losses (Van Helden, 2008). To maintain production and quality standards, intensive fungicide use is often required (Compant et al., 2013). For instance, although the area given to grapes accounts for just 3.5% of the total agricultural area in the European Union (EU), grapes receive around 15% of the synthetic pesticides (as active substances, a.s.) applied to major crops, with an average of 4.7 kg of a.s./ha between 1992 and 2003 (EUROSTAT, 2007). The extensive use of synthetic pesticides and awareness of their negative impact on human health (Loevinsohn, 1987) and the environment (Conway and Pretty, 1991) have led to efforts to limit their use. The use of synthetic fungicides pre-harvest has undergone limitations in order to reduce or eliminate pesticide residues on both grapes and wine (Verger and Boobis, 2013). The EU's Farm to Fork strategy aims for a 50% reduction in the overall use of chemicals and more hazardous active substances by 2030 (European Commission, 2020).

To minimize the negative impacts of pesticides, low-risk, alternative plant protection products are being considered and implemented in Integrated Pest Management and organic farming (Rossi et al., 2012; Pertot et al., 2017). These alternatives include biological control agents (BCAs), which are microorganisms (bacteria, oomycetes, and fungi) that are able to interact, directly or indirectly, with pathogens and to reduce disease development (Kumar and Droby, 2021; Elad and Freeman, 2002; Elmer and Reglinski, 2006). BCAs have minimal impact on human health and the environment, and have limited side effects on non-target organisms (Ritika and Utpal 2014; Parnell et al. 2016; Syed Ab Rahman et al. 2018). They also have short pre-harvest intervals (Marrone, 2007) and can be used in anti-resistant strategies due to their modes of action (MoA) compared to synthetic fungicides (Corkley et al., 2022). Although highly complex and diverse, and often coexisting in the same microorganism, modes of action of BCAs include antibiosis due to the production of enzymes and/or antibiotic metabolites, competition for space and nutrients, hyperparasitism of the pathogen, and induction of resistance in plants (Pertot et al., 2016). However, BCAs often have lower and more variable efficacy compared to synthetic fungicides, especially when applied under field conditions (Tracy, 2014; Pertot et al., 2017; Altieri et al., 2023).

Although several studies have been carried out to elucidate the mode of action of the different BCAs (e.g. Compant et al., 2013; Velivelli et al., 2014; Raymaekers et al., 2020, Hamdi and Alzawi, 2023), research on their effectiveness remains fragmented, making it difficult to draw clear and robust conclusions.

2 Current research on biocontrol of grape diseases: an overview

To estimate the amount of information available on biocontrol of grape disease, a Google search was carried out on 24 July 2023, using the string "biocontrol disease grape", to ensure that all the terms appeared. The search returned over 1.5 million results. When the search was repeated on Google Scholar (using the same string and searching for any type of publication), it yielded 31,200 results. When the search was restricted to scientific papers, this number decreased to approximately 6,400 results. The search results were then segmented by year (Figure 1), revealing that fewer than 50 papers were published in every five-year period until 1985, whereafter the number increased exponentially to more than 2,000 published between 2021 and July 2023, including 523 papers published in the first seven months of 2023. These figures clearly demonstrate a growing interest in the study of BCAs for the control of grape diseases, beginning in the late 1980s.

When the search for scientific papers was further segmented to focus on grape diseases with worldwide distribution (Bois et al., 2017) (Figure 2), the following results were found:

- 55% (2,340 papers) mentioned grey mould, caused by *Botrytis cinerea* (using the string "biocontrol botrytis grape").
- 15.3% (653 papers) mentioned trunk diseases (which include different diseases and causal agents, as outlined in Mondello et al., 2018).

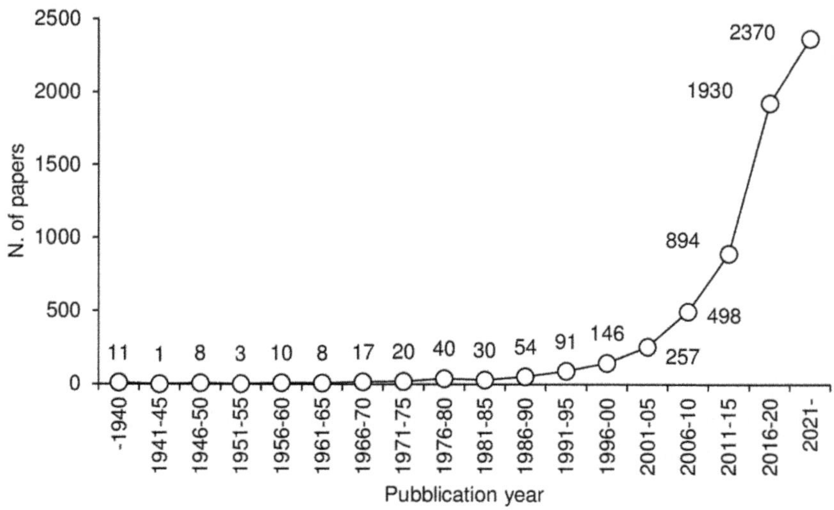

Figure 1 Number of papers published on the biocontrol of grape diseases over time, as obtained through a search on Google Scholar. Excludes references to papers published before 1940 and extrapolates papers for every five-year period until July 2023.

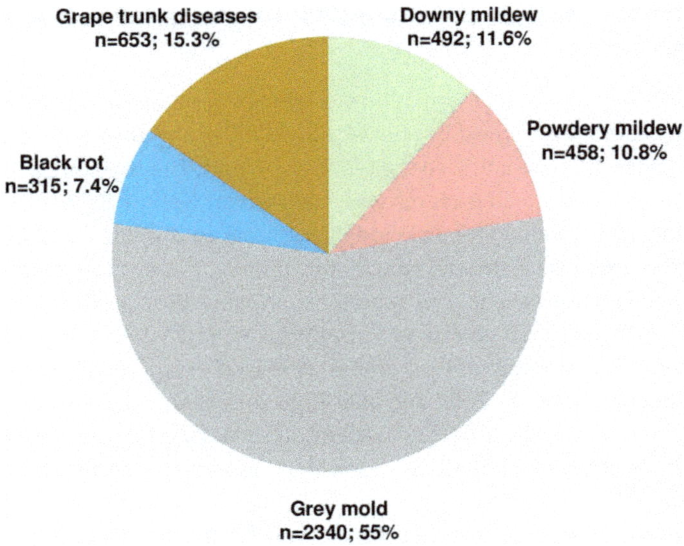

Figure 2 Number of papers published on the biocontrol of major grape diseases up to July 2023, as obtained through a search on Google Scholar.

- 11.6% (492 papers) and 10.8% (458 papers) mentioned downy mildew (caused by *Plasmopara viticola*) and powdery mildew (caused by *Erysiphe necator*), respectively.
- 7.4% (653 papers) mentioned black rot (caused by *Phyllosticta ampelicida*).

Papers mentioning white rot (caused by *Coniella diplodiella*) were few (numbering just 18), unlike those mentioning Aspergillus rot (caused by *Aspergillus* spp.) and anthracnosis (caused by *Elsinoë ampelina*). Surprisingly, 2,270 papers mentioned bunch rots caused by *Aspergillus* (using the search term "biocontrol aspergillus grape"), and 726 papers included both the terms "aspergillus" and "ochratoxin", i.e. the mycotoxin Ochratoxin A (OTA) produced by black *Aspergilla* and particularly *A. carbonarius* (Perrone et al., 2004; Battilani et al., 2006). This result is somewhat skewed by the fact that the term "grape" or the disease of interest was mentioned only in the introduction to the paper, with the actual research focusing on other crops, food products or pathogens. A similar situation was observed for *Colletorichum* rot ("biocontrol colletortichum grape") and sour rot (caused by multiple microorganisms), which were mentioned in 1,670 and 458 papers, respectively.

With the exception of rots caused by *Aspergillus* and *Colletorichum* species (as discussed above), there is consistency between the global relevance of grape diseases and the number of papers on biocontrol, with the exception of grey mould (GM). The extensive research into GM may be attributed to the fact that avoidance of the disease necessitates repeated fungicide applications during berry ripening. There is an urgent requirement for alternative solutions to synthetic fungicides to counter the problems related to the pre-harvest interval (i.e. the minimum number of days between the last pesticide application and the harvest) and to pesticide residues in wine (i.e. the pesticide that may remain on or in berries after they are applied to the crop). *Aspergillus* and *Colletorichum* bunch rots suffer from the same issues as GM (De Curtis et al., 2012; Pedrotti et al., 2020). This is not the case for mildews and black rot, where berries develop so-called ontogenetic resistance (age-related resistance; Gadoury et al., 2003, Kennely et al., 2005; Molitor Berkelmann-Loehnertz, 2011), and no specific pesticide applications are required to protect grape clusters from pre-veraison onwards (Bleyer et al., 2023). Intensive fungicide control is less critical for these diseases compared to GM, and effective alternatives to synthetic pesticides exist, including copper and sulphur (Rantsiou et al., 2020). Growing environmental concerns over the side effects of these inorganic compounds (Conway and Pretty, 1991; Hinckley et al., 2020) and subsequent restrictions in their usage in some countries (e.g. Directive 2009/128 in the European Community), have fueled interest in alternative solutions, including BCAs (Pertot et al., 2017). Interest in biocontrol for trunk diseases stems from the lack (or scarcity) of fungicides able to efficiently prevent plants from being infected by causal agents of wood decay and/or vascular colonization (Mondello et al., 2018). These pathogens typically enter the plant through wounds, especially those made during pruning (Gramaje et al., 2018).

The literature identified via the Google search covers a wide range of topics, including:

- Identification and selection of putative biocontrol microorganisms (e.g. Cordero-Bueso et al., 2017; Bruisson et al., 2019).
- Screening for efficacy against target pathogens in the laboratory or under controlled conditions (e.g. Niem et al. 2020; Zhao et al., 2022).
- Studies into their MoA, such as hyperparasitism (e.g. Angeli et al., 2009, 2012), production of toxic metabolites (e.g. Leifert et al., 1995; Castoria et al., 2001; Toral et al., 2018), competition (e.g. Sipiczki, 2006) and/or induced resistance in plants (e.g. Mohamed et al., 2007; Choudhary and Johri, 2009), using both classical (e.g. Nally et al., 2015) and molecular (e.g. Niem et al, 2023) methods.
- Evaluation of their effectiveness in disease control, either alone or in combination with BCAs and fungicides, under controlled conditions (e.g. Raut et al., 2018; Zhou et al., 2020) and in the vineyard (e.g. Calvo-Garrido et al., 2013; Sawant et al., 2017), using both artificial (e.g. Brown et al., 2021) and natural (e.g. Falk et al., 1996) pathogen inoculum.

This chapter focuses on the efficacy of BCAs, with the goal of critically and quantitatively examining the effectiveness of biocontrol for major grapevine diseases across the following studies:

- Downy mildew (DM);
- Powdery mildew (PM);
- Grey mould (GM);
- Black rot (BR);
- Trunk diseases (TDs; specifically Botryosphaeria dieback, Esca complex and Petri disease, Eutypa dieback, and black foot, as described in Mondello et al., 2018);
- Bunch rots caused by ochratoxigenic fungi (OF), i.e. fungi producing Ochratoxin A (OTA); and
- Sour rot (SR).

The aim is to (i) draw conclusions on the use of BCAs and (ii) identify knowledge gaps to be addressed by future research for more sustained use of biocontrol in vineyards.

3 Assessing the efficacy of biological control agents (BCAs) against grapevine diseases

To review the available literature on the efficacy of biocontrol, a second search was conducted using international bibliographical databases and a systematic literature approach. Systematic literature reviews are an essential aspect of academic research as they build on existing evidence to advance knowledge (Xiao and Watson, 2019). A systematic literature review provides a thorough evaluation of a research topic through a reliable, rigorous and verifiable methodology that involves summarizing, analyzing and synthesizing the related literature (Keele, 2007). A systematic approach reduces errors, limits researcher bias and communication of the information is enhanced (Candel, 2014).

3.1 Systematic literature review

The literature search was conducted across three databases:

- Scopus (https://www.scopus.com; accessed on 1 February 2023).
- Web of Science (https://www.webofscience.com; accessed on 1 February 2023).
- China National Knowledge Infrastructure (https://www.cnki.net/index/; accessed on 2 February 2023).

In order to be included in the database of relevant literature, papers had to satisfy the following pre-defined inclusion and exclusion criteria (Okoli and Schabram, 2010; Candel, 2014):

- The microorganism had to be identified at the taxonomic level and used as a strain or a commercial product.
- The microorganism had to be considered a BCA (i.e. studies focusing on the biology or ecology of the microorganism, or its MoA, were excluded).
- The BCA had to be tested for its ability to control at least one of the targeted diseases/pathogens on grape plants or organs, under controlled or field conditions. Studies conducted in the lab in which the microorganism and/or the pathogen were grown on artificial media were excluded, with the exception of OF.
- The scheduling strategy considered the application of a single BCA only (i.e. strategies that combined different BCAs and/or chemical products were not considered).
- The results needed to be expressed as efficacy relative to untreated control, or the study's data (e.g. disease incidence or severity, OTA concentration) had to allow for the calculation of efficacy.

Based on these criteria, specific queries were formulated by combining the following keywords:

- Biocontrol agent OR biological control agent OR biological control OR biocontrol OR BCA.
- Grape (and its multiple endings) OR vineyard OR *Vitis*.
- The scientific name of the pathogen OR synonyms OR the name of the disease OR other common names, e.g. botrytis bunch rot OR grey mould OR gray mold OR Botrytis bunch rot OR *Botrytis cinerea* OR *Botryotinia fuckeliana* (see disease-specific keywords in Table 1).

The data collection process is schematically described in Figure 3 (Biesbroek et al., 2013; Koricheva and Gurevitch, 2014). Papers obtained from the first search were merged and duplicates excluded; titles and abstracts were then screened to select the papers fulfilling the aim of this chapter, i.e. the *in planta* efficacy of BCAs used for the control of major grapevine diseases. Full texts of papers considered of potential interest were reviewed to ensure relevance. The numbers of screened and selected papers for each disease are reported in Table 1.

The systematic literature review yielded 176 papers in total, conducted between 1988 and 2022, in 32 countries that include Europe, namely Italy (32 studies), Spain (19), France (11) and Greece (11); America, specifically the US

Table 1 Keywords related to the investigated diseases and the corresponding number of papers found and selected for quantitative analysis.

Disease	Keywords	N° hits	N° selected for systematic
Downy mildew	Downy mildew OR *Plasmopara viticola*	212	26
Powdery mildew	Powdery mildew OR White mould OR *Erysiphe necator*	142	13
Black Rot	Black rot OR *Guignardia bidwellii* OR *Phyllosticta ampelicida*	15	0
Grapevine Trunk Diseases	Grape trunk diseases OR GTD OR Esca OR *Phaeoacremonium* OR *Phaeomoniella* OR *Eutypa* OR *Cadophora* OR *Botryospheria* OR *Diplodia* OR *Fomitiporia*	200	37
Grey Mould	Botrytis bunch rot OR grey mould OR gray mould OR grey mold OR gray mold OR *Botrytis cinerea* OR *Botryotinia fuckeliana*	622	83
Sour Rot	Sour rot	28	1
Ochratoxigenic Fungi	*Aspergillus* OR Black mould OR OTA OR OTB OR OTC OR ochratoxin	206	25

(11), Chile (5), Brazil (3) and Canada (2); China (26), India (10) and other Asian or Middle Eastern countries (5); South Africa (7) and Tunisia (5); Australia (3) and New Zealand (2). The selected papers referenced DM (26 papers), PM (13), GM (83), TDs (27), OF (17) and SR (1). No studies on the *in planta* biocontrol efficacy against BR were found.

3.2 Extraction of efficacy data from the literature

Data on BCA efficacy or disease assessments useful for calculating efficacy were extracted from the text, tables, and figures of the collected papers. The GetData Graph Digitizer 2.24 (http://getdata-graph-digitizer.com) was used to obtain precise data from graphs. Measured variables (X; i.e. disease incidence and severity, and OTA concentration) were used to calculate biocontrol efficacy (E) as follows: $E = (X_{NT} - X_T) / X_{NT}$, where NT represents the value of the untreated control and T is the value in any specific BCA-pathogen interaction (e.g. PM disease severity in a specific vineyard and year treated with *Bacillus subtilis*).

The extracted data were organized into a database and used to conduct a quantitative analysis of pooled data, enhancing their relevance compared to individual studies and aiding in summarizing the available scientific knowledge (Noordzij et al., 2009). This method is increasingly being used in plant pathology (Madden and Paul, 2011; Scherm et al., 2014; Ji et al., 2023; Rosace et al., 2023).

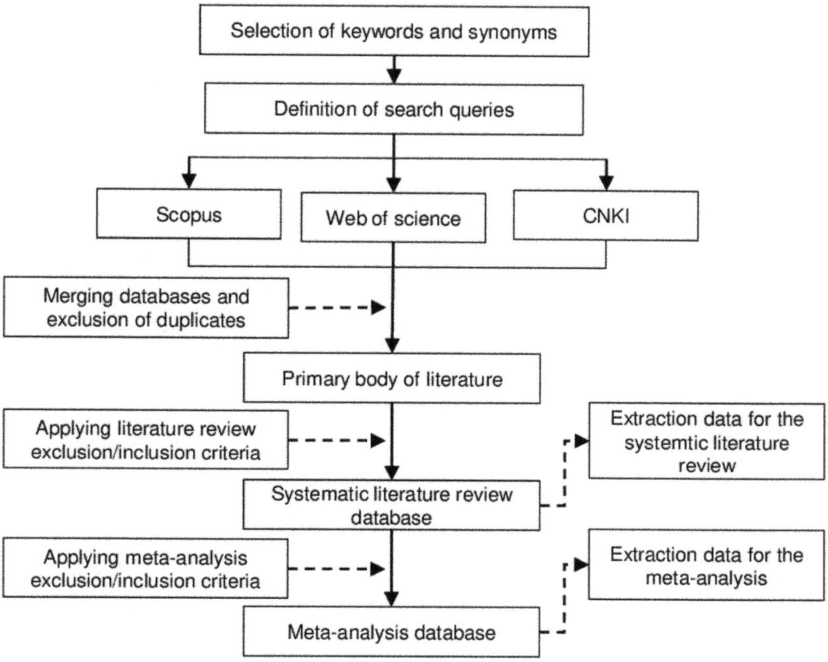

Figure 3 Schematic representation of the systematic literature review and data extraction process. Three international databases were used for the search: Scopus, Web of Science, China National Knowledge Infrastructure (CNKI).

It is necessary to note that the selected papers met the specific criteria outlined earlier (e.g. microorganisms were used as BCAs, efficacy data were available or calculable, BCAs were applied alone and not in combination or alternation with other plant protection products), and were published in the referenced databases. The analysis outlined in the following sections does not include all the papers published to date on the topic.

4 Biocontrol efficacy against downy mildew

A total of 26 papers on biocontrol efficacy on leaves and/or clusters were identified, which provided 138 efficacy figures. In half of these studies (13), BCAs were applied under controlled-environment conditions only, six papers focused on field conditions only, and seven papers included both controlled and field settings. The evaluation of BCA efficacy against DM was largely carried out on leaves, with only a few field studies assessing the effects on clusters.

Under controlled conditions, BCAs were typically applied on potted grapevine plants one to seven days before artificial inoculation with the pathogen. Detached (in laboratory) or attached (in greenhouse) leaves were mostly spray-inoculated with a *P. viticola* sporangia suspension (typically 10^5

sporangia/mL) and incubated at room temperature (20 to 25°C) and high relative humidity (95 to 100% RH) to ensure infection. Following the onset of DM, disease severity was assessed on leaves and/or clusters by using different diagrammatic scales (e.g. Banani et al., 2014; Jaber, 2015). Under field conditions, 9 of the 13 studies were conducted in the presence of natural inoculum; in these conditions, BCAs were applied twice or repeatedly during the season, usually on a calendar basis. Biocontrol efficacy was then assessed seven to 10 days after each application or after the final one (e.g. Falk et al., 1996). In studies where *P. viticola* was artificially inoculated in the field, BCA application, inoculum characteristics, and disease assessment were similar to that used under controlled conditions (e.g. Dagostin et al., 2011).

A total of 24 microorganisms were considered, belonging to 11 genera or singleton species of bacteria, filamentous fungi, and yeasts (Figure 4). Most BCAs were ascomycetous fungi (59% of papers), with the genera *Trichoderma* (39%; mainly *T. harzianum*) and *Fusarium* (35%; mainly *F. proliferatum*) represented more than the singleton species *Beauveria bassiana* (9%), *Saccharomyces cerevisiae* (9%), *Aureobasidium pullulans* (4%), and *Leptosphaerulina australis* (4%). Bacteria accounted for 41% of the papers, with *Bacillus* the most studied genus (56%, mainly *B. subtilis*) followed by *Streptomyces* (16%)

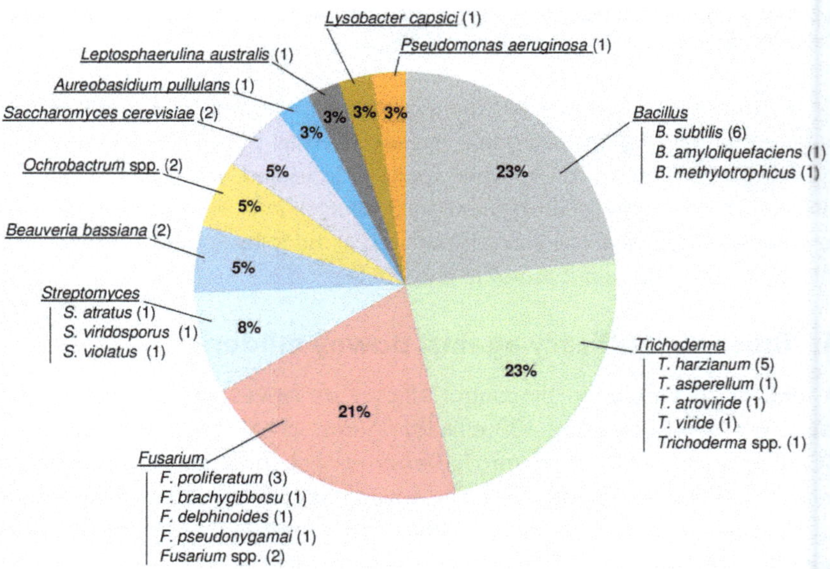

Figure 4 Representation of the microorganisms (singleton species or genera) for which papers on the *in planta* biocontrol efficacy against grape downy mildew have been found through a systematic literature review. Sectors show the amount of papers as a percentage of total papers, while the number of papers for each microorganism is reported in brackets.

and *Ochrobactrum* (13%). The singleton species *Lysobacter capsici* and *Pseudomonas aeruginosa* were studied in only one case.

Figure 5 shows the calculated efficacy of BCAs using disease severity data relative to untreated control, showing significant variability depending on the BCA genera and the test conditions. Under controlled conditions, the highest efficacy was observed for *Streptomyces* (average 92.1%; *n* = 1), *P. aeruginosa* (91.3%; *n* = 1), and *B. bassiana* (80.4 ± 2.8%; *n* = 8). Unfortunately, no field data were available for *P. aeruginosa*, while *Streptomyces* spp. and *B. bassiana* were tested only on leaves in the field. A reduction in efficacy was observed when transitioning from controlled to field conditions: from 92.1% to 68.8% (± 4.2%; *n* = 5) for *Streptomyces*, and from 80.4% to 51.0% (± 8.8%; *n* = 6) for *B. bassiana*. *Bacillus* showed high efficacy on leaves under controlled conditions (69.6 ± 9.5%; *n* = 13) and on clusters in the field (77.0 ± 12.5%; *n* = 3) but performed poorly on leaves in the field (44.5 ± 2.1%; *n* = 17). In contrast, *Trichoderma* exhibited higher efficacy under controlled conditions (59.6 ± 4.5%; *n* = 21) and on leaves in the field (52.4 ± 10.1%; *n* = 5), with negligible biocontrol on clusters in the field (23%; *n* = 1). The efficacy showed by *Fusarium*

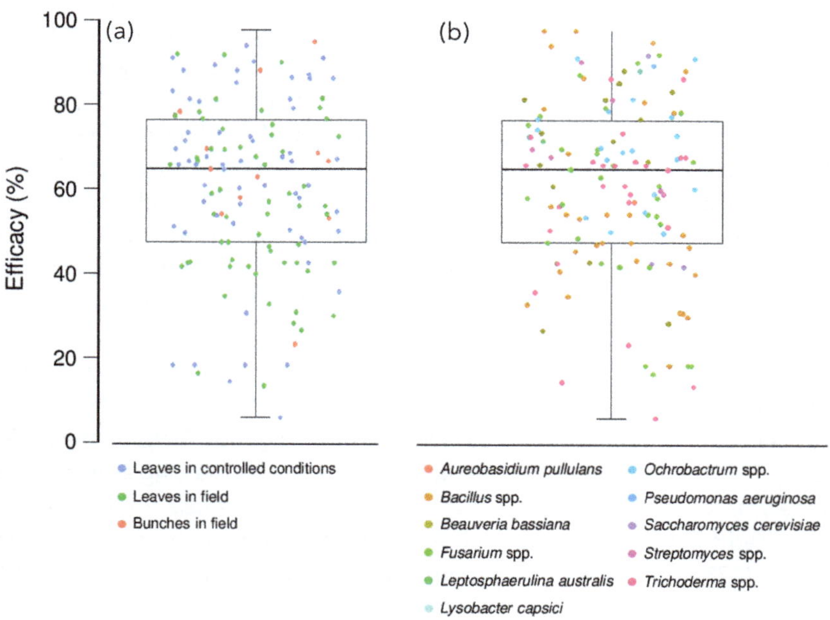

Figure 5 Box plots showing efficacy (%) data on *in planta* downy mildew biocontrol extracted from the papers shown in Figure 4. In panel A, data are grouped based on experimental conditions; in panel B they are based on microorganisms. Efficacy was calculated using disease severity data in relation to an untreated control. The box extends from the 25th to 75th quartile of the data distribution, and the line crossing the box represents the median; whiskers extend to the maximum and minimum.

was slightly higher in the field (59.8 ± 7.8%; *n* = 11, and 66.6 ± 2.3%; *n* = 9, on leaves and clusters, respectively) than in controlled conditions (49.9 ± 7.0%; *n* = 11). The highest efficacy in field was observed for *L. capsici* by Markellou et al. (2022), with 74.0% (*n* = 1) and 90.0% (*n* = 1) on leaves and clusters, respectively; additional studies, however, are needed to confirm the promising performance of this microorganism. The efficacy of *A. pullulans* (*n* = 1), *L. australis* (*n* = 1), *Ochrobactrum* (*n* = 19), and *S. cerevisiae* (*n* = 3) ranged from 58% to 72.9% on leaves in both controlled and field conditions.

5 Biocontrol efficacy against powdery mildew

A review of the literature identified 13 papers on biocontrol efficacy against *E. necator* on leaves and/or clusters, resulting in 203 efficacy figures. Of these studies, nine focused exclusively on field conditions, three on controlled-environment conditions only, and one paper included both controlled and field conditions. The evaluation of BCA efficacy under controlled conditions was carried out exclusively on leaves, while the field studies assessed both the leaves and clusters.

Under controlled conditions, the control provided by BCAs was assessed *in planta*, typically grown in greenhouses. BCAs were applied at least twice, either before or after artificial inoculations were carried out by dusting *E. necator* conidia (Maachia et al., 2015; Konstantinidou-Doltsinis et al., 2007) or via natural inoculum present on greenhouse plants infected by PM (Crisp et al., 2006; Pascale et al., 2017). Disease severity was assessed on leaves at 10 to 20 days after inoculation. In field studies, BCAs were applied at regular intervals to control the disease originating from the natural inoculum. The provided control was assessed in terms of disease severity during the season and/or after the last BCA application (e.g. Ahmed, 2018; Falk et al., 1995; Konstantinidou-Doltsinis et al., 2007).

A total of 22 microorganisms were evaluated, representing five genera or individual species of bacteria and filamentous fungi (Figure 6). Most (59%) of the tested BCAs were ascomycetous fungi, with the genus *Trichoderma* (45%; mainly *T. harzianum*) more frequently considered than the singleton species *Ampelomyces quisqualis* (9%) and *Pseudozyma flocculosa* (5%). Bacteria accounted for 41% of papers, with *Bacillus subtilis* being the most common species (27%), followed by *Pseudomonas fluorescens* and *P. syringae* (14% combined).

Figure 7 illustrates the boxplot of BCAs' efficacy calculated via disease severity data with reference to an untreated control. In panel A, efficacy data are grouped based on the experiment (in lab or field) and the organ (leaves or clusters). In panel B, they are grouped based on the BCA genus. The boxplots

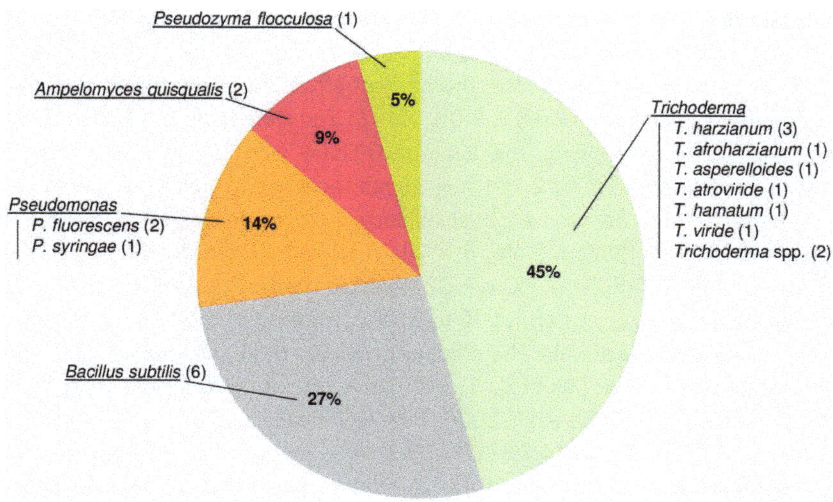

Figure 6 Representation of the microorganisms (singleton species or genera) for which papers on the *in planta* biocontrol efficacy against grape powdery mildew have been found through a systematic literature review. Sectors show the amount of papers as a percentage of total papers, while the number of papers for each microorganism is reported in brackets.

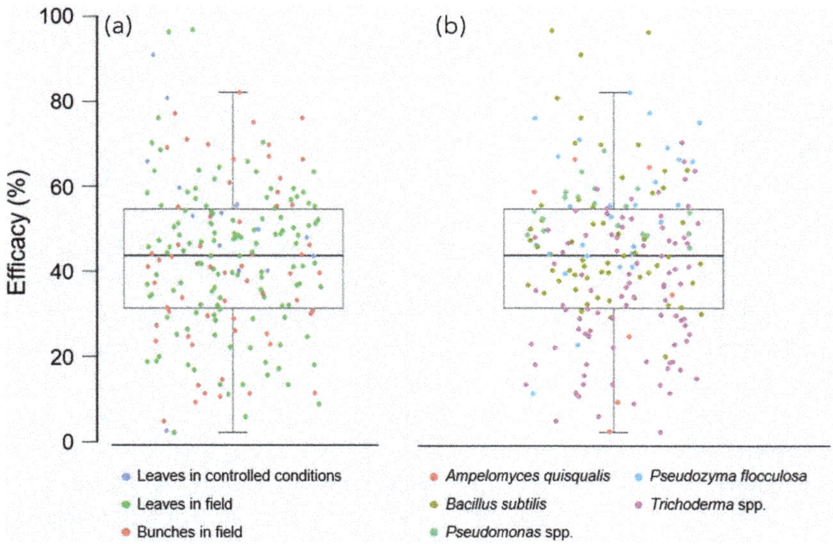

Figure 7 Box plots showing the efficacy (%) data on *in planta* powdery mildew biocontrol extracted from the papers shown in Figure 6. In panel A, data are grouped based on experimental conditions; in panel B they are based on microorganisms. Efficacy was calculated using disease severity data in relation to an untreated control. The box extends from the 25th to 75th quartile of the data distribution, and the line crossing the box represents the median; whiskers extend to the maximum and minimum.

and data dispersion reveal significant variability in the biocontrol efficacy data extracted from the different studies, although some trends can be identified.

Under controlled conditions, the highest efficacy on leaves was observed for *B. subtilis* (average of 60.8 ± 9.0%; n = 6). For *P. flocculosa, Pseudomonas* spp. and *Trichoderma* spp., the average efficacy was 55.2 (± 6.6%; n = 3), 54.0% (n = 1) and 49.1 (± 2.1%; n = 3), respectively. Under field conditions, *Pseudomonas* spp. had the highest efficacy, both on leaves (54.9 ± 1.5%; n = 18) and clusters (55.9 ± 3.4%; n = 18). *B. subtilis* showed a slightly higher efficacy on leaves (49.7 ± 3.1%; n = 32) than on clusters (42.6 ± 2.8%; n = 20), with a reduced efficacy of about 10 to 20% compared to the control obtained under controlled conditions. The efficacy provided by *A. quisqualis* was higher under field conditions (Falk et al., 1995) than in the laboratory experiment (1%; n = 1) conducted by Crisp et al. (2006), with an efficacy on leaves and clusters of 41% (± 12.3%; n = 3) and 44.7% (± 18.5%; n = 3), respectively. In a single study, *Pseudozima flocculosa* exhibited 49.7% efficacy (± 11.2%; n = 6) on clusters under controlled conditions, but no field data were available for leaves. Overall, *Trichoderma* showed poor control, with an average efficacy of 36.3 (±1.9%; n = 73) and 20.3 (±2.3%; n = 16) on leaves and clusters in the field, respectively.

6 Biocontrol efficacy against grey mould

Grey mould is the most studied grape disease for the use of BCAs, with 83 studies on biocontrol efficacy on leaves and/or clusters, contributing 426 efficacy figures. BCAs were applied under controlled-environment conditions in 42 of these papers, in the field in 20 papers, and under both conditions in 17 papers. Evaluation of BCA efficacy against GM was largely carried out on clusters.

Under controlled conditions, studies were conducted on leaves (e.g. Trotel-Aziz et al., 2008) and wounded (e.g. Lemos et al., 2016) and non-wounded berries (e.g. Fedele et al., 2020a) by applying BCAs as preventative (e.g. Maachia et al., 2015) or curative treatments (e.g. Whiteman and Stewart, 1998), a few hours or days before or after the artificial inoculation of a *B. cinerea* conidial suspension. Under field conditions, applications of BCAs were tested on clusters infected by natural (e.g. Calvo-Garrido et al., 2019) or artificial (e.g. Jelenić et al., 2021) inoculum. During these trials, BCAs were applied repeatedly at different timings. GM assessment was carried out by measuring disease incidence and/or severity in both environment-controlled and field conditions, via various assessment diagrams. In the field, assessments were typically carried out prior to harvesting (e.g. Magnin-Robert et al., 2013) and, in a few cases, after incubating field-collected berries with no disease symptoms in moist chambers for a week, to assess latent *B. cinerea* infections (e.g. Fedele et al., 2020b).

A total of 56 microorganisms were studied, belonging to 28 genera or singleton species of bacteria, filamentous fungi, and yeasts. An oomycete was also considered, specifically *Pythium oligandrum* (Figure 8). Most of the tested BCAs were bacteria (51% of studies), with the genera *Bacillus* (27%; mainly *B. subtilis* and *B. amyloliquefaciens*) better represented than *Pantoea* (6%, especially *P. agglomerans*), *Pseudomonas* (5%, mainly *P. fluorescens*), *Enterobacter* (5%), *Paenibacillus* (3%), and singleton species accounting for under 3% each (i.e. *Acinetobacter lwoffii*, *Brevibacillus reuszeri*, *Burkholderia contaminans*, *Lactobacillus plantarum*, and *Serratia liquefaciens*). Yeasts were widely studied (34% of cases), with *Aureobasidium pullulans* (8%), *Candida* spp. (8%, especially *C. sake*), *Metschnikowia* spp. (5%, especially *M. pulcherrima*), and *Hanseniaspora uvarum* (5%) more relevant than other species, which accounted for under 2% each (i.e. *Kloeckera apiculate*, *Saccharomyces cerevisiae*,

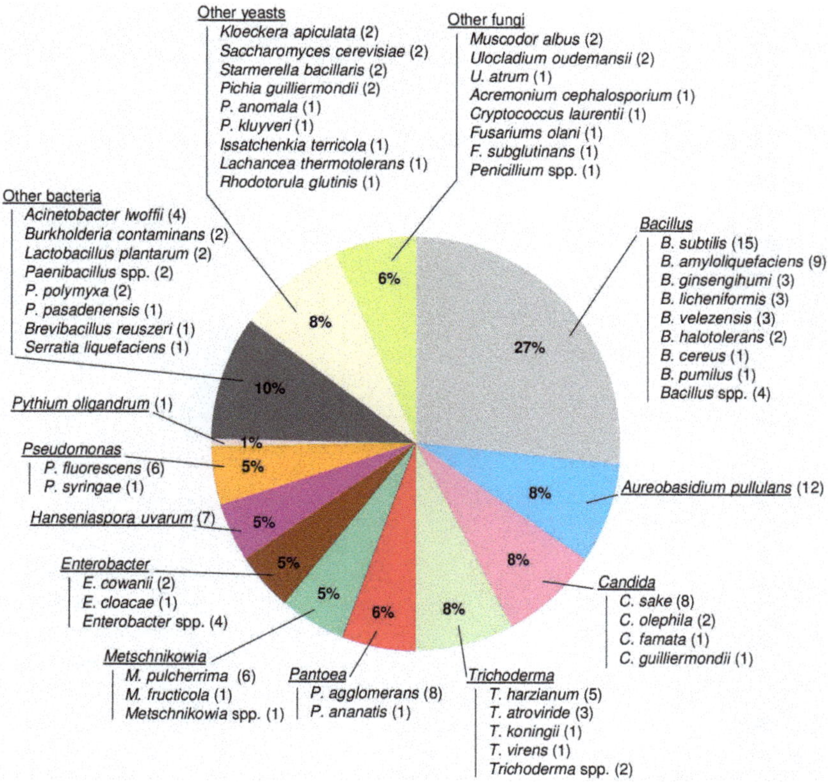

Figure 8 Representation of the microorganisms (singleton species or genera) for which papers on the *in planta* biocontrol efficacy against grey mould have been found through a systematic literature review. Sectors show the amount of papers as a percentage of total papers, while the number of papers for each microorganism is reported in brackets.

Starmerella bacillaris, *Pichia guilliermondii*, *P. anomala*, *P. kluyveri*, *Issatchenkia terricola*, *Lachancea thermotolerans*, and *Rhodotorula glutinis*). Filamentous fungi were studied in 14% of papers, with *Trichoderma* (*T. harzianum* and *T. atroviride*) the most represented (8%). Only one or two papers were found for the genus *Penicillium* and the singleton species *Muscodor albus*, *Ulocladium oudemansii*, *U. atrum*, *Acremonium cephalosporium*, *Cryptococcus laurentii*, *Fusariums olani*, and *F. subglutinans*.

Figure 9 shows a boxplot for BCA efficacy calculated using disease incidence and severity data, with reference to an untreated control. Regarding DM and PM, there was huge variability in the efficacy based on both microorganisms and experimental conditions.

Considering GM on leaves, *B. subtilis* showed an 88% (± 11%; *n* = 2) and 75% (± 9.8%; *n* = 6) reduction in disease incidence and severity, respectively, while *B. amyloliquefaciens* (64%; *n* = 1) and other *Bacillus* species (45.5 ± 7.5%; *n* = 11) showed lower disease severity control. The study by Ronseaux et al.

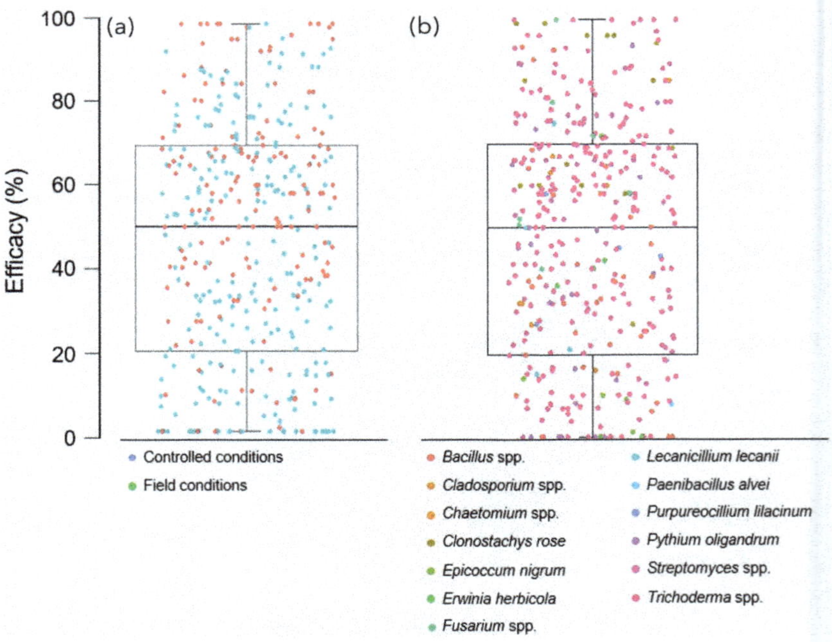

Figure 9 Box plots showing the efficacy (%) data on *in planta* grey mould biocontrol extracted from the papers shown in Figure 8. In panels A and B, disease incidence and severity data, respectively, are grouped based on experimental conditions; in panels C and D they are based on microorganisms. Efficacy was calculated using disease incidence or severity data in relation to an untreated control. The box extends from the 25th to 75th quartile of the data distribution, and the line crossing the box represents the median; whiskers extend to the maximum and minimum.

(2013) showed complete suppression of the disease with *U. atrum*. High control levels, in terms of GM severity on leaves, were observed for *A. lwoffii* (74.7 ± 12.3%; *n* = 3) and *Pseudomonas* spp. (79.8 ± 12.3%; *n* = 3). By contrast, *B. reuszeri* (*n* = 4), *Enterobacter* spp. (*n* = 4), *Paenibacillus* spp. (*n* = 3), *Pantoea* spp. (*n* = 5), and *S. liquefaciens* (*n* = 4) were studied but showed efficacies lower than 50%.

The majority of studies looked at the effect of BCAs on GM on clusters under controlled conditions. The highest control of disease incidence was observed for *C. laurentii* (*n* = 1), *I. terricola* (*n* = 3), *K. apiculata* (*n* = 2), *M. albus* (*n* = 3), and *Penicillium* spp. (*n* = 3), with efficacy greater than 80%. For *C. laurentii*, *I. terricola*, *K. apiculata*, and *Penicillium* spp., disease severity was not evaluated, while *M. albus* (*n* = 2) showed a reduction in disease severity by 97.5% (± 0.5%; *n* = 2). *H. uvarum* reduced both incidence and severity by 73.8% (±7.5%; *n* = 9) and 57.4% (± 11.0%; *n* = 5), respectively. Marsico et al. (2021) and Fedele et al. (2020b) reported reductions in disease severity as high as 100% for *L. thermotolerans* (*n* = 2), *P. oligandrum* (*n* = 2), and *Trichoderma* spp. (*n* = 2). Although disease incidence was not significantly reduced (65±10%; *n* = 2), good efficacy in terms of disease severity was calculated for *L. plantarum* (85%; *n* = 1). Biocontrol efficacy for GM incidence and/or severity ranged from 40% to 70% for *A. cephalosporium* (*n* = /1), *A. pullulans* (*n* = 3/4), *B. contaminans* (*n* = 1/1), *Candida* spp. (n = 4/4), *Enterobacter* spp. (*n* = 5), *Metschnikowia* spp. (*n* = 6/10), *Pantoea* spp. (*n* = /4), *Paenibacillus* spp. (*n* = /3), *Pichia* spp. (*n* = 3/4), and *S. bacillaris* (*n* = 1/26), with less than 30% control observed for *Pseduomonas* spp. (*n* =1/9), *R. glutinis* (*n* = 1/), and *S. cerevisiae* (*n* = 1/3).

Just 14 genera or singleton species were tested under field conditions. The best performances in terms of disease incidence reduction were recorded for *B. subtilis* (59 ± 11.3%; *n* = 3) and *Fusarium* spp. (66.6 ± 10.0%; *n* = 16), followed by *C. sake* (52.2 ± 3.9%; *n* = 18). *Candida sake* confirmed good performances in terms of disease severity, with an average efficacy of 70.8% (± 5.0%; *n* = 19). The ability of *Bacillus* species (*n* = 12) to reduce disease severity, however, was limited, with average efficacies ranging from 26 to 30%. *Trichoderma* spp. (*n* = 14) and *Ulocladium* spp. (*n* = 2) provided an average reduction of disease incidence of about 40%, with higher efficacy for disease severity: 53.7% (± 6.3%; *n* = 14) and 78% (*n* = 1), respectively. Despite less than 30% incidence control being observed for *A. pullulans* (*n* = 28) and *Metschnikowia* spp. (*n* = 1), their efficacy in reducing disease severity was 51.5% (± 8.9%; *n* = 10) and 67% (*n* = 1), respectively. The intermediate GM control observed under controlled conditions for *Enterobacter* spp., *Pantoea* spp., and *Paenibacillus* spp. was reduced in the field, with average efficacy (as disease severity) ranging from 32% to 37%. Low levels of biocontrol were confirmed for *Pseudomonas* spp., with an average efficacy of 11% (± 4.9%; *n* = 10).

7 Biocontrol efficacy against trunk diseases

A total of 27 papers on the biocontrol against the fungi causing trunk diseases were identified, contributing 429 efficacy data points. Whilst 15 studies referred to experiments conducted under controlled conditions, 21 papers focused on field settings, with just one studying both conditions.

A total of 28 microorganisms (oomycetes, fungi, and bacteria) were considered (Figure 10). Ascomycetous fungi were prevalent (78% of papers), with the genus *Trichoderma* (66%; mainly *T. atroviride*) being prevalent over *Fusarium* (5%; mainly *F. lateritium*), and other fungi (8%; with fungal taxa considered in just one paper). Bacteria accounted for 18% of the microorganisms, with *Bacillus* the most frequently considered (13%, mainly *B. subtilis*), followed by *Streptomyces* (3%) and others (3%). *Pythium oligandrum* was the only reported oomycete.

In both controlled and field conditions, BCAs were applied to evaluate their ability to protect detached wood pieces (in controlled conditions; e.g. Silva-Valderrama et al., 2021), wounded nursery cuttings (e.g. Di Marco and Osti, 2007), grafting points (e.g. Pertot et al., 2016), or winter pruning wounds (e.g. Brown, et al., 2021) from tissue colonization by the fungi causing GTDs. In these trials, causal fungi were either artificially inoculated in the form of mycelia

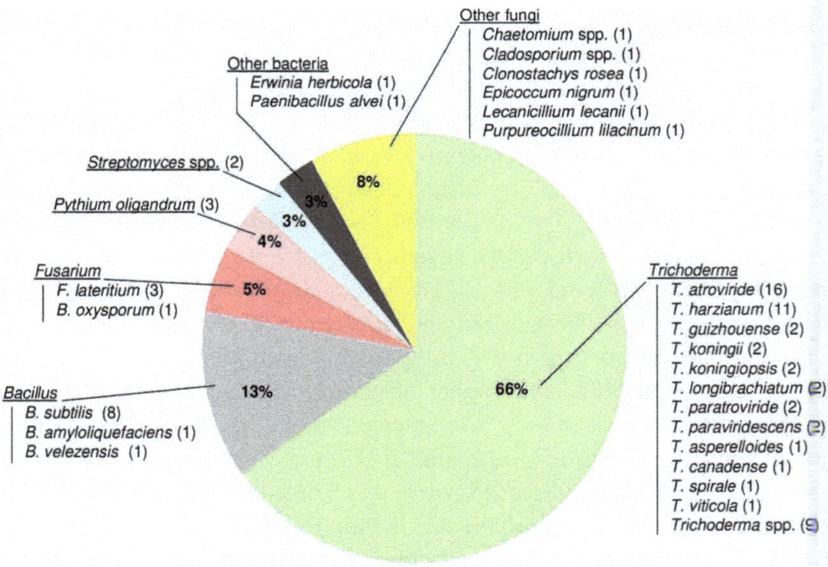

Figure 10 Representation of the microorganisms (singleton species or genera) for which papers on the *in planta* biocontrol efficacy against grape trunk diseases have been found through a systematic literature review. Sectors show the amount of papers as a percentage of total papers, while the number of papers for each microorganism is reported in brackets.

(e.g. Rezgui et al., 2016) or spores (e.g. Mutawila et al., 2015), or the inoculation occurred via natural spread in affected vineyards (e.g. Kotze et al., 2011). The incidence of GTDs was determined through visual assessment of symptoms such as discolouration (e.g. Di Marco et al., 2022) and wood cankers (e.g. Rezgui et al., 2016), as well as by re-isolation of the inoculated fungi (e.g. John et al., 2005) and PCR analysis (e.g. Reis et al., 2021).

Figure 11 illustrates the efficacy of BCAs calculated using disease incidence data relative to an untreated control. For black foot disease, no microorganism tested achieved more than 40% efficacy under either controlled or field conditions. In the field, the most effective BCAs were *Streptomyces* (39.5 ± 8.0% efficacy, n = 18), followed by *T. atroviridae* (31.0 ± 10.5% efficacy, n = 15). For Botryosphaeria dieback, five of the 11 BACs tested under controlled conditions had more than 50% efficacy, with *C. rosea* (79.3 ± 6.8% efficacy, *n* = 12) and *T. harzianum* (74.4% ± 5.3% efficacy, *n* = 10) outperforming the other microorganisms. *Clonostachys rosea* was not tested in the field, and the efficacy of *T. harzianum* was lower (43.2 ± 6.9%, *n* = 12). Other *Trichoderma* spp. showed more than 50% efficacy in the field (*n* = 37), while *Bacillus* spp. reduced the disease by 37.5% (± 7.8% efficacy, *n* = 5). For the Esca complex, BCAs exhibited substantial variation among different taxa of microorganisms, especially under controlled conditions, ranging from 8% efficacy for *Paenibacillus alvei* (*n* = 1)

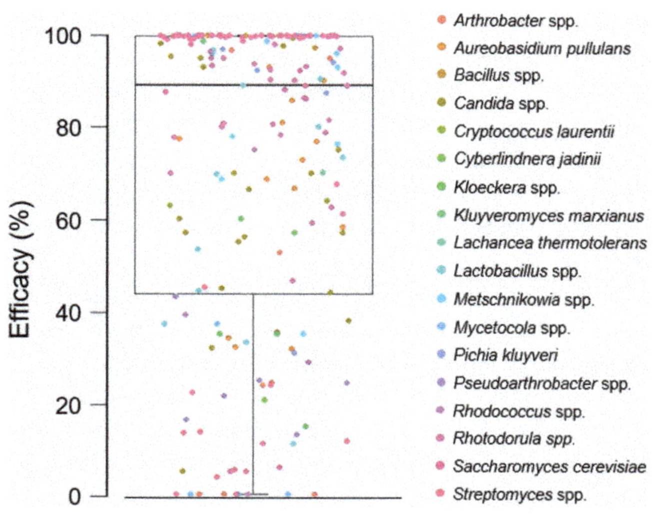

Figure 11 Box plots showing the efficacy (%) data on *in planta* biocontrol of grape trunk diseases extracted from the papers shown in Figure 10. In panel A, data are grouped based on experimental conditions; in panel B they are based on microorganisms. Efficacy was calculated using disease severity data in relation to an untreated control. The box extends from the 25th to 75th quartile of the data distribution, and the line crossing the box represents the median; whiskers extend to the maximum and minimum.

to 78.3% (± 9.6%, *n* = 12) for *T. atroviridae* and 72% (± 23.0%, *n* = 2) for *T. harzianum*. *Clonostachys rosea* also demonstrated high efficacy (74.5 ± 8.3% efficacy, *n* = 6). The efficacy was lower overall for the BCAs tested in the field (average of 53.1 ± 4.4%, *n* = 29) compared to controlled conditions (average of 38.1 ± 0.7%, *n* = 111), with *Trichoderma* spp. (average of 36.3 ± 0.4%, *n* = 70), *Streptomyces* spp. (47.2 ± 5.9%, n = 24) and *Bacillus* spp. (45.2 ± 9.0%, *n* = 5) exhibiting higher efficacy than *P. oligandrum* (27.2 ± 9.3%, *n* = 12). For Eutypa dieback, experiments under controlled conditions ranged from no efficacy (Lecanicillium lecanii, *n* = 1) to complete (*Fusarium* spp., *n* = 2), with the remainder ranging from 32% (± 3.7%, *n* = 7, for *Bacillus* spp.) to 58.3% (± 8.7%, *n* = 3, for *T. harzianum*). When tested in the field, all BCAs showed more than 50% (*T. harzianum*, *T. atroviridae*, *Bacillus* spp.) or 60% (other *Trichoderma* and *Fusarium* spp.) efficacy.

8 Biocontrol efficacy against ochratoxigenic fungi and sour rot

A total of 17 studies evaluated the biocontrol efficacy against OF, providing 223 efficacy figures. Most studies were conducted under controlled environmental conditions (15 of 17) with the remainder in the field (De Felice et al., 2008; Dimakopoulou et al., 2008).

Under controlled conditions, BCAs and *Aspergillus* spp. were co-cultured on synthetic media (e.g. Zhu et al., 2015), or on berries (e.g. Lappa et al., 2018) or were applied to media previously contaminated with the fungus (e.g. El Khoury et al., 2017). The OTA concentration in the medium was quantified with liquid chromatography (HPLC). In the field studies (De Felice et al., 2008; Dimakopoulou et al., 2008), a cell suspension of *A. pullulans* was applied at pre-bunch closure, early veraison, and/or before harvest, and the OTA contamination produced by natural infection by *Aspergillus* spp. was determined at harvest through HPLC.

A total of 33 microorganisms from 18 genera were considered, with 30% being bacteria and 70% yeast. No filamentous fungi were tested as BCAs against ochratoxigenic fungi. Most tested BCAs belonged to *Candida* (18%, comprising six different species), followed by 18% focused on "other yeasts" (Figure 12), and 12% on *Saccharomyces cerevisiae*. Among bacteria, 18% were represented by *Streptomyces* spp., *Lactobacillus*, and *Bacillus* species, while "other bacteria" made up 12%.

Figure 13 displays the boxplot of BCA efficacy calculated using OTA concentration data with reference to an untreated control. The highest efficacy (>80%) in terms of reducing ochratoxin A (OTA) production by *A. carbonarius* was observed for the bacteria *Bacillus* (*n* = 7) and *Rhodococcus* (*n* = 3), with OTA reductions exceeding 90%. Among yeasts, *S. cerevisiae* (*n* = 43) and

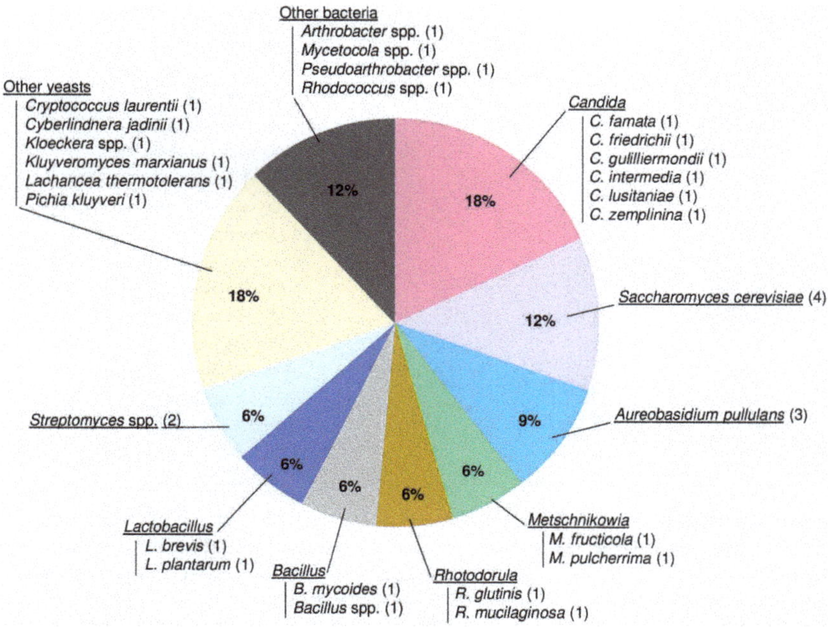

Figure 12 Representation of the microorganisms (singleton species or genera) for which papers on the biocontrol efficacy against the production of ochratoxin A (OTA) by *Aspergilla* have been found through a systematic literature review. Sectors show the amount of papers as a percentage of total papers, while the number of papers for each microorganisms is reported in brackets.

Kluyveromyces marxianus (*n* = 1) also showed more than 90% efficacy in reducing OTA production by *A. carbonarius*. Conversely, the lowest efficacy was observed for *Arthrobacter* and *Mycetocola*, with 8.2% (± 8.1; *n* = 3) and 12.4% (± 12.4; *n* = 3) efficacy, respectively, in controlling OTA production by *A. carbonarius*. *Bacillus* also exhibited complete efficacy (100 ± 0%; *n* = 3) in controlling OTA production by *A. niger*. In addition, *Candida* (*n* = 4), *Pichia kluyveri* (*n* = 4), and *S. cerevisiae* (*n* = 10) showed promising results (>88%) in reducing OTA production by *A. ochraceus*.

8.1 Biocontrol efficacy against sour rot (SR)

Only one paper evaluated BCA efficacy in controlling SR. Specifically, Calvo-Garrido et al. (2013) investigated the efficacy of the yeast *Candida sake* under field conditions on both disease incidence and severity on clusters. The study found that SR incidence at harvest was reduced by 27.3% and severity by 66.6%.

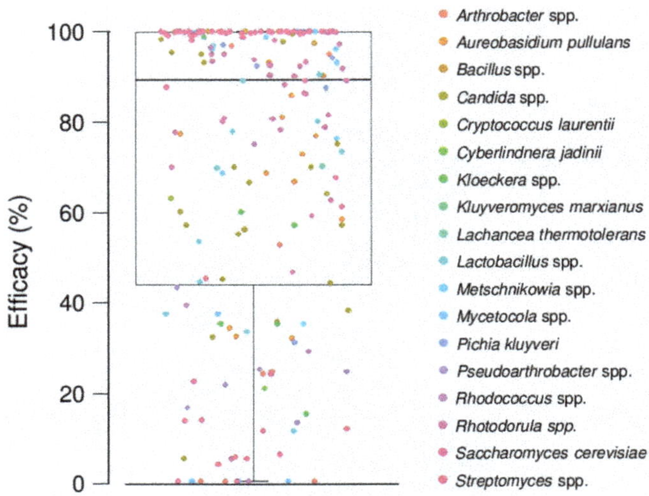

Figure 13 Box plots showing the efficacy (%) data on biocontrol against the production of ochratoxin A (OTA) by *Aspergilla* provided by different microorganisms; data were extracted from the papers shown in Figure 12. Efficacy was calculated using disease OTA concentration in relation to an untreated control. The box extends from the 25th to 75th quartile of the data distribution, and the line crossing the box represents the median; whiskers extend to the maximum and minimum.

9 General lessons from the literature on efficacy

The biocontrol of grape diseases has been studied extensively, as demonstrated by the large number of studies and the variety of microorganisms tested. These microorganisms belong to different kingdoms and have been evaluated for their ability to control one or multiple target pathogens. They display various MoAs and have diverse nutritional and environmental requirements. Only some of these studies, however, provide meaningful information on the efficacy of BCA application on plants (either under controlled conditions or in the vineyard).

The significant differences (clearly shown by the boxplots in Figures 5, 7, 9, and 11) among the experiments carried out at various locations (across continents in both the southern and northern hemispheres) and in different years, in which the disease pressure ranged from mild to extremely severe, and the environmental conditions were highly diverse, made synthesis of the results difficult. Variations in experimental settings (e.g. controlled vs field conditions, pre- vs post-infection timing, distribution methods, etc.) and methods for assessing BCA efficacy also increased this difficulty. In addition, some BCAs have been classified at the genus level only, or when the classification is provided for the species, so no information is available for the specific strain. Even when the strain details are supplied, they are rarely deposited in a mycotheque, making replication by other researchers problematic.

As a general trend, there is a clear reduction in BCA efficacy when moving from controlled environments to field studies, suggesting that the former studies primarily provide information about the potential effect of microorganisms, which must then be confirmed through repeated field experiments. Unfortunately, this secondary check is often not available. For instance, *L. thermotolerans* provided full protection against *B. cinerea* on berries in controlled conditions, but, to the best of our knowledge, the microorganism has not been tested in the vineyard. Similarly, much of the information available on ochratoxigenic fungi comes from lab experiments only, which may limit the generalizability of these findings.

Only a few of the microorganisms tested in the aforementioned studies have become commercial plant protection products. As an example, 85 of the 176 studies included in this analysis were conducted in Europe, taking into consideration 61 microorganisms. However, only 14 of these microorganisms have been authorized to be used in viticulture in Europe, namely *Ampelomyces quisqualis* AQ10 against powdery mildew; *Bacillus pumilus* QST2808 against powdery mildew; *B. amyloliquefaciens* MBI600 against powdery mildew, grey mould, and sour rot; FZB24 against grey mould and sour rot; *Aureobasidium pullulans*, *B. amyloliquefaciens* subsp. *plantarum* D747, *B. subtilis* QST713, *Metschnikowia fructicola* NRRLY-27328, *Saccaromyces cerviasiae* LASO2, *Pythium oligandrum* M1 against grey mould; *Trichoderma asperellum* ICC012, *T. gamsii* ICC080 against grey mould, sour rot, and trunk diseases; *T. atroviride* I-1237 against trunk diseases; and *T. atroviride* SC1 against grey mould and trunk diseases. No biocontrol products are currently authorized for downy mildew or black rot.

Since robust evidence on biocontrol efficacy is essential for both the authorization of BCA-based plant protection products and their adoption by farmers, a different approach to the study of BCAs is urgently required. The following sections outline two potential approaches: (i) using meta-analysis to better utilize literature data and draw robust conclusions about biocontrol efficacy; and (ii) developing BCA-based disease control tactics based on studying the interactions between the microorganism, the status of the pathogen, the plant, and the environmental conditions.

10 Understanding and improving biological control efficacy through network meta-analysis: powdery mildew, trunk diseases, and mycotoxigenic fungi

Previous discussions have highlighted the need to critically and quantitatively evaluate the factors influencing the success or failure of BCAs in disease management. Network meta-analysis can enable the integration and analysis of results obtained from different sources of information to determine the overall effectiveness of biocontrol in relation to biocontrol agents and application-oriented factors (e.g. number of applications). Meta-analysis operates on the principle that overall knowledge is based on combining and synthesizing results from individual studies and observations (Madden and Paul, 2011; Borenstein et al., 2009; Cooper et al., 2009; Hunter and Schmidt, 2004).

In this chapter, three case studies were selected and investigated through network meta-analysis – namely powdery mildew, trunk diseases, and mycotoxigenic fungi. The case studies were selected because they impact different organs, have simple to complex aetiology, occur at varying points of the grape-growing season, and present different challenges with regard to fungicide availability for disease control.

Network meta-analyses were carried out following the method described by González-Domínguez et al. (2019). Briefly, literature records were used to extract data on BCA treatments, chemical treatments, and untreated controls. The response variables considered in the statistical analysis included: (i) disease severity on both leaves and clusters for PM; (ii) disease incidence on pruning wounds for TDs; and (iii) mycotoxin (OTA) concentration (expressed in g/kg). Disease severity and incidence, as well as OTA concentration, were directly retrieved from tables and figures in the selected papers as described above. To be included in the meta-analysis, studies had to present data on the within-study variance, as indicated in González-Domínguez et al. (2019). The result of an individual study included in a meta-analysis is generally referred to as the estimated effect size (Madden and Paul, 2011).

The meta-analysis was conducted using R software (v. 4.1.1; package "metafor"; Viechtbauer et al., 2010) by applying a multivariate random effects model to log-transformed values of disease severity, disease incidence, or OTA concentration. To verify that the variability in the estimated effect was due to heterogeneity (i.e. variance between studies), residual heterogeneity and consistency were assessed (Donegan et al., 2013; Higgins and Thompson, 2002; Jackson et al., 2012). The percentage reduction in disease due to treatment was calculated relative to the untreated control (Gonzalez-Dominguez et al., 2019; Ngugi et al., 2011), and differences in disease control were evaluated in pairwise combinations (Gonzalez-Dominguez et al., 2019).

10.1 Biocontrol reduces powdery mildew compared to untreated controls but is less effective than chemical treatments, and efficacy increases with the number of applications

Data on PM severity on leaves and clusters observed in untreated and BCA-treated plots were extracted from 64 individual field studies documented in nine papers, in which *Ampelomyces quisqualis*, *Bacillus* spp. (mainly *B. subtilis*,

94% of studies with *Bacillus*), *Pseudomonas fluorescens*, and *Trichoderma* spp. were used. In 55 of the 64 studies, control provided by BCAs was compared to both untreated and chemical control (mainly with sulphur and/or synthetic fungicides like azoxystrobin, flusilazole, and triadimefon). PM epidemics varied widely across the studies, with an average disease severity of 37.8% (a range of 4.4% to 82.1% on leaves) and 36.7% (a range of 10.6% to 50.4% on clusters) in the untreated control. Disease severity also showed high variability among plots treated with BCAs, with an average disease severity of 20% (range 1.3% to 60%) and 20.2% (range 3.1% to 35.9%) on leaves and clusters, respectively. Lower disease severity was observed in fungicide-treated plots, with an average disease severity of 8.7% (a range of 0% to 35%) and 7.7% (a range of 0% to 29.4%) on leaves and clusters, respectively (Figure 14).

The meta-analysis facilitated a significant and robust comparison of PM control (Figure 15). The severity of PM was found to be significantly lower in plots treated with BCAs or fungicides than in untreated plots, and higher disease control was observed for fungicides than for BCAs. The average estimated efficacy was 48.7% (confidence interval, CI, 44.8 – 52.3%; *n* = 200) for BCAs (as a category including all the microorganisms tested) and 74.8% (CI 71.4 – 77.8%; *n* = 96) for the chemical control. All BCAs significantly reduced PM severity compared to the untreated control. *A. quisqualis* showed the highest variability in estimated

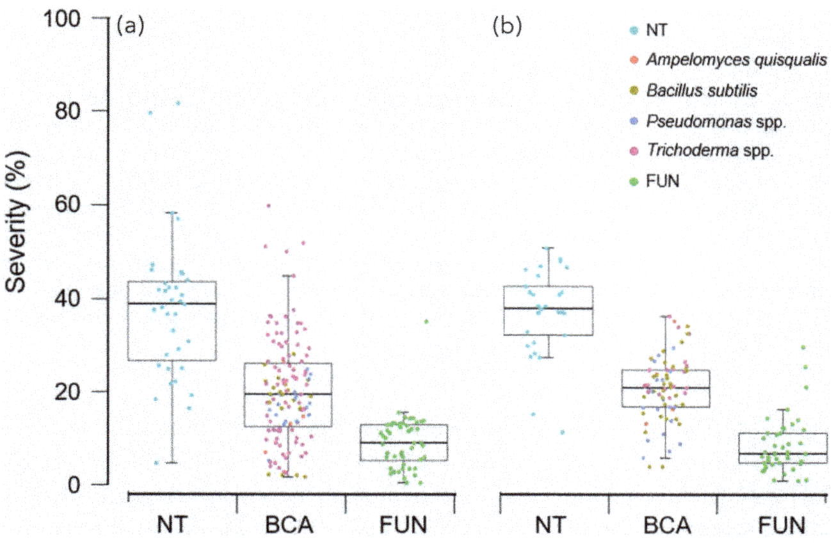

Figure 14 Powdery mildew severity (%) on leaves (A) and bunches (B) observed in field plots untreated (NT), treated with a biological control agent (BCA), or a chemical fungicide (FUN). The box extends from the 25th to 75th quartile of the data distribution, and the line crossing the box represents the median; whiskers extend to the maximum and minimum. Dot colours represent different microorganisms.

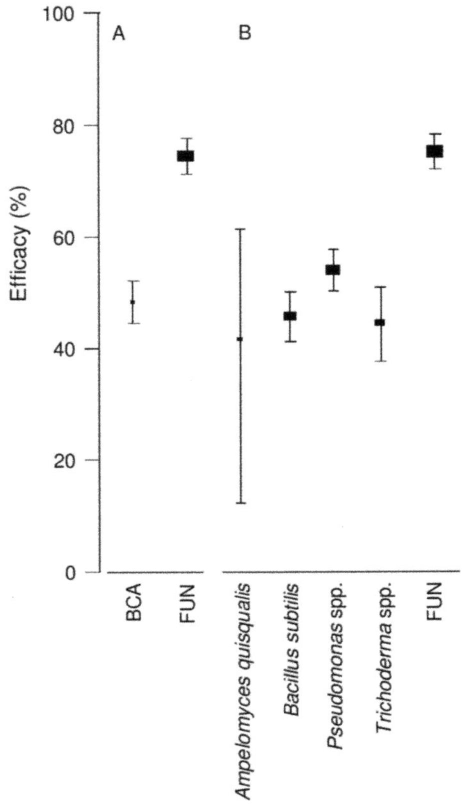

Figure 15 Forest plots showing the efficacy (%) of different treatments for the control of powdery mildew, as estimated through meta-analysis. Efficacy was calculated using disease severity data in relation to an untreated control, which is represented by the x-axis (i.e. no efficacy). Dots represent the average efficacy of treatments, and dot size increases with the precision of estimates; whiskers show a 95% confidence interval. The robustness of meta-analysis was estimated as residual heterogeneity (Q_E), degree of freedom (df), *p*-value (*P*), and % of total variation in treatment effect estimates due to heterogeneity between studies (I^2). In panel A, biological control agents (BCA) vs fungicides (FUN): QE = 3,552,924; df = 357; $P < 0.0001$; $I^2 > 99\%$. In panel B, single microorganisms vs FUN: QE = 3,543,447; df = 357; $P < 0.0001$; $I^2 > 95\%$.

efficacy (41.7%, CI; 12.3 – 61.3%; *n* = 6) due to the low number of observations. A similar efficacy was estimated for *Bacillus* spp. and *Trichoderma* spp. (45.8%, CI 41.2 – 50.0%, *n* = 54 and 44.6%, CI 37.6 – 50.8%, *n* = 104, respectively). The highest average efficacy (54.1%, CI 50.2 – 57.7%, *n* = 36) was estimated for *P. fluorescens*. Pairwise comparison showed that estimated effects for *A. quisqualis*, *Bacillus* spp., and *Trichoderma* spp. were not significantly different ($P > 0.7$) to one another, but all were different from *P. fluorescens* ($P < 0.05$), indicating that *P. fluorescens* provided superior PM control compared to the other BCAs.

The number of BCA applications in the selected studies varied from one to five per season, with seven or eight applications tested in two of the 64 studies. A single application significantly reduced PM severity compared to the untreated control. No significant ($P = 0.08$) differences in PM severity were observed between one and two applications, while a significant ($P < 0.001$) increase in disease control was noted for each additional application (data not shown).

10.2 Microorganisms exhibit varying levels of control over Botryosphaeria dieback, Black foot, Eutypa dieback, and Esca complex when compared to untreated plots and chemicals treatments

Data on the incidence of trunk diseases observed in untreated and BCA-treated plots were extracted from 91 individual field studies reported in 16 papers, in which *Bacillus* spp., *Fusarium* spp., *P. oligandrum*, *Streptomyces* spp., and *Trichoderma* spp. were evaluated. In 52 of the 91 studies, control provided by BCAs was compared to fungicides (mostly benomyl, in 71% of studies). Significant variability was observed in TD incidence among the studies, with an

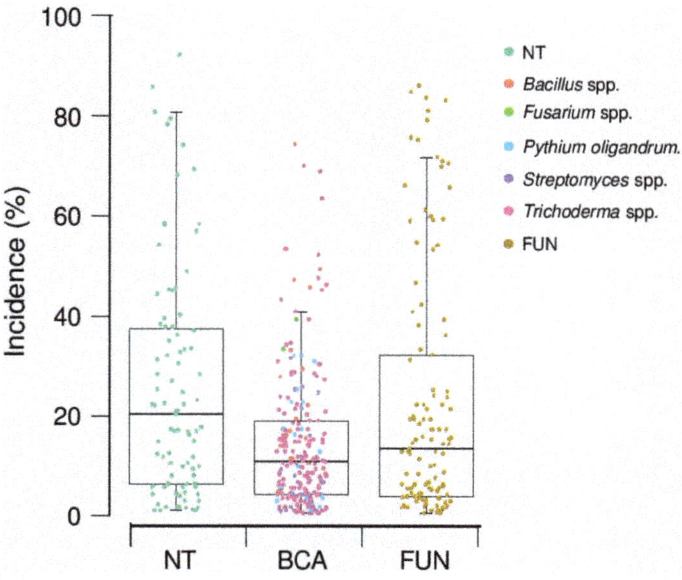

Figure 16 Incidence (%) of grape trunk diseases observed in field plots untreated (NT), treated with a biological control agent (BCA), or a chemical fungicide (FUN). The box extends from the 25th to 75th quartile of the data distribution, and the line crossing the box represents the median; whiskers extend to the maximum and minimum. Dot colours represent different microorganisms.

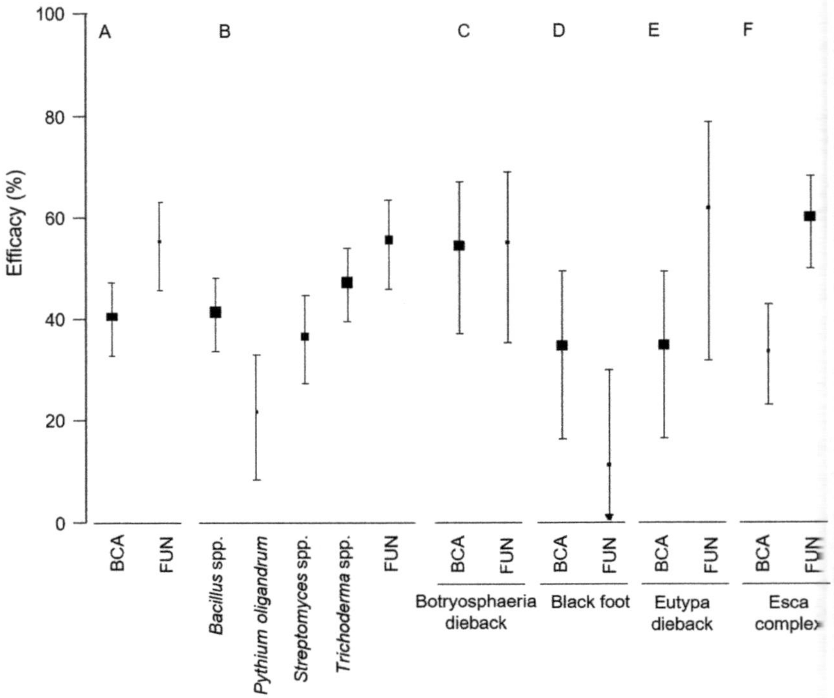

Figure 17 Forest plots showing the efficacy (%) of different treatments for the control of grape trunk diseases, as estimated through meta-analysis. Efficacy was calculated using disease severity data in relation to an untreated control, which is represented by the x-axis (i.e. no efficacy). Dots represent the average efficacy of treatments, and dot size increases with the precision of estimates; whiskers show a 95% confidence interval. Robustness of meta-analysis was estimated as residual heterogeneity (Q_E), degree of freedom (df), p-value (*P*), and % of total variation in treatment effect estimates due to heterogeneity between studies (I^2). In panel A, biological control agents (BCA) vs fungicides (FUN): QE = 60,352; df = 430; $P < 0.0001$; $I^2 >95\%$. In panel B, single microorganisms vs FUN: QE = 59,810; df = 426; $P < 0.0001$; $I^2 >79\%$. In panel C, BCAs vs FUN against Botryosphaeria dieback: QE = 4,225; df = 109; $P < 0.0001$; $I^2 >80\%$. In panel D, BCAs vs FUN against Black foot disease: QE = 2,183; df = 79; $P < 0.0001$; $I^2 >80\%$. In panel E, BCAs vs FUN against Eutypa dieback: QE = 20,208; df = 72; $P < 0.0001$; $I^2 >80\%$. In panel F, BCA vs FUN against Esca complex: QE = 11,555; df = 176; $P < 0.0001$; $I^2 >80\%$.

average disease incidence of 25.1% (a range of 0.5% to 92%) in the untreated control. Disease incidence also showed high variability among plots treated with BCAs and fungicides, with an average disease incidence of 14.1% (a range of 0% to 74%) and 22.5% (a range of 0% to 85.9%), respectively (Figure 16).

The meta-analysis results were statistically robust (see Figure 17). The incidence of TDs (as a whole) was significantly (P<0.0001) lower in plots treated with BCAs or fungicides than in untreated plots. The estimated efficacy of

BCAs and fungicides averaged 40.4% (CI 32.7% – 47.2%; *n* = 221) and 55.3% (CI 45.8% – 63.1%; *n* = 121), respectively (Figure 17A). When the efficacy against TDs (as a whole) of individual BCAs was investigated (Figure 17B), *Bacillus* spp. showed the lowest estimated efficacy, which averaged 21.6% (CI 8.4% – 32.9%; *n* = 23), followed by *P. oligandrum* (average 37.7%, CI 50.2% – 57.7%; *n* = 24), *Trichoderma* spp. (average 43.0%, CI 35.2% – 49.8%; *n* = 148), and *Streptomyces* spp. (average 48.6%, CI 41.0% – 55.3%; *n* = 24), the latter being not significantly (P = 0.23) different from chemical control. *Fusarium* spp. was excluded from this comparison because there were only two studies available.

For Botryosphaeria dieback (Figure 17C), the estimated efficacy of BCAs (average of 54.3%, CI 36.9% – 66.9%; *n* = 48) and fungicides (average of 55%, CI 35.1% – 68.7%; *n* = 45) were similar to the control. For Black foot (Figure 17D), the disease in fungicide-treated plots did not differ from the untreated control (P = 0.31), while disease in the BCA-treated plot was significantly ($P < 0.05$) different from the untreated plot, with estimated efficacy of 34.6% (CI 16.2% – 49.0%; *n* = 55). In the case of Eutypa dieback (Figure 17E), plots treated with fungicides or BCAs had significantly less disease than the untreated control $P < 0.01$), but they did not differ (P = 0.07) from each other due to the great variability in estimated efficacies (CI 16.1% – 49.0%, *n* = 38, and CI 31.6% – 78.6%, *n* = 19, for BCAs and fungicides, respectively). For Esca complex (Figure 17F), plots treated with fungicides or BCAs had lower levels of disease than the untreated control, with estimated efficacy for fungicides (average 59.9%, CI 49.6% – 68.0%, *n* = 49) being significantly ($P < 0.001$) higher than for BCAs (average 33.4%, CI 22.7% – 42.6%, *n* = 90).

10.3 Both bacteria and fungi reduce ochratoxin production

Data on BCA efficacy on OTA production by *Aspergillus* were extracted from 39 individual studies reported in 13 papers. In these studies, BCAs included bacteria (of genera *Arthrobacter*, *Streptomyces*, *Bacillus*, *Lactobacillus*, *Mycetocola*, *Pseudoarthrobacter*, *Rhodococcus*) and fungi (as yeasts or yeast-like fungi, i.e. *A. pullulans*, *C. laurentii*, *C. jadinii*, *K. marxianus*, *L. thermotolerans*, *Metschnikowia* spp., *Saccharomyces* spp.). OTA concentrations varied widely across the studies, from 0 to more than 2,000 μg/kg, regardless of treatment (Figure 18).

Results of the meta-analysis were robust (Figure 19). BCAs significantly ($P < 0.0001$) lowered the estimated OTA concentration by 71% (CI 53.8 – 81.8%; *n* = 104), with no differences ($P > 0.12$) between bacteria (69.3% efficacy, CI 58.0 – 77.6%; *n* = 77) and fungi (81.2%, CI 64.0 – 90.1%; *n* = 27).

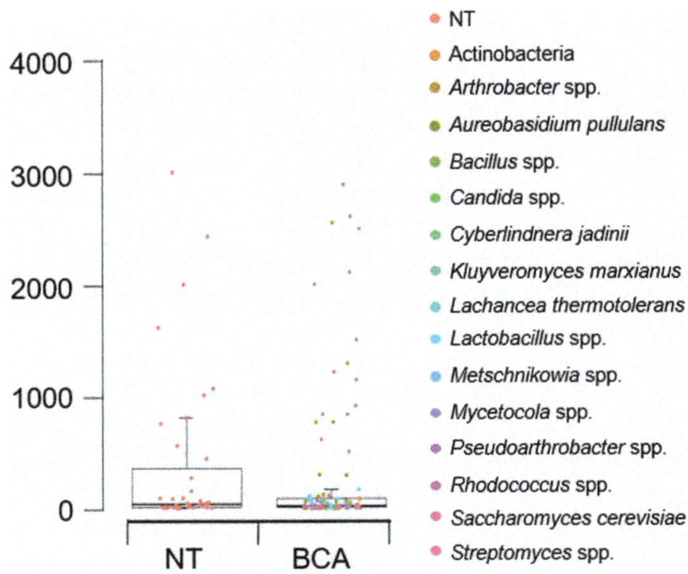

Figure 18 Concentration of ochratoxin A (OTA) (μg/kg) produced by *Aspergilla* on untreated artificial media (NT) and treated with an BCA. The box extends from the 25th to 75th quartile of the data distribution, and the line crossing the box represents the median; whiskers extend to the maximum and minimum. Dot colours represent different microorganisms.

11 Improving biocontrol efficacy through understanding interactions between pathogen, plant and environmental conditions: *Plasmopora viticola*, *Erysiphe necator*, and *Botrytis cinerea*

As shown above, meta-analysis can improve our understanding of the literature data published to date and allow robust insight into the practical use of BCAs. Unfortunately, not much of the current literature can be used for this kind of statistical analysis, because many published papers lack the necessary information. Meta-analysis cannot account for variations in the experimental protocols used across studies, including the timing of BCA applications in vineyards with respect to grape phenology, the epidemiology of the target pathogen, as well as the environmental conditions that the BCA faces following its application. These factors are of paramount importance and are responsible for the huge variability that any BCA exhibits across studies, as demonstrated above. A scenario study carried out by Fedele et al. (2020d), using a mathematical model that incorporates the MoA of BCAs, demonstrated that factors such as timing with respect to the pathogen (preventative vs curative intervention), temperature, and moisture requirements for BCA growth

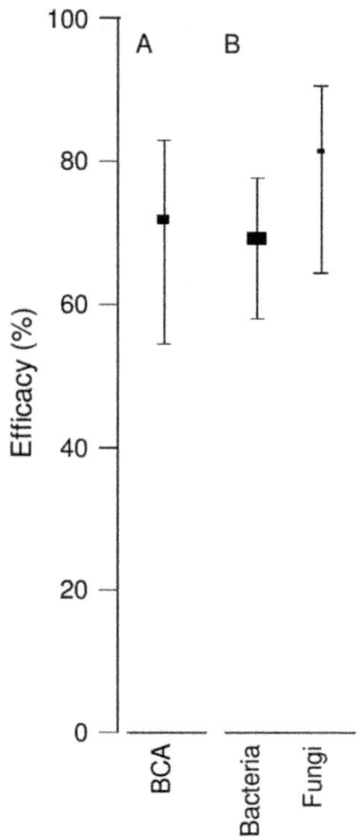

Figure 19 Forest plots showing the efficacy (%) of different treatments for the control of ochratoxin A (OTA) (µg/kg) production by *Aspergilla*, as estimated through meta-analysis. Efficacy was calculated using OTA concentration in an untreated control, which is represented by the x-axis (i.e. no efficacy). Dots represent the average efficacy of treatments, and dot size increases with the precision of estimates; whiskers show the 95% confidence interval. Robustness of meta-analysis was estimated as residual heterogeneity (Q_E), degree of freedom (df), p-value (*P*), and % of total variation in treatment effect estimates due to heterogeneity between studies (I^2). In panel A, biological control agents (BCA): QE = 215,351; df = 141; $P < 0.0001$; $I^2 > 99\%$. In panel B, biocontrol by bacteria vs yeasts: QE = 191,500; df = 141; $P < 0.0001$; $I^2 > 99\%$.

and survival, all significantly influenced biocontrol efficacy. Temperature and moisture accounted for more than 90% of the variability in biocontrol outcomes.

Since the unpredictability of the control efficacy following a BCA application is one of the main reasons for the limited adoption of biocontrol by farmers (He et al., 2021), a different approach is necessary, where BCA applications are based on the status of the pathogen, the plant, and the environmental conditions (Fedele et al., 2020d; Altieri et al., 2023). The following paragraphs offer insights for downy and powdery mildews, and for grey mould.

11.1 Breaking the life cycle of Plasmopora viticola

Plasmopara viticola survives the winter as oospores in the leaf residue above the soil or in soil. Oospores break dormancy in the year following their formation or in the next few years (Caffi et al., 2011). They repeatedly germinate in spring (typically between bud break and late July), producing sporangia that release bi-flagellated motile zoospores in water. Zoospores survive in water until they are dispersed by rain to the vegetation; otherwise, they die. Upon reaching the vegetation, zoospores swim to the stomata, encyst, and germinate, and the germ tubes penetrate the stomata opening, causing (primary) infection that results in the typical DM lesions found on leaves and clusters. Asexual sporangia are then produced on DM lesions, spread through air currents, and cause secondary infections that further progress the disease. Oospores are primarily responsible for the establishment of disease in the vineyard and for maintaining its presence throughout the season (Gessler et al, 2011). Prevention of primary infections is key for successful DM control, currently achieved by repeated fungicide applications.

BCAs can contribute to the control of *P. viticola* through various mechanisms: (i) parasitizationu of oospores and consequent reduction of the oospore pool; (ii) killing of zoospores or interference with their normal swimming behaviour in water above the ground (i.e. before their dispersal to plants) or on the plant surfaces (i.e. before their gemination and penetration); and (iii) suppression of the production or germination of secondary sporangia, and consequent prevention of secondary infections.

Concerning (i), oospore parasitization in soils has been documented for other oomycetes, such as *Phytophthora erythroseptica*, *Ph. megasperma* var *sojae*, *Ph. cactorum*, and *Pytium* spp. by oomycetes, hyphomycetes, chytridiomycetes, actinomycetes, and bacteria (Sneh et al., 1977; Wynn and Epton, 1979; Hoch and Abawi, 1979; Sutherland and Lockwood, 1984; Khan et al., 1993) but not for *P. viticola*. Some microorganisms were able to prevent leaf infection in bioassays where oospores were kept in water and in the presence of floating leaf discs. Disease suppression on leaf discs, however, was inconsistent over repeated experiments (Dagostin et al., 2006; Vecchione et al., 2007).

Concerning (ii), a *Paenibacillus* sp. isolated from the rhizosphere of *Sorghum bicolor*, and its peptide paenimyxin, inhibited the mobility of *P. viticola* zoospores and decreased infection sites on leaf discs when added to the sporangial suspension used for artificial inoculation (Hao et al., 2017). Secondary metabolites produced by a soil-isolated *Streptomyces* sp., an endophytic *Fusarium* sp. (from *Piper guinense*), and a marine *Streptomyces* sp. also exhibited motility inhibitory activity (up to 100%) and lytic activity (up to >90%) against zoospores (Abdalla et al., 2011; Kongue Tatong et al., 2014;

Dame et al., 2016). The effectiveness of these microorganisms under natural conditions has not yet been investigated; we can speculate that they could survive better in soil rather than on plant surfaces, thus inhibiting zoospores originating from macrosporangia, but this needs to be confirmed by additional research.

Concerning (iii), an *Epicoccum nigrum* strain isolated from DM-affected leaves in the field was found to encase sporangia with abundant mycelium, preventing their dispersal. Any antagonistic effect, however, was not observed and its ability to colonize leaves remains questionable (Kortekamp, 1997). A strain of *Fusarium proliferatum,* isolated from atypical DM lesions, was effective in reducing sporangial production on leaf discs by 97%. Its hyphae were seen to consistently coil around and inside sporangiophores (Falk et al., 1996). Similar results were also observed with other *Fusarium* spp. (Ghule et al., 2018). Strains of *Aletrnaria alternata* from unusual DM symptoms completely inhibited sporulation on leaf discs and caused ultrastructural changes in *P. viticola* hyphae and haustoria, due to the production of toxic metabolites (Musetti et al., 2006). In a wide-ranging study, 47 of 1,700 organisms applied at infection time of the leaf discs, together with *P. viticola,* were able to reduce or prevent sporulation (Vecchione et al., 2004). *Acremonium* spp. isolated from asymptomatic grapevine leaves were reported as hyperparasite of *P. viticola* in leaves, with occasional formation of sporangiophores and sporangia in the leaves colonized by the endophyte, or with deformed sporangiophores (Burruano et al. 2008). These fungi also produce secondary metabolites that inhibit sporangial germination – as the ratio between germinated (empty) and non-germinated (degenerated) sporangia – with different inhibition degrees based on species and strains of the same species (Assante et al., 2005; Arnone et al., 2008, 2009; Lo Piccolo et al., 2015). Inhibition of sporangial germination was also observed for various unspecified bacteria (seven of 17 tested) (Vecchione et al., 2004). To our knowledge, the field activity of the abovementioned microorganisms has only been tested by Falk et al. (1996), who found that certain fungi reduced DM severity on clusters and leaves by 53 to 99% following weekly applications. However, none of these microorganisms have been further developed as a commercial product.

11.2 Understanding the epi-season of Erysiphe necator

The life cycle of *E. necator* spans two viticultural seasons, a concept known as the "epi-season" (Magarey and Moyer, 2010), in which the current season leaves a "legacy" for the next. Sexual fruiting bodies (the chasmothecia) form from late summer to autumn on PM colonies, especially after harvest, when the interruption of fungicidal treatments gives the pathogen the opportunity to develop. Mature chasmothecia (black in colour) are dispersed to the trunks of the vines by rainfall, where they remain trapped in the bark and remain viable until the following spring. Chasmothecia that fall to the ground typically lose their vitality during winter. Ascospores are released by chasmothecia in spring (normally between bud break and fruit set; Rossi et al., 2010). If environmental conditions are favourable, they can cause primary (or ascosporic) infections. These infections have symptoms that are atypical and often go unnoticed. However, ascosporic infections play a key role for the establishment of PM in the vineyard and for the initial multiplication of the asexual inoculum, the conidia, which is responsible for secondary infections and, therefore, the polycyclic and "explosive" nature of PM epidemics.

In the efficacy studies mentioned above, the microbial products were used in late spring and/or summer to control the disease in its epidemic phase. In this period, PM colonies produce huge quantities of conidia under a wide range of environmental conditions, which are dispersed by air currents and cause repeated infection cycles that sporulate in turn following a latent period of five to six days. The control of PM in these conditions is challenging, and even sulphur and synthetic fungicides do not enable full disease control (Legler et al., 2011). During summer, the air temperature is high and free moisture in the grapevine canopy is low. These environmental conditions do not limit the development of *E. necator* (Gadoury et al., 2012) but may prevent sustained growth of other microorganisms, including BCAs.

Given the life cycle of *E. necator* and the concept of the epi-season, PM control should be anticipated and targeted to the sexual phase of the fungus. This means: (i) controlling chasmothecia in the development phase, between late summer and early autumn; (ii) killing mature chasmothecia on the vine bark in winter before bud break; and (iii) preventing ascosporic infections during the following spring. In epidemiological terms, this approach delays the epidemic phase of the disease (Vanderplank, 1963) in a period in which clusters have already acquired ontogenic resistance to *E. necator* (Gadoury et al., 2003).

The suppression of chasmothecia production in (i) can be achieved by using *A. quisqualis*, a hyperparasite of *E. necator* that naturally colonizes mycelia, conidia, and chasmothecia when they are young, i.e. they are yellow to orange (Kiss et al., 2004; Angeli et al., 2009; Legler et al., 2012). Parasitized chasmothecia are typically dull, fawn-coloured, and flaccid (Falk et al., 1995). In controlled conditions, *A. quisqualis* was able to suppress chasmothecia formation by 41%, while field application of *A. quisqualis* during the pre- and post-harvest delayed disease onset and epidemic development in the following season by one to four weeks, and lowered both disease incidence and severity on clusters in the following season by approximately 50% compared to an untreated control (Caffi et al. 2013; Legler et al. 2016). In an organic strategy based on fall application of *A. quisqualis* against chasmothecia and sulphur treatments targeted to prevent ascosporic infections in spring, the average disease control was >95%, and similar to the control obtained by using synthetic fungicides (Caffi et al., 2013). The trunk sanitation in (ii) is commonly obtained with lime sulphur (Gadoury et al., 1994), fungicidal oils (Thiessen et al., 2018), or Bordeaux mixture (Todaro and Miles, 2018). No BCAs are yet known to parasitize overwintered chasmothecia.

Concerning (iii), application of *Bacillus*-based BCAs against ascospric infection appears promising. These products have different MoAs (Cawoy et. al., 2011; Tran et al., 2022) that inhibit spore germination and hyphal growth. Their efficacy has been rated as good and reliable under low to medium disease pressure, with reduced efficacy under high disease pressure (Adaskaveg et al., 2022). Since the PM pressure is lower for ascospric than for conidial infections, these treatments are typically recommended between bud break and full bloom (Adaskaveg et al., 2022). It may also be considered that ascosporic and conidial infections have different needs that also suggest using BCAs against the former. Ascosporic infections are triggered by rain and require water on plant surfaces, with a temperature range of 5–28°C (optimum 20–25°C). Conidial infections are hindered by rain and the presence of water but are favored by high relative humidity, with an optimal temperature of 20–25°C (but infection is possible up to 35°C).

11.3 Target the multiple infection pathways of Botrytis cinerea

The life cycle of *B. cinerea* is complex, and multiple infection pathways exist (Elmer and Michailides, 2007; González-Domınguez et al., 2015). At vine flowering, latent infections occur, which will remain latent until berry ripening, when they can result in rotten berries (infection pathways I, IIa, and IIb of Elmer and Michailides, 2007). From fruit set to veraison, *B. cinera* saprophytically colonizes bunch trash (i.e. calyptras, dead stamens, aborted flowers and berries, and tendrils), and produces conidia that may serve for conidial infection of berries (infection pathway III and IV). From veraison onwards, berries can be infected by conidia (infection pathway Va) or by the mycelium from adjacent affected berries (the berry-to-berry infection; pathway Vb; González-Domínguez et al., 2015).

This complexity requires multiple interventions during the season, specifically at flowering (known as intervention in A), pre-bunch closure (B), veraison (C), and ripening (D) to target the different infection pathways (González-Domínguez et al., 2019). Biocontrol of GM may then comply with this complexity. From pre-flowering to fruit set, BCAs may be able to prevent both conidial germination on inflorescences and the establishment of latent infection of berries. From fruit-set to pre-bunch closure, they may competitively colonize bunch trash as a saprophyte and prevent sporulation by *B. cinerea*. From veraison until harvest, BCAs may colonize the berry surface and prevent infection by conidia and aerial mycelium (berry-to-berry infection).

At the abovementioned timings, BCAs may express different MoAs and develop on different plant substrates. Intervention in A requires the capability of BCA to inhibit conidial germination through the production of bioactive metabolites and/or to induce resistance in clusters to prevent latent infections. Intervention in B implies the BCA's capability to compete with *B. cinerea* for the saprophytic growth on bunch trash and inhibit its sporulation. Interventions in C and D require the BCA to compete with the pathogen for the infection sites (i.e. microcracks and wounds on the berry skin), to inhibit conidial germination and hyphal growth, and/or to induce resistance, e.g. by increasing the production of phytoalexins in the outer berry layers. Mechanisms of action of microbial biocontrol agents against *B. cinerea* have recently been reviewed by Roca-Couso et al. (2021). To add complexity, the different plant substrates (flowers and flower parts, bunch trash, and berries at different ripening stages) display different biochemical characteristics (e.g. pH, nutrient types and availability, water potential) that also vary during the season (Fedele et al., 2020b, 2020c). For instance, *Bacillus amyloliquefaciens* grow more profusely on berries at veraison and berry softening than at earlier or later stages; *Aureobasidium pullulans* grow more at veraison, the growth of *Metschnikowia fructicola* increases as berry ripening progresses, and *Trichoderma atroviride* show the opposite (Altieri et al., 2022).

Disease control interventions typically start at flowering and continue until pre-harvest, which means that BCAs need to be adapted to the varying weather conditions from spring to late summer. Temperature, humidity, and rainfall have been shown to influence the growth and efficacy of different BCAs in various ways (Fedele et al., 2020a, Altieri et al., 2022, 2023). Unfortunately, little information is available to date regarding the environmental requirements of the different BCAs (Fedele et al., 2020a). The environmental niche approach proposed by Fedele et al. (2020a) – which focuses on the environmental conditions necessary for the presence of a species and the maintenance of its population – could aid the selection of the appropriate BCA to be used in field conditions for a specific intervention, as well as the screening of new BCAs.

A multi-criteria approach is essential for effective biocontrol of GM, in which different BCAs are chosen based on the targeted infection pathway, the required MoA, and the environmental conditions at the time of intervention and in the days following it (Pertot et al., 2017; Fedele et al., 2020d).

12 Conclusions

Despite the substantial number of studies published on the biocontrol of grape diseases, the availability of commercial products remains limited (with obvious variations among countries) and acceptance by farmers remains low.

The reasons to why BCAs are less used than anticipated have been widely discussed (e.g. Velivelli et al., 2014; Singh et al., 2020; Lahlali et al., 2022). In a recent paper, Collinge et al. (2022) discussed what has been achieved to date and outlined the steps required for BCAs to become a standard part of the disease management toolkit, as disease resistance and pesticides are. The focus of this paper was on microorganism selection, development of a commercial product (from production to formulation), delivery systems, and regulatory and industrial approval. This chapter, however, specifically addresses BCA efficacy when applied in vineyards.

Two main limitations arise from the literature analysis: (i) BCAs typically provide lower disease control compared to chemical fungicides, particularly under high disease pressure; and (ii) there is significant variability in biocontrol efficacy across experiments, which undermines famer trust. The huge variability among microorganisms (species and strains) used, application timings and frequency, methods for efficacy assessment, and varying environmental conditions make comparison among the literature studies difficult and inconsistent so that it is virtually impossible to create practical guidelines for BCA use against grape disease. The situation is further complicated by the lack of detailed experiment information and incomplete reporting of experimental errors, limiting the potential for formal meta-analysis. These limitations should be considered by researchers when planning experiments and writing papers.

Another notable issue from the literature analysis is that BCA application timing is often adopted from traditional fungicide scheduling, resulting in calendar-based applications during the typical disease season. This approach overlooks the fact that biocontrol relies on living microorganisms and their interaction with the pathogen and the plant (the plant site that is challenged by the pathogen), in a variable environment. Much progress is needed in this area.

An innovative, step-by-step approach is proposed here. The first step is to clearly define the pathogen stage for which biocontrol is required (e.g. oospores of *P. viticola* on the vineyard floor, developing chasmothecia of *E. necator* of the vegetation in late season, or saprophytic colonization of bunch trash to prevent growth and sporulation of *B. cinerea*). The second step is to define the MoA that is required to control the pathogen at a specific stage (e.g. parasitism of oospores or chasmothecia, or competition for the bunch trash material). The third step is to create a list of microorganisms responding to the following criteria: (i) possess the required MoA; (ii) ability to grow on the substrate of interest (e.g. leaf residues and soil where the *P. viticola* oospores survive, the leaves and clusters where the PM colonies develop chasmothecia, or the bunch trash that is colonized by *B. cinerea*); and (iii) share the same environmental niche as the pathogen at the target stage, so that the BCA may grow when the pathogen grows. The fourth step is to perform efficacy experiments across multiple environments and seasons to confirm the efficacy of the selected BCAs against the target pathogen and stage. By following this approach, positive experimental outcomes can result in practical guidelines for farmers.

13 Data availability statement

The list of published papers used in data collection and supporting the conclusions of this chapter can be made available by the authors upon request.

14 Conflict of interest

The authors declare that their research was conducted in the absence of any commercial or financial relationships that could be construed as a potential conflict of interest.

15 Author Contributions

RV and SI mainly contributed to the conception and design of the study. AV, CLM, JT, and SI contributed to data collection. AV, RV, and SI carried out the analysis of results. AV, CLM, JT, RV, and SI contributed to writing the original draft preparation. BP and FG contributed to reviewing and editing the original draft. All authors wrote the manuscript, contributed to the article, and approved the submitted version.

16 Acknowledgments

AV conducted this study within the Doctoral School on the Agro-Food System (AgriSystem) at the Università Cattolica del Sacro Cuore, Piacenza, Italy. SI was partially supported by the "National Research Centre for Agricultural Technologies (Agritech)" CN00000022. PNRR-M4C2, Investment 1.4 - Funded by the European Union - NextGenerationEU" - CUP: J33C22001160007.

17 References

Abdalla, M. A., et al. (2011). Khatmiamycin, a motility inhibitor and zoosporicide against the grapevine downy mildew pathogen *Plasmopara viticola* from Streptomyces sp. ANK313. *The Journal of Antibiotics*, *64*(10), 655–659.

Ab Rahman, S. F. S., et al. (2018). Emerging microbial biocontrol strategies for plant pathogens. *Plant Science*, *267*, 102–111.

Adaskaveg, J. E., et al. (2022). *Fungicides, bactericides, biocontrols, and natural products for deciduous tree fruit and nut, citrus, strawberry, and vine crops in California 2022*. Riverside: University of California. 74 p.

Ahmed, M. F. A. (2018). Evaluation of some biocontrol agents to control Thompson seedless grapevine powdery mildew disease. *Egyptian Journal of Biological Pest Control*, *28*, 1–7.

Altieri, V., et al. (2022). Biocontrol of *Botrytis cinerea* as influenced by grapevine growth stages and environmental conditions. *XXVII Congress of the Italian Phytopathological Society*. September 21st–23rd, 2022.

Altieri, V., et al. (2023). Efficacy of preharvest application of biocontrol agents against gray mold in grapevine. *Frontiers in Plant Science*, *14*, 1154370. https://doi.org/10.3389/fpls.2023.1154370

Angeli, D., et all. (2009). Occurrence of *Erysiphe necator* chasmothecia and their natural parasitism by *Ampelomyces quisqualis*. *Phytopathology*, *99*(6), 704–710.

Angeli, D., et al. (2012). Existence of different physiological forms within genetically diverse strains of *Ampelomyces quisqualis*. *Phytoparasitica*, *40*, 37–51.

Arnone, A., et al. (2008). Acremine G, dimeric metabolite from cultures of *Acremonium byssoides* A20. *Journal of Natural Products*, *71*(1), 146–149.

Arnone, A., et al. (2009). Acremines H–N, novel prenylated polyketide metabolites produced by a strain of *Acremonium byssoides*. *Tetrahedron*, *65*(4), 786–791.

Assante, G., et al. (2005). Acremines AF, novel secondary metabolites produced by an endophytic strain of *Acremonium* sp., isolated in grape leaves infected by *Plasmopara viticola*. *Tetrahedron*, *61*(32), 7686–7692.

Banani, H., et al. (2014). Characterization of resistance mechanisms activated by *Trichoderma harzianum* T39 and benzothiadiazole to downy mildew in different grapevine cultivars. *Plant Pathology*, *63*(2), 334–343.

Battilani, P., et al. (2006). European research on ochratoxin A in grapes and wine. *International Journal of Food Microbiology*, *111*, S2–S4.

Biesbroek, G. R., et al. (2013). On the nature of barriers to climate change adaptation. *Regional Environmental Change*, *13*, 1119–1129. https://doi.org/10.1007/s10113-013-0421-y

Bleyer, K., et al. (2023). Using ontogenetic resistance of grapevine for fungicide reduction strategies. *European Journal of Plant Pathology*, *165*(1), 115–124.

Bois, B., et al. (2017). Climate vs grapevine pests and diseases worldwide: the first results of a global survey. OENO *One*, *51*(2), 133–139.

Borenstein, M., et al. (2009). *Introduction to meta-analysis*. Chichester, UK: John Wiley & Sons.

Brown, A. A., et al. (2021). Pruning-wound protectants for trunk-disease management in California table grapes. *Crop Protection*, *141*, 105490.

Bruisson, S., et al. (2019). Endophytes and epiphytes from the grapevine leaf microbiome as potential biocontrol agents against phytopathogens. *Frontiers in Microbiology*, *10*, 2726.

Burruano, S., et al. (2008). Interaction between Acremonium byssoides and *Plasmopara viticola* in *Vitis vinifera*. *Phytopathologia Mediterranea*, *47*(2), 122–131.

Caffi, T., et al. (2011). Long-term survival of *Plasmopara viticola* oospores. *IOBC/WPRS Bulletin*, *66*, 111–114.

Caffi, T., et al. (2013). Combining sanitation and disease modelling for control of grapevine powdery mildew. *European Journal of Plant Pathology*, *135*, 817–829.

Calvo-Garrido, C., et al. (2013). Biological control of botrytis bunch rot in organic wine grapes with the yeast antagonist *Candida sake* CPA-1. *Plant Pathology*, *62*(3), 510–519.

Calvo-Garrido, C., et al. (2013). *Candida sake* CPA-1 and other biologically based products as potential control strategies to reduce sour rot of grapes. *Letters in Applied Microbiology*, *57*(4), 356–361.

Calvo-Garrido, C., et al. (2019). Microbial antagonism toward Botrytis bunch rot of grapes in multiple field tests using one *Bacillus ginsengihumi* strain and formulated biological control products. *Frontiers in Plant Science*, *10*, 105.

Candel, J. J. (2014). Food security governance: A systematic literature review. *Food Security*, *6*, 585–601.

Castoria, R., et al. (2001). *Aureobasidium pullulans* (LS-30) an antagonist of postharvest pathogens of fruits: study on its modes of action. *Postharvest Biology and Technology*, *22*, 7e17.

Cawoy, H., et al. (2011). *Bacillus*-based biological control of plant diseases. In M. Stoytcheva (Ed.), *Pesticides in the modern world-pesticides use and management* (pp. 273–302). InTech Academic Press.

Choudhary, D. K. & Johri, B. N. (2009). Interactions of *Bacillus* spp. and plants–with special reference to induced systemic resistance (ISR). *Microbiological Research*, *164*(5), 493–513.

Collinge, D. B., et al. (2022). Biological control of plant diseases –What has been achieved and what is the direction? *Plant Pathology*, *71*(5), 1024–1047.

Compant, S., et al. (2013). Use of beneficial bacteria and their secondary metabolites to control grapevine pathogen diseases. *BioControl*, *58*, 435–455.

Conway, G. R. & Pretty, J. N. (1991). *Unwelcome harvest: agriculture and pollution*. London: Earthscan Publishing.

Cooper, H. & Hedges, L. V. (2009). Research synthesis as a scientific process. In A. H. Cooper et al. (Eds.), *The handbook of research rynthesis and meta-analysis* (2nd ed., pp. 3–16). Thousand Oaks, CA: Sage Publications.

Cordero-Bueso, G., et al. (2017). Wild grape-associated yeasts as promising biocontrol agents against *Vitis vinifera* fungal pathogens. *Frontiers in Microbiology*, *8*, 2025.

Corkley, I., et al. (2022). Fungicide resistance management: Maximizing the effective life of plant protection products. *Plant Pathology*, *71*(1), 150–169.

Crisp, P., et al. (2006). An evaluation of biological and abiotic controls for grapevine powdery mildew. 1. Greenhouse studies. *Australian Journal of Grape and Wine Research*, *12*(3), 192–202.

Dagostin, S., et al. (2006). Potential use of biocontrol agents to prevent *Plasmopara viticola* oospore germination. *IOBC/WPRS Bulletin*, *29*(11), 43.

Dagostin, S., et al. (2011). Are there alternatives to copper for controlling grapevine downy mildew in organic viticulture? *Crop Protection*, *30*(7), 776–788.

Dame, Z. T., et al. (2016). Oligomycins and pamamycin homologs impair motility and induce lysis of zoospores of the grapevine downy mildew pathogen, Plasmopara viticola. *FEMS Microbiology Letters*, *363*(16), fnw167.

De Curtis, F., et al. (2012). Environmental factors affect the activity of biocontrol agents against ochratoxigenic *Aspergillus carbonarius* on wine grape. *International Journal of Food Microbiology*, *159*(1), 17–24.

De Felice, D. V., et al. (2008). Strains of *Aureobasidium pullulans* can lower ochratoxin A contamination in wine grapes. *Phytopathology*, *98*(12), 1261–1270.

Dimakopoulou, M., et al. (2008). Phyllosphere grapevine yeast *Aureobasidium pullulans* reduces *Aspergillus carbonarius* (sour rot) incidence in wine-producing vineyards in Greece. *Biological control*, *46*(2), 158–165.

Di Marco, S. & Osti, F. (2007). Applications of Trichoderma to prevent *Phaeomoniella chlamydospora* infections in organic nurseries. *Phytopathologia Mediterranea*, *46*(1), 73–83.

Di Marco, S., et al. (2022). Activity of *Trichoderma asperellum* strain ICC 012 and *Trichoderma gamsii* strain ICC 080 toward diseases of esca complex and associated pathogens. *Frontiers in Microbiology*, *12*, 813410.

Donegan, S., et al. (2013). Assessing key assumptions of network meta-analysis: a review of methods. *Research Synthesis Methods*, *4*(4), 291–323.

El Khoury, R., et al. (2017). Ability of soil isolated actinobacterial strains to prevent, bind and biodegrade ochratoxin A. *Toxins*, *9*(7), 222.

Elad, Y. & Freeman, S. (2002). Biological control of fungal plant pathogens. In *Agricultural applications* (pp. 93–109). Berlin, Heidelberg: Springer.

Elmer, P. A. G. & Michailides, T. J. (2007). Epidemiology of *botrytis cinerea* in orchard and vine crops. In Y. Elad, et al. (Eds.), *Botrytis: Biology, pathology and control* (pp. 243–272). Dordrecht: Springer Netherlands. https://doi.org/10.1007/978-1-4020-2626-3_14

Elmer, P. A. G. & Reglinski, T. (2006). Biosuppression of *Botrytis cinerea* in grapes. *Plant Pathology*, *55*(2), 155–177.

European Commission. (2020). *A farm to fork strategy for a fair, healthy and environmentally-friendly food system*. COM(2020) 381 Final. Brussels: European Commission. https://eur-lex.europa.eu/legal-content/EN/TXT/?uri=CELEX%3A52020DC0381.

EUROSTAT. (2007). The use of plant protection products in the European Union. Edition. http://ec.europa.eu/eurostat (accessed 30 June 2023).

Falk, S. P., et al. (1995). Partial control of grape powdery mildew by the mycoparasite *Ampelomyces quisqualis*. *Plant Disease*, *79*(5), 483–490.

Falk, S. P., et al. (1996). *Fusarium proliferatum* as a biocontrol agent against grape downy mildew. *Phytopathology*, *86*(10), 1010–1017.

Fedele, G., et al. (2020a). Biocontrol of *Botrytis cinerea* on grape berries as influenced by temperature and humidity. *Frontiers in Plant Science*, *11*, 1–11. https://doi.org/10.3389/fpls.2020.01232

Fedele, G., et al. (2020b). Reduction of *Botrytis cinerea* colonization of and sporulation on bunch trash. *Plant Disease*, *104*(3), 808–816.

Fedele, G., et al. (2020c). Influence of environment on the biocontrol of *Botrytis cinerea*: a systematic literature review. In A. De Cal, et al. (Eds.), *How research can stimulate the development of commercial biological control against plant diseases* (pp. 61–82). Cham: Springer. https://doi.org/10.1007/978-3-030-53238-3_5

Fedele, G., et al. (2020d). A generic model accounting for the interactions among pathogens, host plants, biocontrol agents, and the environment, with parametrization for *Botrytis cinerea* on grapevines. *Agronomy*, *10*(2), 222. https://doi.org/10.3390/agronomy10020222

Gadoury, D. M., et al. (2012). Grapevine powdery mildew (*Erysiphe necator*): a fascinating system for the study of the biology, ecology and epidemiology of an obligate biotroph. *Molecular Plant Pathology*, *13*(1), 1–16.

Gadoury, D. M., et al. (1994). Reduction of powdery mildew and other diseases by over-the-trellis applications of lime sulfur to dormant grapevines. *Plant Disease*, *78*(1), 83–87.
Gadoury, D. M., et al. (2003). Ontogenic resistance to powdery mildew in grape berries. *Phytopathology*, *93*(5), 547–555.
Gessler, C., et al. (2011). *Plasmopara viticola*: a review of knowledge on downy mildew of grapevine and effective disease management. *Phytopathologia Mediterranea*, *50*(1), 3–44.
Ghule, M. R., et al. (2018). Identification of *Fusarium* species as putative mycoparasites of *Plasmopara viticola* causing downy mildew in grapevines. *Australasian Plant Disease Notes*, *13*, 1–6.
González-Domínguez, E., et al. (2015). A mechanistic model of *Botrytis cinerea* on grapevines that includes weather, vine growth stage, and the main infection pathways. *PloS One*, *10*, 1–23. https://doi.org/10.1371/journal.pone.0140444
González-Domínguez, E., et al. (2019). A network meta-analysis provides new insight into fungicide scheduling for the control of *Botrytis cinerea* in vineyards. *Pest Management Science*, *75*(2), 324–332.
Gouadec, D. & Blouin, J. (2007). *Les Parasites de la Vigne: Stratégies de Protection Raisonnée Stratégies de Protection Raisonnée*. Paris, France: Dunod.
Gramaje, D., et al. (2018). Managing grapevine trunk diseases with respect to etiology and epidemiology: current strategies and future prospects. *Plant Disease*, *102*, 12–39. https://doi.org/10.1094/PDIS-04-17-0512-FE
Hamdi, R. F. & Alzawi, K. S. (2023). Using fungi and bacteria as biological control agents of fungal plant diseases an article. Journal of University of Anbar for Pure Science, *17*(1), 50–61.
Hao, Z., et al. (2017). Biocontrol of grapevine aerial and root pathogens by *Paenibacillus* sp. strain B2 and paenimyxin in vitro and in planta. *Biological Control*, *109*, 42–50.
He, D. C., et al. (2021). Biological control of plant diseases: An evolutionary and eco-economic consideration. *Pathogens*, *10*(10), 1311.
Higgins, J. P. T. & Thompson, S. G. (2002). Quantifying heterogeneity in a meta-analysis. *Statistics in Medicine*, *21*, 1539–1558.
Hinckley, E. L. S., et al. (2020). A shift in sulfur-cycle manipulation from atmospheric emissions to agricultural additions. *Nature Geoscience*, *13*(9), 597–604.
Hoch, H. C. & Abawi, G. S. (1979). Mycoparasitism of oospores of *Pythium ultimum* by *Fusarium merismoides*. *Mycologia*, *71*(3), 621–625.
Hunter, J. E. & Schmidt, F. L. (2004). *Methods of meta-analysis: Correcting error and bias in research findings* (2nd ed.) Thousand Oaks, CA: Sage Publications Inc.
Jaber, L. R. (2015). Grapevine leaf tissue colonization by the fungal entomopathogen *Beauveria bassiana* sl and its effect against downy mildew. *BioControl*, *60*, 103–112.
Jackson, D., et al. (2012). Quantifying the impact of between-study heterogeneity in multivariate meta-analyses. *Statistics in Medicine*, *31*, 3805–3820.
Jelenić, J., et al. (2021). Endophytic *Fusarium* spp. affect yield, but can they be a solution for *Botrytis* control in vineyard production? *Fungal Ecology*, *54*, 101114.
Ji, T., et al. (2023). Temperature-dependent growth and spore germination of fungi causing grapevine trunk diseases: Quantitative analysis of literature data. *Plant Disease*, *107*(5), 1386–1398.
John, S., et al. (2005). Protection of grapevine pruning wounds from infection by *Eutypa lata* using *Trichoderma harzianum* and *Fusarium lateritium*. *Australasian Plant Pathology*, *34*, 569–575.

Keele, S. (2007). Guidelines for performing systematic literature reviews in software engineering. Technical Report, EBSE Technical Report EBSE-2007-01, 1-57.

Kennelly, M. M., et al. (2005). Seasonal development of ontogenic resistance to downy mildew in grape berries and rachises. *Phytopathology*, *95*(12), 1445–1452.

Khan, N. I., et al. (1993). Parasitism of oospores of *Pythium* spp. by strains of Actinoplanes spp. *Canadian Journal of Microbiology*, *39*(10), 964–972.

Kiss, L., et al. (2004). Biology and biocontrol potential of *Ampelomyces mycoparasites*, natural antagonists of powdery mildew fungi. *Biocontrol Science and Technology*, *14*(7), 635–651.

Kongue Tatong, M. D., et al. (2014). Banchromene and other secondary metabolites from the endophytic fungus Fusarium sp. obtained from Piper guineense inhibit the motility of phytopathogenic *Plasmopara viticola* zoospores. *Tetrahedron Letters*, *55*(30), 4057–4061.

Konstantinidou-Doltsinis, S., et al. (2007). Control of powdery mildew of grape in Greece using Sporodex® L and Milsana®. *Journal of Plant Diseases and Protection*, *114*, 256–262.

Koricheva, J. & Gurevitch, J. (2014). Uses and misuses of meta-analysis in plant ecology. *Journal of Ecology*, *102*(4), 828–844.

Kortekamp, A. (1997). *Epicoccum nigrum* LINK: a biological control agent of *Plasmopara viticola* (BERK. et CURT.). *Vitis*, *36*(4), 215–216.

Kotze, C., et al. (2011). Evaluation of biocontrol agents for grapevine pruning wound protection against trunk pathogen infection. *Phytopathologia Mediterranea*, *50*, S247–S263.

Kumar, A. & Droby, S. (2021). *Microbial management of plant stresses: current trends, application and challenges* (First eds. A. Kumar and S. Droby). Cambridge: Woodhead Publishing.

Lahlali, R., et al. (2022). Biological control of plant pathogens: A global perspective. *Microorganisms*, *10*(3), 596.

Lappa, I. K., et al. (2018). Control of *Aspergillus carbonarius* in grape berries by *Lactobacillus plantarum*: a phenotypic and gene transcription study. *International Journal of Food Microbiology*, *275*, 56–65.

Legler, S. E., et al. (2011). Effect of different plant protection products on the sexual stage of grapevine powdery mildew. In *Proceedings of the IOBC/WPRS European meeting of the working group" Integrated Protection and Production in Viticulture"* (pp. 2–2). N/A.

Legler, S. E., et al. (2012). A nonlinear model for temperature-dependent development of *Erysiphe necator* chasmothecia on grapevine leaves. *Plant Pathology*, *61*(1), 96–105.

Legler, S. E., et al. (2016). Sporulation rate in culture and mycoparasitic activity, but not mycohost specificity, are the key factors for selecting *Ampelomyces* strains for biocontrol of grapevine powdery mildew (Erysiphe necator). *European Journal of Plant Pathology*, *144*, 723–736.

Leifert, C., et al. (1995). Antibiotic production and biocontrol activity by *Bacillus subtilis* CL27 and *Bacillus pumilus* CL45. *Journal of Applied Microbiology*, *78*(2), 97–108.

Lemos Junior, W. J. F., et al. (2016). Biocontrol ability and action mechanism of Starmerella bacillaris (synonym Candida zemplinina) isolated from wine musts against gray mold disease agent Botrytis cinerea on grape and their effects on alcoholic fermentation. *Frontiers in Microbiology*, *7*, 1249.

Lo Piccolo, S., et al. (2015). Identification of *Acremonium* isolates from grapevines and evaluation of their antagonism towards Plasmopara viticola. *Annals of Microbiology*, *65*, 2393–2403.

Loevinsohn, M. (1987). Insecticide use and increased mortality in rural Central Luzon, Philippines. *The Lancet*, *329*(8546), 1359–1362.

Maachia, B. S., et al. (2015). Biological control of the grapevine diseases 'grey mold' and 'powdery mildew' by *Bacillus* B27 and B29 strains.

Madden, L. V., and Paul, P. A. (2011). Meta-analysis for evidence synthesis in plant pathology: An overview. *Phytopathology*, 101(1), 16–30.

Magarey, P. & Moyer, M. (2010, July). Towards establishing slow input regimes in Australian viticulture 3: Use of "epi-season" and" lag phase control" in applying epidemiological knowledge of grapevine powdery mildew, to reduce the number of sprays and inoculum reservoirs for long-term control. In *Proceedings of the 6th International Workshop on Grapevine Downy and Powdery Mildew, Bordeaux, France* (pp. 4–9).

Magnin-Robert, M., et al. (2013). Differential induction of grapevine resistance and defense reactions against *Botrytis cinerea* by bacterial mixtures in vineyards. *BioControl*, *58*, 117–131.

Markellou, E., et al. (2022). Evaluation of plant protection efficacy in field conditions and side effects of Lysobacter capsici AZ78, a biocontrol agent of *Plasmopara viticola*. *Biocontrol Science and Technology*, *32*(8), 930–951.

Marrone, P. G. (2007). Barriers to adoption of biological control agents and biological pesticides. *CABI Reviews*, *2007*, 12-pp.

Marsico, A. D., et al. (2021). Native vineyard non-*Saccharomyces* yeasts used for biological control of *Botrytis cinerea* in stored table grape. *Microorganisms*, *9*(2), 457.

Mohamed, N., et al. (2007). Defense responses in grapevine leaves against *Botrytis cinerea* induced by application of a *Pythium oligandrum* strain or its elicitin, oligandrin, to roots. *Phytopathology*, *97*(5), 611–620.

Molitor, D. & Berkelmann-Loehnertz, B. (2011). Simulating the susceptibility of clusters to grape black rot infections depending on their phenological development. *Crop Protection*, *30*(12), 1649–1654.

Mondello, V., et al. (2018). Grapevine trunk diseases: a review of fifteen years of trials for their control with chemicals and biocontrol agents. *Plant Disease*, *102*(7), 1189–1217.

Musetti, R., et al. (2006). Inhibition of sporulation and ultrastructural alterations of grapevine downy mildew by the endophytic fungus *Alternaria alternata*. *Phytopathology*, *96*(7), 689–698.

Mutawila, C., et al. (2015). Development of benzimidazole resistant *Trichoderma* strains for the integration of chemical and biocontrol methods of grapevine pruning wound protection. *BioControl*, *60*, 387–399.

Nally, M. C., et al. (2015). Antifungal modes of action of *Saccharomyces* and other biocontrol yeasts against fungi isolated from sour and grey rots. *International Journal of Food Microbiology*, *204*, 91–100.

Ngugi, H. K., et al. (2011). Meta-analysis to determine the effects of plant disease management measures: review and case studies on soybean and apple. *Phytopathology*, *101*(1), 31–41.

Niem, J. M., et al. (2023). Biocontrol potential of an endophytic *Pseudomonas poae* strain against the grapevine trunk disease pathogen *Neofusicoccum luteum* and its mechanism of action. *Plants*, *12*(11), 2132.

Niem, J. M., et al. (2020). Diversity profiling of grapevine microbial endosphere and antagonistic potential of endophytic *Pseudomonas* against grapevine trunk diseases. *Frontiers in Microbiology*, *11*, 477.

Noordzij, M., et al. (2009). Systematic reviews and meta-analyses: when they are useful and when to be careful. *Kidney international*, *76*(11), 1130–1136.

Okoli, C., and Schabram, K. (2010). A guide to conducting a systematic literature review of information systems research. *Papers Inf. Syst*. 10, 1–46. doi: 10.2139/ssrn.1954824

Parnell, J. J., et al. (2016). From the lab to the farm: an industrial perspective of plant beneficial microorganisms. *Frontiers in Plant Science*, *7*, 1110.

Pascale, A., et al. (2017). *Trichoderma* and its secondary metabolites improve yield and quality of grapes. *Crop protection*, *92*, 176–181.

Pedrotti, C., et al. (2020). Essential oil as sustainable alternative for diseases management of grapes in postharvest and in vineyard and its influence on wine. *Journal of Environmental Science and Health, Part B*, *56*(1), 73–81.

Perrone, G., et al. (2004). PCR assay for identification of *Aspergillus carbonarius* and *Aspergillus japonicus*. *Molecular Diversity and PCR-detection of Toxigenic Fusarium Species and Ochratoxigenic Fungi: Under the aegis of COST Action 835 'Agriculturally Important Toxigenic Fungi 1998–2003', EU project (QLK1-CT-1998--01380) and the ISPP 'Fusarium Committee'*, 641–649.

Pertot, I., et al. (2016). *Trichoderma atroviride* SC1 prevents *Phaeomoniella chlamydospora* and *Phaeoacremonium aleophilum* infection of grapevine plants during the grafting process in nurseries. *BioControl*, *61*, 257–267.

Pertot, I., et al. (2017). A critical review of plant protection tools for reducing pesticide use on grapevine and new perspectives for the implementation of IPM in viticulture. *Crop Protection*, *97*, 70–84.

Pertot, I., et al. (2017). Combining biocontrol agents with different mechanisms of action in a strategy to control *Botrytis cinerea* on grapevine. *Crop Protection*, *97*, 85–93.

Rantsiou, K., et al. (2020). Impact of chemical and alternative fungicides applied to grapevine cv Nebbiolo on microbial ecology and chemical-physical grape characteristics at harvest. *Frontiers in Plant Science*, *11*, 700.

Raut, V. M., et al. (2018). Evaluation of Milastin-K (*Bacillus subtilis* KTSB 1015 1.5% AS) as a potential biocontrol agent for management of downy and powdery mildew in grape vines. *Pestology*, *42*(10), 15–24.

Raymaekers, K., et al. (2020). Screening for novel biocontrol agents applicable in plant disease management–a review. *Biological Control*, *144*, 104240.

Reis, P., et al. (2021). Combining an HA+ Cu (II) site-targeted copper-based product with a pruning wound protection program to prevent infection with *Lasiodiplodia* spp. in grapevine. *Plants*, *10*(11), 2376.

Rezgui, A., et al. (2016). Endophytic bacteria with antagonistic traits inhabit the wood tissues of grapevines from Tunisian vineyards. *Biological Control*, *99*, 28–37.

Ritika, B. & Utpal, D. (2014). An overview of fungal and bacterial biopesticides to control plant pathogens/diseases. *African Journal of Microbiology Research*, *8*(17), 1749–1762.

Roca, P. (2022, April). State of the world vine and wine sector. OIV *Press Conference*. https://doi.org/org/10.1787/9789264244047-44-en

Roca-Couso, R., et al. (2021). Mechanisms of action of microbial biocontrol agents against *Botrytis cinerea*. *Journal of Fungi*, *7*(12), 1045.

Ronseaux, S., et al. (2013). Interaction of *Ulocladium atrum*, a potential biological control agent, with Botrytis cinerea and grapevine plantlets. *Agronomy*, *3*(4), 632–647.

Rosace, M. C., et al. (2023). Susceptibility of pruning wounds to grapevine trunk diseases: A quantitative analysis of literature data. *Frontiers in Plant Science*, *14*, 1063932.

Rossi, V., et al. (2010). Dynamics of ascospore maturation and discharge in Erysiphe necator, the causal agent of grape powdery mildew. *Phytopathology*, *100*, 1321–1329.

Rossi, V., et al. (2012). Helping farmers face the increasing complexity of decision-making for crop protection. *Phytopathologia Mediterranea*, *51*, 457–479.

Sawant, I. S., et al. (2017). Enhanced biological control of powdery mildew in vineyards by integrating a strain of Trichoderma afroharzianum with sulphur. *Biological Control*, *114*, 133–143.

Scherm, H., et al. (2014). Meta-analysis and other approaches for synthesizing structured and unstructured data in plant pathology. *Annual Review of Phytopathology*, 52(1), 453–476.

Silva-Valderrama, I., et al. (2021). Biocontrol potential of grapevine endophytic and rhizospheric fungi against trunk pathogens. *Frontiers in microbiology*, *11*, 614620.

Singh, A., et al. (2020). Biological control of plant diseases: opportunities and limitations. In: Varma, A., et al. (Eds), *Plant Microbiome Paradigm*, 121–146. https://doi.org/10.1007/978-3-030-50395-6_7

Sipiczki, M. (2006). Metschnikowia strains isolated from botrytized grapes antagonize fungal and bacterial growth by iron depletion. *Applied and Environmental Microbiology*, *72*(10), 6716–6724.

Sneh, B., et al. (1977). Parasitism of oospores of *Phytophthora megasperma* var. *sojae*, *P. cactorum*, *Pythium* sp., and *Aphanomyces euteiches* in soil by oomycetes, chytridiomycetes, hyphomycetes, actinomycetes, and bacteria. *Phytopathology*, *67*(5), 622–628.

Sutherland, E. D. & Lockwood, J. L. (1984). Hyperparasitism of oospores of some peronosporales by *Actinoplanes missouriensis* and *Humicola fuscoatra* and other Actinomycetes and fungi. *Canadian Journal of Plant Pathology*, *6*(2), 139–145.

Thiessen, L. D., et al. (2018). Interruption and reduction of *Erysiphe necator* chasmothecia development utilizing fungicidal oil. *Plant Health Progress*, *19*(2), 153–155.

Todaro T. & Miles T. (2018). Early season vineyard disease management. https://www.canr.msu.edu/news/early_season_vineyard_disease_management (accessed 20 July 2023).

Toral, L., et al. (2018). Antifungal activity of lipopeptides from *Bacillus* XT1 CECT 8661 against *Botrytis cinerea*. *Frontiers in Microbiology*, *9*, 1315.

Tracy, E. F. (2014). The promise of biological control for sustainable agriculture: a stakeholder-based analysis. *Journal of Science Policy & Governance*, *5*(1), 1–13.

Tran, C., et al. (2022). Antimicrobial *Bacillus*: metabolites and their mode of action. *Antibiotics*, *11*(1), 88.

Trotel-Aziz, P., et al. (2008). Characterization of new bacterial biocontrol agents *Acinetobacter*, *Bacillus*, *Pantoea* and *Pseudomonas* spp. mediating grapevine resistance against *Botrytis cinerea*. *Environmental and Experimental Botany*, *64*(1), 21–32.

Van Helden, M. (2008). Protection intégrée: La protection intégrée vis-à-vis des ravageurs de la vigne. *Ravageurs de la vigne*. Féret, France, 321–335.
Vanderplank, J. E. (1963). *Plant diseases*. New York: Elsevier Science.
Vecchione, A., et al. (2004). Screening and identification of potential biocontrol agents against grapevine downy mildew considering an integrated control strategy of the disease. *S. Michele all'Adige, Italy, 27*(8), 85–88.
Vecchione, A., et al. (2007). Biological control of *Plasmopara viticola*: a multisite approach. *Acta Horticulturae, 754*, 361–366. https://doi.org/10.17660/ActaHortic.2007.754.47
Velivelli, S. L., et al. (2014). Biological control agents: from field to market, problems, and challenges. *Trends in Biotechnology, 32*(10), 493–496.
Verger, P. J. P. and Boobis, A. R. (2013). Reevaluate pesticides for food security and safety. *Science, 341*, 717–718. https://doi.org/10.1126/science.1241572
Viechtbauer, W. (2010). Conductingmeta-analyses in R with the metafor package. *Journal of Statistical Software, 36*, 1–48.
Whiteman, S. A. & Stewart, A. (1998). *Suppression of Botrytis cinerea sporulation on irradiated grape leaf tissue by the antagonistic bacterium Serratia liquefaciens*. Netherlands: Springer.
Wynn, A. R. & Epton, H. A. S. (1979). Parasitism of oospores of *Phytophthora erythroseptica* in soil. *Transactions of the British Mycological Society, 73*(2), 255–259.
Xiao, Y. & Watson, M. (2019). Guidance on conducting a systematic literature review. *Journal of Planning Education and Research, 39*(1), 93–112.
Zhao, H., et al. (2022). Cell-free supernatant of *Bacillus velezensis* suppresses mycelial growth and reduces virulence of *Botrytis cinerea* by inducing oxidative stress. *Frontiers in Microbiology, 13*, 980022.
Zhou, Q., et al. (2020). Application of antagonist *Bacillus amyloliquefaciens* NCPSJ7 against *Botrytis cinerea* in postharvest Red Globe grapes. *Food Science & Nutrition, 8*(3), 1499–1508.
Zhu, C., et al. (2015). Inhibition of the growth and ochratoxin A production by Aspergillus carbonarius and *Aspergillus ochraceus* in vitro and in vivo through antagonistic yeasts. *Food Control, 50*, 125–132.

Chapter 8

Disease management in cereals using biocontrol agents

Alexander Schouten, Wageningen University, The Netherlands

1 Introduction

Cereals, such as rice, wheat, maize, rye, oats, barley and millet, are pivotal staple crops at a global scale and, when processed properly, they can be a rich source of vitamins, antioxidants, minerals, fats, oils and proteins as well (Poutanen et al., 2022). Yet, the large-scale growing of cereals is not without problems as various fungal and bacterial pathogens can challenge these valuable crop plants, reducing yield and, in several cases, quality, sometimes by accumulating toxins in infected seeds (Andrade and Caldas, 2015; Anonymous, 2023b; Kayim et al., 2022; Kumar et al., 2022; Pereira et al., 2022). Depending on the crop and region, the impact of fungal and bacterial diseases on crop losses can vary significantly (Savary et al., 2019). In wheat (*Triticum aestivum*), which is the second most important food crop in the world (Poutanen et al., 2022), Septoria tritici blotch (STB) caused by the ascomycete fungus *Zymoseptoria tritici* (syn. *Mycosphaerella graminicola*), can be a devastating foliar disease, particularly in Northwest Europe and Sub-Saharan Africa, resulting in yield losses of 5 to 7%

http://dx.doi.org/10.19103/AS.2024.0142.19

(Savary et al., 2019). The *Fusarium graminearum* species complex (FGSC), which is the causal agent of Fusarium head blight (FHB), not only affects yield but also accumulates mycotoxins, such as deoxynivalenol (DON), in the infected grains. This disease seems to be particularly problematic in China with yield losses reaching almost 9% (Savary et al., 2019). Stem rust is most prominent in Sub-Saharan Africa with reported losses reaching almost 9% (Savary et al., 2019).

Due to the application of fungicides, resistant cultivars and growing and sanitation strategies, most diseases could be contained to acceptable levels. However, the emergence of fungicide resistance, races overcoming introgressed resistance genes in novel cultivars, the alteration of the growth environment is changing that for the worse. To illustrate this, *Puccinia graminis, P. striiformis, P. triticina* and *P. hordei* are causal agents of various types of cereal rusts and used to be most destructive pathogens in cereals globally (Cock et al., 2021). In the past century, these pathogens were effectively contained by eradicating the alternating host, i.e. barberry *(Berberis),* growing fast ripening wheat cultivars, growing cultivars containing stem or stripe rust resistance (Sr) or other resistance genes, such as *Rph,* and applying fungicides (Carmona et al., 2020; Dinh et al., 2022). In the past decades, these pathogens are now reemerging in various cereal growing areas, due to the proliferation of the (Sr) resistance-breaking race, Ug99 (Lewis et al., 2018; Saunders et al., 2019), the reintroduction of barbery because of nature conservation concerns (Lewis et al., 2018; Saunders et al., 2019), the emergence fungicide resistance (Cock et al., 2021) and the banning of particular fungicides due to new regulations, e.g. in the EU (Anonymous, 2023a).

The excessive use of fungicides, which are still allowed in cereal cultivation, increase the risk of resistance or tolerance development in the fungal pathogens, including human pathogens, thus even jeopardizing drug treatment of humans (Fisher et al., 2022). The various azole fungicides, which essentially all target CYP51 that plays an important role in the ergosterol biosynthetic pathway, can all be rendered inactive or less active by mutations in the *CYP51* gene. Such mutations have occurred not only in natural *F. graminearum* populations but also in *Zymoseptoria tritici, Pseudocercospora fijjensis,* the causal agent of black Sigatoka in banana, and *Aspegillus fumigatus,* the causal agent of aspergilloses in immunosuppressed people (Fisher et al., 2022). This indicates that different organisms can independently find similar or even almost identical solutions upon selection pressure. However, there are other solutions to reduce the toxicity of a fungicide, such as increasing efflux mechanisms and fungicide degradation (Yin et al., 2023).

Claviceps species, the causal agents of ergot in both grasses and cereals, infect the ovary resulting in the development of sclerotia, i.e. the fungal resting structures, containing various ergotoxins in the spikes, which can lead to convulsive ergotism, gangrenous ergotism, or both and even death in animals and humans when consumed. In historical writings and paintings, ample evidence can be found that this intoxication was a serious issue for humans, particularly because the cause of the associated illnesses was not known (Lee, 2009). There are no fungicides available against *Claviceps* species and various species have a broad host range, meaning that they can survive on grasses in the absence of cereal cultivation. Nowadays, cereal harvests are critically monitored for the presence of ergot sclerotia, which should not exceed 0.2 g per kg of unprocessed cereals according to current European law (Anonymous, 2021). The overwintering sclerotia remain viable for approximately one year and by removing these from the soil surface, e.g. by plowing them into the soil, the inoculum can be significantly reduced (Anonymous, 2023b).

Next to fungi, cereals can also be challenged by pathogenic bacteria, such as *Xanthomonas translucens* pv. *undulosa, Pseudomonas syringae* pv. *syringae, P. syringae* pv. *atrofaciens, Pseudomonas fuscovaginae, Erwinia rhapontici, Pseudomonas cichorii* and *Clavibacter tessellarius,* which are the causal agents of the leaf streak and black chaff disease, bacterial leaf blight, basal glume rot, bacterial brown sheath, pink seed, stem melanosis bacterial mosaic in wheat, respectively (Tambong, 2022), and *Pseudomonas avenae* in millet (Claflin et al., 1989) and *P. syringae pv. panici, P. avenae,* the causal agents of bacterial brown stripe, and *Xanthomonas oryzae* pv. *oryzae,* the causal agent of bacterial blight in rice (Shakya and Chung, 1983) (Liu et al., 2012; Shakya et al., 1986). The development of resistance toward fungicides and the reoccurring breaking of plant resistance by pathogens is currently a wider and ever-increasing problem in agriculture, which is further aggravated by the limited availability of chemical replacements, making the need for alternative approaches to control fungal plant diseases overall even more pressing.

Twenty-five native North American grass species harbored *Fusarium graminearum* species complex (FGSC) isolates, which all grew asymptomatically inside the plant with little or no trichothecene accumulation, although the infecting isolates were capable of causing disease and producing mycotoxins in wheat. The coexistence of North American grasses and isolates from FGSC is supposedly very ancient, indicating that evolutionary processes modified the host-fungus relationship toward a benign and possibly mutualistic interaction (Lofgren et al., 2018). This example illustrates that the cereal germplasm may contain elements that can be useful in pacifying or antagonizing plant pathogens. One can think of plant-synthesized secondary metabolites and the more recently discovered antimicrobial peptides (Campos et al., 2018; Tian et al., 2022).

Particular associated microorganisms may have a profound influence on reducing cereal diseases as well. Plants are not an isolated operational entity when dealing with adverse biotic circumstances and most microorganisms in the plant's environment are generally not harmful. Plants can even manipulate this beneficial microbial population for their own benefit. This non-pathogenic microbial population residing in the endosphere, the phyllosphere, the bulk soil and the rhizosphere, can be a crucial element in biotic and abiotic stress management by plants, e.g. by priming the plant or inducing defense responses, by directly attacking pathogen, by blocking the colonizing pathogenic microorganism, by transferring pathogen-specific viruses (hypovirulence), by producing specific antibiotics or altering plant synthesized antimicrobial secondary metabolites into more effective ones. In this way, beneficial microorganisms may thus support plants to fend off pathogens. Soils may also harbor suitable bacteriophages, targeting plant pathogenic bacteria. By better understanding these beneficial biological entities and plant manipulation mechanisms, the development of cultivars that will be more resilient toward pathogens seriously threatening cereal cultivation will become simplified. This chapter therefore focuses on the thus far described biological entities in and around cereals that can increase their resilience toward diseases and the underlying mechanisms responsible for this.

2 Antifungal compounds synthesized by microorganisms

Microorganisms can produce anti-fungal compounds and, in this way, particularly when present in the endosphere, rhizosphere or phyllosphere, protect the plant from being colonized and cereals have been at the basis of this type of biocontrol. In the 1960s, the take-all decline (TAD) phenomenon was essentially discovered in *Poaceae,* in which take-all disease, caused by *Gaeumannomyces graminis,* is significantly reduced during prolonged growth of the host plant. It was observed that in several important growing areas, such as the Inland Pacific Northwest of the United States, the Netherlands and the UK, both wheat and barley could be effectively cultivated as a monoculture for many decades without the emergence of take-all disease (Baker and Cook, 1974; Gerlagh, 1968; Shipton, 1972). When wheat or barley cultivation on a particular plot was initiated, take-all could inflict serious biological and economic damage, but only in the first three to five years of consecutive cultivation. After that, take-all disease all but vanished from that particular plot. Remarkably, take-all disease could reappear when the monoculture was interrupted, e.g. by fallow or crop rotation (Baker and Cook, 1974; Cook, 2007; Shipton, 1972). It soon became clear that the soil was an important factor, changing from disease-conducive to disease-suppressive over time. Disease-suppressive soils could be transferred to take-all conducive plots, accelerating the development of suppressiveness toward take-all disease (Baker and Cook, 1974; Cook, 2007; Cook et al., 1995). The observation that suppressive soils became conducive again after these soils had been pasteurized showed that, in fact, the soil microbiome was involved (Cook and Rovira, 1976; Shipton, 1972) (Raaijmakers and Weller, 1998). The uninterrupted wheat cultivation caused the accumulation of particular antibiotic-producing pseudomonads in the rhizosphere of wheat and barley, thereby preventing *G. graminis* from proliferating (Cook, 2007). Further research led to the insight that, depending on the growing area, individual soil-borne *Pseudomonas* species or species isolates can mediate TAD by producing different unrelated compounds with antifungal activity, of which phloroglucinol, various phenazine derivatives, pyoluteorin, pyrrolnitrin and cyclic lipopeptides are currently the best described. Several of these compounds, like phenazine-1-carboxilic acid, pyrrolnitrin and cyclic lipopeptides, have antifungal activities against other fungal pathogens as well, both in the rhizosphere and phyllosphere. For more thorough reviews on TAD and the mechanisms behind this phenomenon, see Cook (2007), Duffy et al. (2003) and Schouten (2018).

A similar observation was made for the decline of *Rhizoctonia* barepatch disease in wheat, caused by the basidiomycete *Rhizoctonia solani*. Here, too, a monoculture of wheat initiated a decline in bare patch disease incidence to undetectable levels (Roget, 1995) and this suppressiveness could be relocated to non-suppressive soils and abolished by pasteurization and, also, radioactive irradiation (Wiseman et al., 1996). The microorganism(s) and mechanism(s) involved in the decline of Rhizoctonia barepatch are not fully understood to date but the involvement of *Trichoderma* spp., parasitic toward *Rhizoctonia,* was suggested. Nevertheless, antibiosis seems to play a role as well. A *Bacillus* sp., strain L324-92, could suppress root diseases caused by, *G. graminis* var. *tritici, R. solani* AG-8, and several *Pythium* species when applied as a seed inoculation in wheat (Kim et al., 1997). Although the mode of action by which the L324-92 repressed wheat root diseases has thus far not been resolved, recently characterized cyclic lipopetides, produced by *Bacillus amyloliquefaciens* S76-3, isolated from wheat spikes, showed antagonistic activity toward *F. graminearum* (Gong et al., 2015). Another study showed that phenazine-producing recombinant derivatives from *Pseudomonas fluorescens* strain Q8r1-96 were significantly more effective against *R. solani* AG-8 root rot than the parental strain (Huang et al., 2004).

From the rice rhizosphere, *Bacillus atrophaeus* FA12 and *B. cabrialesii* FA26 were isolated, both showing broad-range antifungal and antibacterial activities against various plant pathogens, including *Xanthomonas oryzae* pv. *oryzae* (*Xoo*), the causal agent of bacterial blight disease in rice. In pot experiments, both FA12 and FA26, applied as seed inoculants, significantly enhanced the vigor of rice seedlings by 79% and 109%, and suppressed bacterial blight by 60% and 55%, respectively. As iturins and fengycin, both cyclic lipopetides, were the major antimicrobial constituents encountered, it was postulated that these compounds were primarily involved in defending rice against *Xoo*

(Rajer et al., 2022). The types of cyclic lipopeptides that can be produced by the various microorganisms, including *Bacillus* and *Pseudomonas* species, are mesmerizing. Depending on the amino acid and fatty acid composition, their function and, therefore, the suspected ecological role may vary and are far from being fully elucidated. Nevertheless, they can facilitate motility (i.e. swarming), adhesion to (plant) surfaces and can play a role in microbial antagonism by disrupting the membranes of fungi and oomycetes. Zoospores of oomycetes seem to be particularly sensitive as they can disintegrate within a split second, i.e. explode, upon contact with various types of cyclic lipopeptides. Plant cell membranes are insensitive to cyclic lipopeptides due to the presence of sterols. Nevertheless, next to the curative abilities, cyclic lipopeptides may prevent disease indirectly, by triggering induced systemic resistance responses (ISR) toward pathogens in the host plants (Ongena and Jacques, 2008; Raaijmakers et al., 2010).

In the phyllosphere as well, bacteria may be important allies in mediating resistance. Spikes inoculated with *Pseudomonas chlororaphis* subsp. *aurantiaca* Pa40 showed a reduction in wheat sharp eyespot disease (*Rhizoctonia cerealis*), similar to chemical treatment with validamycin A, in a greenhouse experiment and Pa40 harbored genes coding for enzymes involved in the biosynthesis of both pyrrolnitrin and 2-hydroxylated phenazine (Jiao et al., 2013).

3 Physically preventing pathogens to invade

Plants have the ability to physically stop pathogens from entering. Pathogens can be recognized by the plant cells and upon contact, particularly with pathogenic fungi, cell walls can be reinforced by defensive substances such as callose, glycoproteins such as extensin, which are rich in the amino acid hydroxyproline, complex phenolic compounds, including lignin and suberin, and minerals such as silicon and calcium (Kashyap et al., 2021; Xin and Herburger, 2021). Furthermore, the xylem, which can serve as super-highway for pathogens for dispersal inside a host, can be blocked through tyloses (Kashyap et al., 2021). In resistant tomato plants, a combination of these physiochemical barriers were shown to limit the colonization and invasion of the bacterial pathogen, *Ralstonia solanacearum* (Planas-Marquès et al., 2020). However, as the authors recognize, breeding for these inducible structural defenses may be difficult as they are regulated by intricate polygenic traits.

Physical barriers may also be erected by plant-associated microorganisms. Finger millet *(Eleusine coracana)*, a traditional African crop plant, has a broad resistance to potential pathogens, such as the mycotoxin-producing *Fusarium graminearum*. This cereal can have an extraordinary symbiosis with a root-inhabiting endophytic *Enterobacter* sp., designated M6. M6 is also able to form a biofilm on the root hairs and can swarm towards *Fusarium*, forming a physical barrier by trapping or blocking the entry of the pathogen, after which the fungus is killed (Mousa et al., 2016). Mutagenesis studies indicated that the M6 killing abilities require c-di-GMP-dependent signaling, the synthesis of diverse fungicidal constituents by M6 and resistance toward *Fusarium*-produced antibiotics. This intricate defense mechanism of chemotaxis and blocking is thus followed by antibiosis, suggesting a longstanding host-endophyte-pathogen co-evolution. It was suggested that M6 may be effectively transferred to other cereals, such as maize and wheat (Mousa et al., 2016).

4 Cereal secondary metabolites for the defense against fungal pathogens: true active compounds or just an intermediate?

Cereals can constitutively express or *de novo* synthesize secondary metabolites (Meyer et al., 2016), such as alkaloids, phenolic compounds, like hydroxybenzoic and hydroxycinnamic acids, tannins, proanthocyanidins, phenolic lipids, (prenylated) flavonoids and coumarins (Cushnie and Lamb, 2005; Cushnie and Lamb, 2011). In this respect, the well-studied cereal, barley, can produce various compounds that can be active against powdery mildew (*Erisiphe graminis* f. sp. *hordei;* gramine, putresceine, hordatine A and B and spermidine and p-coumaroyl-hydroxy agmatine), foot rot (*Helmithosporium* sativus; spermidine hordatine A and B) and brown rust (*P. hordei;* spermidine) (Kremer and Ben-Hammouda, 2009). For most cereals, including wheat, the knowledge concerning antifungal compounds is relatively scarce (Ahuja et al., 2012). Similar to the wild barley (*Hordeum lechleri*), wheat, maize *(Zea mays)* and rye (*Secale cereale*), can synthesize various benzoxazinoids. Resistance was in several cases associated with the accumulation of particular (glucosylated) benzoxazinoids against the causal agents of FHB in wheat (Niemeyer, 1988; Søltoft et al., 2008) and against the causal agent of northern corn leaf blight, *Setosphaeria turtica* (anamorph: *Exserohilum turcicum*) in maize (Tanwir et al., 2017; Yang et al., 2019). Direct antifungal activity was reported toward *Microdochium nivale* (currently *Fusarium nivale*) (Virtanen and Hietala, 1955; Wahlroos and Virtanen, 1959), *Helminthosporium turcicum* (Couture et al., 1971), *Parastagonospora nodorum* (Baker and Smith, 1977) (Du Fall and Solomon, 2013) and *Rhizoctonia solani* (Song et al., 2011). *In vitro* assays indicated that DIBOA (2,4-dihydroxy-1,4-benzoxazin-3-one) proved to be the most potent growth inhibitor of *Gaeumannomyces graminis* var. *tritici* (*Ggt*) strains, followed by DIMBOA (2,4-dihydroxy-7-methoxy-1,4-benzoxazin-3-one), while their respective conversion products, BOA (1,4-benzoxazin-3-one) and MBOA (7-methoxy-1,4-benzoxazin-3-one), were significantly less potent (Wilkes et al., 1999). Benzoxazinoid accumulation can also trigger other defense mechanisms or prime the plant for defense (Ahmad et al., 2011; Song et al., 2011).

Next to benzoxazinoids, wheat can accumulate serotonin and the derivatives N-feruloyltryptamine and N-feruloylserotonin, which all interfered with the sporulation of *P. nodorum* (Du Fall and Solomon, 2013) and *Cochliobolus sativus* (anamorph: *Bipolaris sorokiniana)* (Ishihara et al., 2017). Another secondary metabolite that has caught attention with respect to resistance toward fungal infections is ferulic acid (4-hydroxy-3-methoxycinnamic acid), which is generally abundantly present in plants, particularly cereals, although primarily conjugated with mono-, oligo- and polysaccharides, polyamines and lipids (Kumar and Pruthi, 2014; Ou and Kwok, 2004). As a crosslinker between polysaccharides and lignin, ferulic acid is pivotal for the lignocellulose, which confers rigidity to plant cell walls and serves as a physical barrier for insects and microorganisms as well. In itself, FA has a broad range of biological activities too, such as antioxidant, being antimicrobial and providing resistance in barley toward the aphid, *Schizaphis graminum* (Cabrera et al., 1995), and in wheat against the orange wheat blossom midge (*Sitodiplosis mosellana*) in a dose-dependent manner (Abdel-Aal et al., 2001).

In vitro, antifungal activity of ferulic acid was reported toward *Sclerotinia sclerotiorum, Fusarium oxysporum, Alternaria* sp., *Botrytis cinerea,* and *Penicillium digitatum* (Lattanzio et al., 1994; Ou and Kwok, 2004). When seeds of two spring wheat cultivars were treated with chitosan the germination and the synthesis of phenolic acids, especially ferulic acid, and lignin in the primary leaf, was promoted. At the same time, this not only tempered seed-borne *F. graminearum* infection but also repressed the proliferation of the fungus to the primary roots of the germinating seedlings (Reddy et al., 1999). The accumulation of ferulic acid synthesis in the period from anthesis until approximately 20 days after anthesis correlated with resistance toward *Fusarium* (McKeehen et al., 1999) and wheat genotypes with the highest ferulic acid contents in wheat spikes showed the highest resistance toward FHB disease in a field situation (Martin et al., 2017).

Added to disease resistance, the antifungal secondary metabolites may also affect the *in planta* accumulation of mycotoxins in a direct way. In *F. culmorum* and *F. graminearum* strains, ferulic acid negatively affected *Tri-gene* transcription in a dose-dependent manner, thus inhibiting the biosynthesis of type B trichothecene biosynthesis, which include deoxynivalenol (DON) and its acetylated forms 3-acetyl-4-deoxynivalenol and 15-acetyl-4-deoxynivalenol (3- and 15-ADON), and nivalenol (NIV) and its acetylated form 4-acetylnivalenol or fusarenone X (FX) (Boutigny et al., 2009). Conversely, when *P. nodorum* was exposed to serotonin, an increase in the accumulation of the mycotoxin altenariol was observed (Du Fall and Solomon, 2013). Through the action of other fungal effectors, e.g. SnTox3, the SnToxA-induced serotonin defense pathway in wheat may be suppressed again (Du Fall and Solomon, 2013), thus illustrating the ability of a pathogen to effectively outmanoeuvre plant defence strategies.

Nevertheless, there may be some negative side effects to consider when aiming for increasing ferulic acid levels. Relatively low concentrations ferulic acid stimulated sorghum growth, but relatively high concentrations negatively affected both seed germination and seedling growth of sorghum (Rasmussen and Einhellig, 1977). It was suggested that, after being released into the soil, ferulic acid may be converted by bacterial activity into phytotoxic derivatives, such as 2-methoxy-4-ethenylphenol (Liebl and Worsham, 1983; Nair et al., 1988), thus potentially harming crop growth in the following year. Whether or not such conversions are also taking place *in planta* by endophytic bacteria is currently not known.

The large number and sometimes transient accumulation of secondary metabolites and their derivatives, some of which still may have to be discovered, in cereals, together with the various fungal invaders, all with their specific sensitivity (resistance) towards these metabolites, makes it difficult to determine the actual role of a particular metabolite in plant resistance towards plant fungi. For example, the colonization by the beneficial mycorrhiza, *Glomus intraradicis,* of barley *(Hordeum vulgare)* resulted in an early and transient accumulation of antifungal hydroxycinnamic acid amides, suggesting only a brief defense response upon penetration, although there was a continuous accumulation of cyclohexanone derivatives, particularly blumenin, during further establishment of the plant-fungus association (Peipp et al., 1997).

Some plant-synthesized metabolites may be further derivatized by microorganisms, leading to the actual antifungal compound. A compelling example of this is the benzoxazinoid, BOA, which can be converted by the activity of the soil microbiome into the more stable aminophenoxazine, APO (Figure 1) (Niemeyer, 2009). APO was demonstrated to be toxic to *Fusarium verticillioides*. However, the fungus itself can modify BOA, via the synthesis of the 2-aminophenol intermediate, into N-(2-hydroxyphenyl) malonamic acid (HPMA) and 2-acetamidophenol (BOA-X), both of which are less toxic for the fungus when compared to APO. Thus, by quickly channeling BOA into a different biosynthesis pathway, the accumulation of the toxic APO can be reduced or even abolished by the fungus (Figure 1) (Glenn et al., 2003). In turn, the APO insensitive endophytic bacterium, *Bacillus mojavensis,* can stay ahead of this fungal metabolization strategy by channeling BOA, well before secretion into the rhizosphere, through the 2-aminophenol intermediate, into its synthesis pathway of APO (Figure 1). In this way, APO can still accumulate to levels that reduce *Fusarium verticillioides* proliferation and the associated mycotoxin accumulation in maize (Bacon et al., 2007).

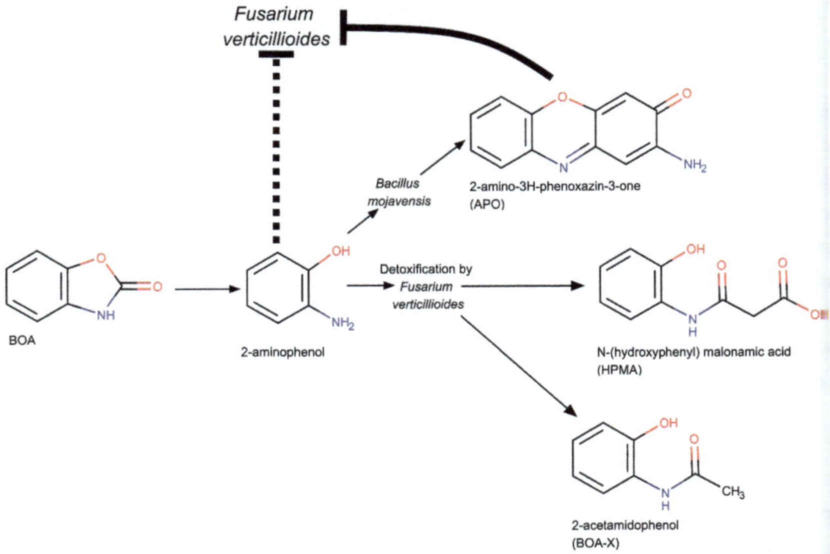

Figure 1 The plant-synthesized benzoxazinoid BOA (1,4-benzoxazin-3-one) is converted by microbial activity into, for *Fusarium verticillioides* potentially toxic, 2-aminphenol. The fungus has nevertheless the ability to convert 2-aminophenol into the less toxic compounds HPMA and BOA-X. The endophytic bacterium *Bacillus mojavensis* can undercut these detoxification mechanisms by converting 2-aminophenol into APO, which is toxic for the fungus as well, thus protecting the plant against *F. verticillioides* infection (figure modified from Glenn et al., 2003).

5 Induced defense responses in cereals

Beneficial endophytic and rhizosphere microorganisms can induce resistance responses in the host plant toward pathogens. Generally, components, present on or released by microbes, called microbial-associated molecular patterns (MAMPs), can be recognized by the plant, setting off an intricate cascade of (signaling) events that may lead to the accumulation of hormones to two independent immune responses, salicylic acid-induced systemic resistance (ISR) or jasmonic acid-induced systemic acquired resistance (SAR). The direction of the immune response can be further fine-tuned by other hormones, such as auxin, ethylene, abscisic acid, cytokinins, brassinosteroids and gibberellin. Flagelin monomer, fgl22, which is one of the major building blocks of the bacterial flagellum, and chitin, an important element in the fungal cell wall are only two examples of the many MAMPS that have been characterized (De Kesel et al., 2021; Yu et al., 2019). Next to immediately activating clear defense responses, this induction can also be latent, thus merely priming the plant and not showing an apparent accumulation of defense-related transcripts, such as those coding for phenyl ammonia lyase (PAL) and pathogenesis-related (PR) proteins. Primed plants are able to respond much more quickly and dramatically, accumulating defense-related transcripts, when triggered by biotic or abiotic stresses, such as root-associated microorganisms, particular synthetic compounds, pathogens and herbivores (Conrath et al., 2006; De Kesel et al., 2021; Frost et al., 2008; Heil and Silva Bueno, 2007; Pineda et al., 2010; Yu et al., 2019). Next to MAMPs, several fungi and bacteria can produce plant hormones, such as auxins, cytokinins and gibberellins, and thus can influence the overall direction the immune response will go, although other mechanisms may be involved as well (Aly et al., 2011; De Kesel et al., 2021; Li et al., 2006; Pieterse et al., 2012; Redman et al., 2011; Schouten, 2019; Tan and Zou, 2001; Yu et al., 2019).

Colonization of barley roots by the endophytic *Pochonia chlamydosporia* resulted in an enrichment of transcripts associated with abiotic stress responses, primarily those coding for heat shock proteins, plant hormone biosynthesis, such as auxin, ethylene and jasmonic acid (JA), and effector-triggered immunity (ETI) and pattern-triggered immunity (PTI) (Larriba et al., 2015). Several endophytes, isolated from clover and positive for the production of IAA, ACC deaminase and could be transferred to sterile rice seedlings. Although endophytic colonization was cultivar dependent, when established, the endophytes led to growth promotion and an increased resistance against pathogens in rice. The increase in resistance was attributed to the induction of defense responses by the plant (Etesami et al., 2014). Although being a rather aggressive root colonizer, even killing cells in the cortex, the endophyte *Serendipita indica* (formerly *Piriformaspora indica*) is considered a mutualist by promoting plant growth and supporting the plant in resisting biotic and abiotic stress elements (Deshmukh et al., 2006; Jacobs et al., 2011; Qiang et al., 2012). This endophyte shows a large host range (Qiang et al., 2012) and significantly reduced disease symptom severity of a typical leaf (*Blumeria graminis* f. sp. *tritici*), stem base (*Pseudocercosporella herpotrichoides*) and root *(F. culmorum)* pathogen in wheat in the greenhouse. However, subsequent field experiments with *S. indica* reduced the disease severity caused by *P. herpotrichoides,* whereas the leaf pathogen, *B graminis* f. sp. *tritici* was not hindered at all. The endophyte either triggered systemic defence responses or primed the host plant (Serfling et al., 2007). Studies in barley implied that roots are partly protected from necrotrophic pathogen *Fusarium culmorum* through mobilizing the plant's antioxidant capabilities (Harrach et al., 2013). However, *S. indica* originates from a sub-tropical arid region (Verma et al., 1998) and may therefore be ill-adapted to the temperate climatic conditions found in Europe (Rabiey et al., 2017; Serfling et al., 2007).

6 *Trichoderma:* a multi-role biocontrol agent

Species of the genus *Trichoderma* can be found globally in soils and in and on plant roots, wood and decaying plant debris, oomycetes and fungi (Druzhinina et al., 2011). They have been recognized as potent competitors or pathogens toward other microorganisms, even nematodes, in many environments and, although some species can be plant-pathogenic, several species or isolates within species have been considered important (potential) antagonists, which can be used in biological control approaches (Druzhinina et al., 2011; Jaklitsch and Voglmayr, 2015; Kubicek et al., 2008; Sood et al., 2020). Beneficial *Trichoderma* isolates can be very versatile in their biology and activity; asymptomatically colonizing plants, supporting plants regarding abiotic stress conditions, improving photosynthesis efficiency and nitrogen management, and fending off pathogens by direct attack, i.e. mycoparasitism, competitive inhibition or antibiosis, or inducing plant defense responses or enhanced plant growth (Druzhinina et al., 2011; Ferreira and Musumeci, 2021; Manzar et al., 2021; Sood et al., 2020; Zhang et al., 2021).

The genome of *Trichoderma virens* was shown to hold genes coding for non-ribosomal peptide synthetases (NRPSs) and polyketide synthase/NRPS (PKS/NRPS) hybrid enzymes. Mutating one of PKS/NRPS hybrid genes resulted in an impaired ability of *T. virens* to induce the defense responses through PAL (phenylalanine ammonia lyase), whereas the induction of another defense response, i.e. allene oxide synthese remained (Djonovic et al., 2007; Mastouri et al., 2010; Mukherjee et al., 2012; Shoresh and Harman, 2008; Shoresh et al., 2010; Vargas et al., 2008; Viterbo et al., 2005). As mentioned in Section 2, a *Trichoderma* sp. was, next to other antagonists, suggested to be involved in controlling *Rhizoctonia* barepatch disease in wheat (Wiseman et al., 1996). *T. koningii,* significantly reduced take-all disease in wheat, caused by by *G. graminis* var. *tritici,* in growth-chamber experiments, and somewhat improved wheat growth in the absence of *G. graminis* var. *tritici* in field trials (Duffy et al., 1996). Root colonization by *T. harzianum,* present in the biopreparation Supresivit, promoted root growth and enhanced nutrient uptake of wheat and other cereals, potentially enhancing the overall health and productivity of the plants (Hysek et al., 2002; Sharma et al., 2012). A *T. viride* isolate could also be used as biofertilizer in wheat (Mahato et al., 2018). More recently, a *T. afroharzianum* isolate was shown to effectively control FHB, Fusarium crown rot and damping off in wheat. This was attributed to the synthesis of particular enzymes and secondary metabolites, thus inhibiting fungal proliferation and, at the same time, reducing mycotoxin accumulation in wheat grains (Bouanaka et al., 2021). Five strains of *T. harzianum* and two strains of *T. koningii* significantly reduced tan spot disease severity in wheat, which is caused by *Pyrenophora tritici-repentis,* in greenhouse trials (Perello et al., 2003). A combination of *T. harzianum* and the bacterium *Streptomyces viridosporus* elevated resistance against the wheat rust, *Puccinia triticina,* in two susceptible genotypes Morocco and Sids-1 (El-Sharkawy et al., 2015). A combination of *T. harzianum,* isolate HL1, *T. viride,* isolate HL5, and an arbuscular mycorrhizal fungus effectively affected uredospores germination of *Puccinia graminis* Pers. f. sp. *tritici* (El-Sharkawy et al., 2018). *T. asperellum*, isolate T3, could be a potential agent for managing anthracnose, caused by *Colletotrichum graminicola*, in sorghum (Manzar et al. 2021).

Generally, *Trichoderma* seems a very promising antagonist toward plant pathogenic fungi, but they still are not effective enough in a biocontrol setting. To solve this, efforts are being made to improve this, e.g. by increasing their chitinase and glucanase activity (Hassan, 2014).

Another problem is that, despite being antagonistic, *Trichoderma* species can become pathogenic. This was demonstrated for *Trichioderma afroharzianum* in maize, causing ear rot disease, leading to massive infections on maize cobs, in Germany, France and Italy (Pfordt et al., 2020). This disease could also be found in wheat and barley but not in sorghum (Pfordt et al., 2023). The application of a *Trichoderma* biocontrol agent may therefore not be without risks, although the development of hypovirulent strains may help solve this problem.

7 Hypovirulence

Fungi can be infected with viruses and, in the case of plant pathogenic fungi, these mycoviruses can have a profound effect on virulence. The best-described example of this is *Cryphonectria parasitica*, the causal agent of chestnut blight, a disease that has been ravaging chesnut on a massive scale, particularly on the East Coast of the United States. In the 1960s, a hypovirulent strain, carrying a double-stranded RNA (dsRNA) virus, was found in Italy. This strain was still capable of colonizing chestnut, but behaved much more benign, resembling an endophyte, thus not killing the tree. Through anastomosis, this virus could be transferred to the virulent strain, thereby significantly reducing the occurrence of chestnut blight. Unfortunately, this approach could not be applied in the United States, as the hypovirulent strain did not anastomose with the *C. parasitica* strains encountered there. Instead, a heavily melanized barrier was formed, indicating different vegetative compatibility groups, and the virus could not be transferred (Milgroom and Cortesi, 1999). It has become clear that, in various species of endophytic fungi of grasses, mycoviruses are omnipresent (Herrero et al., 2009). A mycovirus in the endophyte *Curvularia protuberata* was necessary for providing thermotolerance of panic grass, *Dichanthelium lanuginosum* (Marquez et al., 2007), indicating that mycoviruses can also play a role in coping with abiotic stress. Although the potential of hypovirulence was established, particularly under practical conditions, there are still several major challenges regarding the actual application of hypovirulence-inducing mycoviruses as extensively reviewed by Kyrychenko et al. (2018), Garcia-Pedrajas et al. (2019) and Myers and James (2022), such as how their (artificial) transmission to and their subsequent loss by plant pathogenic fungi, their biology and the responses by the fungal host as a whole. Furthermore, not all mycoviruses, even when related, cause hypovirulence. Yet, the knowledge on these aspects is also rapidly expanding, which may indeed lead to applications of mycoviruses in biological control other than the *C. parasitica* case in Italy. This also applies to hypovirulence-inducing mycoviruses in cereal pathogens. For *F. graminearum* several mycoviruses were discovered over the past two decades. FgV1, now also called FgV1-DK21, reduced virulence and has been therefore been scrutinized as a biocontrol agent for some time (Stack et al., 2003). Remarkably, a strongly related viral strain, FgV1-ch, showing 96% sequence identity with FgV1- DK21, did not reduce fungal virulence. Although this may sound disappointing, the comparison of the two mycoviral genomes may help to determine the factors that determine hypovirulence (Zhang et al., 2020b). *F. graminearum* isolate China-9 harbors a mycovirus (FgV-ch9), which affected colony and mycelial morphology, reduced the forming of conidia and virulence in wheat and maize plants, especially at relatively high viral dsRNA levels. When the purified virus-like particles from China-9 were transferred to the pathogenic *F. graminearum* isolate PH-1, similar phenotypes in this isolate were observed as well, including reduced virulence in wheat and maize. Unfortunately, PH-1 was able to cure itself from the introduced mycovirus through the next generation of conidia, thus restoring virulence (Darissa et al., 2012). This curing ability was already described for *Rosellinia necatrix* infected with another mycovirus, RnPV1-W8 (Sasaki et al., 2006), and seems to be a phenomenon to take into account when searching for biological control approaches through the introduction of hypovirulence inducing mycoviruses. Subsequent research on FgV-ch9 has nevertheless brought important new insights into a mechanism by which mycoviruses cause hypovirulence, which is the down-regulation of expression of a fungal host gene, *vrl,* by means of a protein, P3, encoded on the FgV-ch9 ds-RNA (Bormann et al., 2018). Studies on other *F. graninearum* dsRNA hypoviruses, FgHV1- and FgHV2, have fully resolved their genome sequences and potentially revealed a strategy for avoiding RNA-silencing mechanisms responsible for host immunity (Wang et al., 2016). The *F. graminearum,* strain Fg-4-2, contained a FgV1 mycovirus strain, FgV1-SD4, caused hypovirulence and eliminated the accumulation of the mycotoxin, deoxynivalenol (Paudel et al., 2022). This study also indicated that mycoviruses may not only be transferred by anastomosis, thus requiring identical vegetative compatibility groups, but also by virions, both under lab and natural conditions. The downside of applying hypovirulence strains in cereals is the speed by which FHB generally develops, meaning that serious crop damage has been done even before the spread of the hypovirulence in the virulent strain and the formation of new spores (Garda-Pedrajas et al., 2019).

The phenomena of hypovirulence may however still become relevant in an alternative approach, which is based on a combination of hypovirulence and microbial antagonism. To protect wheat, rice, barley, maize and oats, against several *Fusarium* diseases, FHB, stripe rust and rice blast (Tian et al., 2020), *Sclerotinia sclerotiorum* could be used as an endophytic antagonist, thereby simultaneously reducing mycotoxin accumulation and increasing plant growth and crop yield. Unfortunately, *S. sclerotiorum* is a necrotrophic pathogen in dicot plants, being destructive on a global scale and, as a result, causing significant economic losses annually. This split nutritional strategy, also termed schizotrophism, i.e. being destructively pathogenic or mutualistically endophytic (Tian et al., 2020), would render *S. sclerotiorum* unsuitable as a biocontrol agent in practice. However, in the same research, it was shown that the mycovirus-infected hypovirulent *S. sclerotiorum* strain, DT-8, previously isolated from Rapeseed (Yu et al., 2010; Zhang et al., 2020a), could asymptomatically infect wheat roots, increase yield and consistently reduced *Fusarium* disease by 40-60% in field trials, results which were as good as the virus-free DT-8 derivative, DT-8VF.

8 Bacteriophages to control pathogenic bacteria

Bacteriophages, i.e. viruses that are specialized in infecting and killing bacteria, are considered the most abundant biological entities on the planet, outnumbering the bacteria by an estimated factor of 10, and can be found in all places where bacteria are residing, such as aquatic environments, animals (intestines), soils, including the rhizosphere and phyllosphere (Forero-Junco et al., 2022; Salmond and Fineran, 2015). In marine environments, it has been estimated that 20% of the bacteria are being lysed daily (Suttle, 1994) and in other environments, this value is thought to be similar (Meaden et al., 2015). Genome sizes of bacteriophages, generally being a linear double-stranded DNA molecule, can vary significantly, ranging from several kilobases to almost 500 kb, with the latter being called jumbo phages (Kim et al., 2019; Ongenae et al., 2021; Yuan and Gao, 2017). Although different in shapes and sizes, 96% of all known bacteriophages belong to the tailed Caudovirus order and within this order there, too, is a tremendous variation in structure, virion size and DNA content (Ackermann, 2009; Stone et al., 2019; Yuan and Gao, 2017).

Bacteriophages have two infection strategies, being lytic or lysogenic. Lytic bacteriophages immediately start replicating after the DNA has been injected into the cell, thereby rapidly killing the bacterium. Lysogenic bacteriophages, also called temperate bacteriophages, may not start replication immediately, but instead the injected DNA is internalized into the chromosome or residing extrachromosomically as a plasmid and called a prophage. This prophage can provide the bacterium with enhanced fitness, e.g. resistance toward antibiotics and lytic bacteriophages, and virulence, e.g. through phage-encoded toxins, and the infected bacteria can proliferate normally, whereby the phage DNA is transferred upon cell division as well (Brussow et al., 2004; Davies et al., 2016). However, upon biotic or abiotic stress or imminent cell death, which can be detected by the prophage as a so-called SOS-signal, the lytic phase is initiated (Brady et al., 2021). Genomically, lytic and lysogenic bacteriophage can be rather similar, whereby the lysogenic bacteriophage carries particular genes that are involved in controlling lysogeny (Gillis and Mahillon, 2014).

Bacteriophages are generally rather specialized, targeting only specific bacterial species or strains within species (Comeau et al., 2008; Stone et al., 2019). Because they parasitize bacteria at a large scale, bacteriophages can rapidly shape the composition of microbial populations (Comeau et al., 2008), even under *in vitro* conditions (Borin et al., 2023), making them potentially powerful biocontrol agents, e.g. to fight multi-drug resistant bacteria in medicine (Monteiro et al., 2019).

Already in 1935, thus not long after the discovery of bacteriophages, a bacteriophage was used to reduce Stewart's wilt, caused by *Pantoea stewartii,* in maize (Thomas, 1935), but only the past decade additional efforts were made to find and apply bacteriophages to control bacterial diseases in plants (Clavijo-Coppens et al., 2021; Clavijo-Coppens et al., 2022; Svircev et al., 2018; Vu and Oh, 2020). This has led to several interesting phage formulations, sometimes comprising a cocktail of different bacteriophages, against cereal pathogens, such as *Xanthomonas oryzae pv. oryzae* and *Burkholderia glumae and B. plantarii* in rice and *Clavibacter nebraskensis* in maize (Adachi et al., 2012; Chae et al., 2014; Dong et al., 2018; Kimmelshue et al., 2019; Riber et al., 2023), although these are still in an experimental stage. The efficacy ranged from 20 to 100% and the plant's developmental stage seems to have an influence, because, in one study, the treatment of seeds reduced disease incidence by 58 to 76% whereas the treatment of seedlings reduced disease incidence by only 20% (Kimmelshue et al., 2019). Recently, a significant variety of bacteriophages were encountered in the phyllosphere of wheat (Dougherty et al., 2023; Forero-Junco et al., 2022; Riber et al., 2023), which would suggest that, despite being UV-light sensitive, bacteriophages may act in controlling bacterial leaf pathogens as well.

9 Conclusions

Integrated pest management (IPM) strategies, including using microorganisms and viruses to control plant pathogens, are by no means novel and ideas on this were already summarized by Baker and Cook in 1974. In fact, such biocontrol practices were sometimes already in place well before the introduction of chemical pesticides and the examples given in this review, such as the TAD phenomena and phage application, illustrate that cereal cropping systems have been at the forefront in recognizing the value of biological control strategies. Baker and Cook (1974) also recognized that chemical pesticides were too profitable, too effective and too simple to ignore, for both industry and farmers. But things have been changing. Due to their increasing ineffectivity and the banning of pesticides, due to human health and environmental concerns, more emphasis is put on IPM approaches, making it relevant to rejuvenate old and acquire new knowledge on biological approaches to control pests and diseases. However, as biological entities are involved, with their sometimes sophisticated and intricate biology, survival strategies and interactions with other biological entities, the complexities of such approaches are significantly greater than the application of a chemical. This may be one of the reasons why many biocontrol applications still have not gone beyond greenhouse experiments. At the same time, due to the enormous advances in molecular techniques, biochemistry, biology, ecology, next-generation sequencing and bioinformatics and other computational sciences in the past decades, we are currently much more capable of studying such intricate interactions between plants, plant pathogens and biocontrol agents than approximately 60 years ago, the period in which the development chemical pesticides was on the march. By scrutinizing such interactions and biocontrol mechanisms, completely new discoveries are made as well, such as the recently characterized single-stranded DNA-type mycoviruses (Myers and James, 2022), which may offer new opportunities within biocontrol strategies as well.

Next to the risks concerns regarding pesticides, pesticide resistance development in pathogens and the relatively limited and time-consuming options regarding resistance breeding in crops forces societies and scientists to search for additional and more sustainable approaches. It may very well be that because of their complexity, biocontrol agents are the sustainable answers to controlling pests and diseases, as is illustrated by the TAD phenomenon. Of course, there are still many issues to address, also with respect to the protection of cereals against the various fungal and bacterial diseases, such as the speed by which a biocontrol approach is effective, the potential schizotrophic nature of a biocontrol agent, how well the host plant can accommodate the biocontrol agent in the rhizosphere, phyllosphere or as an endophyte, how competitive the biocontrol agent is toward other microorganisms in the agricultural environment, whether or not a monoculture, a rotation system or intercropping is a suitable approach, how a biocontrol agent can be applied and sustain itself effectively over time and what the abiotic factors are that guarantee success of a biocontrol agent. Also, there is the question of how pathogens may adapt to coping with antagonism or induced plant immunity in the long run (Duffy et al., 2003; Mazzola et al., 1995; Schouten, 2018; Schouten et al., 2004; Yu et al., 2019). To address these aspects, a strong multidisciplinary approach, in which biological ecological, molecular and biochemical knowledge are combined, is imperative. The industry, applying these approaches in an integrated fashion for product development, is still relatively small, but growing rapidly and the chemical industry is trying to make a turn toward biocontrol product development as well. Cereals comprise major crops, such as wheat, maize and rice, and biological control solutions will therefore have a major impact on sustainability and food security, including mycotoxin accumulation in our food chain.

10 Where to look for further information

Due to the multidisciplinary research that is needed to understand and apply biological control strategies against pests and diseases, there are many places one can look for further information. The International Organisation for Biological Control (IOBC) (http://www.iobc-global.org) has for many years been promoting environmentally safe methods of pest and disease control. It is a global voluntary organization, bringing together scientists studying biological control from various angles, communicating through workshops, newsletters and its official journal, *BioControl*. The International Society for Microbial Ecology (https://www.isme-microbes.org/) supports research and education of microbial ecologists and the wider community, e.g. through *The ISME Journal*. Both IOBC and ISME can serve as a good starting point to look for further information.

The expanding research on and application of microorganisms for controlling pests and diseases in an IPM approach is considered a potentially accessible and cost-effective strategy, particularly for poor farmers in third-world countries. The CGIAR (formerly the Consultative Group for International Agricultural Research) comprises a consortium of 15 research centers, known as the CGIAR Consortium of International Agricultural Research Centres, and is funded by national governments, multilateral funding and development agencies and private foundations (http://www.cgiar.org). Among them are the International Rice Research Institute (IRRI) in the Philippines and the International Maize and Wheat Improvement Centre, better known by its Spanish acronym CIMMYT for Centro Internacional de Mejoramiento de Mafz y Trigo (http://www.cimmyt.org). Several CGIAR centers and their partners have joined further through initiating the Systemwide Program on Integrated Pest Management (SP-IPM), which aims at more productive and healthy agro-ecosystems by accelerating the development of innovative and adapted technologies in order to improve pest and disease management (http://www.spipm.cgiar.org and http://www.spipm.cgiar.org/members/centers).

As illustrated in this review, there is a plethora of journals with papers and reviews on microbial antagonism. A number of journals focus on fundamental and applied research in the field of biological control. Several books, reviewing biocontrol approaches and mechanisms, can be found as well. Natural antifungal compounds in cereals were more thoroughly reviewed previously, together with the antibiotics related to TAD (Schouten, 2018).

Last, but not least, there are several companies working on research and product development with the ultimate aim to apply microbes or viruses to control plant pathogens and pests.

11 References

Abdel-Aal, E.S.M., et al. 2001. Screening spring wheat for midge resistance in relation to ferulic acid content. *J. Agricult. Food Chem.* 49:3559-3566.

Ackermann, H.W. 2009. Phage classification and characterization. *Methods in Molecular Biology* 501:127–140.

Adachi, N., et al. 2012. Control of bacterial seedling rot and seedling blight of rice by bacteriophage. *Plant Dis.* 96:1033–1036.

Ahmad, S., et al. 2011. Benzoxazinoid metabolites regulate innate immunity against aphids and fungi in maize. *Plant Physiol.* 157:317–327.

Ahuja, I., et al. 2012. Phytoalexins in defense against pathogens. *Trends Plant Sci.* 17:73–90.

Aly, A.H., et al. 2011. Fungal endophytes: unique plant inhabitants with great promises. *Appl. Microbiol. Biot.* 90:1829–1845.

Andrade, P.D. and Caldas, E.D. 2015. Aflatoxins in cereals: worldwide occurrence and dietary risk assessment. *World Mycotoxin J.* 8:415–431.

Anonymous. 2021. Commision Regulation (EU) 2021/1399 on Amending Regulation (EC) No 1881/2006 as regards maximum levels of ergot sclerotia and ergot alkaloids in certain foodstuffs. *Official Journal of the European Union.* L 301/:1–3.

Anonymous. 2023a. Articles 191 to 193 of the Treaty on the Functioning of the European Union (TFEU).

Anonymous. 2023b. *The encyclopaedia of cereal diseases*. A.C. Oilseeds, editor. AHDB Cereals & Oilseeds, Coventry, UK.

Bacon, C.W., et al. 2007. Interactions of *Bacillus mojavensis* and *Fusarium verticillioides* with a benzoxazolinone (BOA) and its transformation product, APO. *J. Chem. Ecol.* 33:1885–1897.

Baker, E.A. and Smith, I.M. 1977. Antifungal compounds in winter-wheat resistant and susceptible to *Septoria nodorum*. *Ann. Appl. Biol.* 87:67–73.

Baker, K.F. and Cook, R.J. 1974. *Biological control of plant pathogens*. W.H. Freeman and Company, San Francisco.

Borin, J.M., et al. 2023. Comparison of bacterial suppression by phage cocktails, dual-receptor generalists, and coevolutionarily trained phages. *Evol. Appl.* 16:152162.

Bormann, J., et al. 2018. Expression of a structural protein of the mycovirus FgV-ch9 negatively affects the transcript level of a novel symptom alleviation factor and causes virus infection-like symptoms in *Fusarium graminearum*. *J. Virol.* 92:17.

Bouanaka, H., et al. 2021. On the biocontrol by *Trichoderma afroharzianum* against *Fusarium culmorum* responsible of fusarium head blight and crown rot of wheat in Algeria. *Egypt. J. Biol. Pest Control.* 31:13.

Boutigny, A.L., et al. 2009. Ferulic acid, an efficient inhibitor of type B trichothecene biosynthesis and *Tri* gene expression in *Fusarium* liquid cultures. *Mycol. Res.* 113:746–753.

Brady, A., et al. 2021. Molecular basis of lysis-lysogeny decisions in Gram-positive phages. In *Annual review of microbiology*, Vol. 75. S. Gottesman, editor. Annual Reviews, Palo Alto, 563–581.

Brussow, H., et al. 2004. Phages and the evolution of bacterial pathogens: From genomic rearrangements to lysogenic conversion. *Microbiol. Mol. Biol. Rev.* 68:560-+.

Cabrera, H.M., et al. 1995. Changes in ferulic acid and lipid-content in aphid-infested barley. *Phytochem.* 39:1023–1026.

Campos, M.L., et al. 2018. The role of antimicrobial peptides in plant immunity. *J. Exp. Bot.* 69:4997–5011.

Carmona, M., et al. 2020. Role of fungicide applications on the integrated management of wheat stripe rust. *Frontiers in Plant Science.* 11:22.

Chae, J.C., et al. 2014. Diversity of bacteriophages infecting *Xanthomonas oryzae* pv. *oryzae* in paddy fields and its potential to control bacterial leaf blight of rice. *J. Microbiol. Biotechnol.* 24:740–747.

Claflin, L.E., et al. 1989. *Pseudomonas avenae,* causal agent of bacterial leaf stripe of pearl-millet. *Plant Dis.* 73:1010–1014.

Clavijo-Coppens, F., et al. 2021. Novel virulent bacteriophages infecting Mediterranean isolates of the plant pest *Xylella fastidiosa* and *Xanthomonas albilineans. Viruses-Basel.* 13:21.

Clavijo-Coppens, F., et al. 2022. Phage-mediated biocontrol against plant pathogenic bacteria. In *Biocontrol of plant disease: recent advances and prospects in plant protection*. C. Prigent-Combaret and B. Dumas, editors. John Wiley & Sons, Inc., London, UK, 173–216.

Comeau, A.M., et al. 2008. Exploring the prokaryotic virosphere. *Res. Microbiol.* 159:306–313.

Conrath, U., et al. 2006. Priming: Getting ready for battle. *Mol. Plant-Microbe Interact.* 19:1062–1071.

Cook, N.M., et al. 2021. High frequency of fungicide resistance-associated mutations in the wheat yellow rust pathogen *Puccinia striiformis* f. sp. *tritici. Pest Managem. Sci.* 77:3358–3371.

Cook, R.J. 2007. Take-all decline: Model system in the science of biological control and clue to the success of intensive cropping. In *Biological control: a global perspective.* C. Vincent, M. Goettel, and G. Lazarovits, editors. CABI, Wallingford, 399–414.

Cook, R.J. and Rovira, A.D. 1976. Role of bacteria in biological-control of *Gaeumannomyces graminis* by suppressive soils. *Soil Biol. Biochem.* 8:269–273.

Cook, R.J., et al. 1995. Molecular mechanisms of defense by rhizobacteria against root disease. *Proc. Natl. Acad. Sci. USA.* 92:4197–4201.

Couture, R.M., et al. 1971. Role of cyclic hydroxamic acids in monogenic resistance of maize to *Helminthosporium turcicum. Physiol. Plant Pathol.* 1:515–521.

Cushnie, T.P.T. and Lamb, A.J. 2005. Antimicrobial activity of flavonoids. *Int. J. Antimicrob. Agents.* 26:343–356.

Cushnie, T.P.T. and Lamb, A.J. 2011. Recent advances in understanding the antibacterial properties of flavonoids. *Int. J. Antimicrob. Agents.* 38:99–107.

Darissa, O., et al. 2012. A dsRNA mycovirus causes hypovirulence of *Fusarium graminearum* to wheat and maize. *Eur. J. Plant Pathol.* 134:181–189.

Davies, E.V., et al. 2016. The role of temperate bacteriophages in bacterial infection. *FEMS Microbiol. Lett.* 363:10.

De Kesel, J., et al. 2021. The Induced Resistance Lexicon: Do's and Don'ts. *Trends Plant Sci.* 26:685–691.

Deshmukh, S., et al. 2006. The root endophytic fungus *Piriformospora indica* requires host cell death for proliferation during mutualistic symbiosis with barley. *Proc. Natl. Acad. Sci. USA.* 103:1845018457.

Dinh, H.X., et al. 2022. The barley leaf rust resistance gene *Rph3* encodes a predicted membrane protein and is induced upon infection by avirulent pathotypes of *Puccinia hordei. Nat. Commun.* 13:13.

Djonovic, S., et al. 2007. A proteinaceous elicitor Sm1 from the beneficial fungus *Trichoderma virens* is required for induced systemic resistance in maize. *Plant Physiol.* 145:875–889.

Dong, Z.X., et al. 2018. Isolation and characterization of a novel phage Xoo-sp2 that infects *Xanthomonas oryzae* pv. *oryzae. Journal of General Virology.* 99:1453–1462.

Dougherty, P.E., et al. 2023. Widespread and largely unknown prophage activity, diversity, and function in two genera of wheat phyllosphere bacteria. *The ISME Journal:1–11* (published online).

Druzhinina, I.S., et al. 2011. *Trichoderma:* the genomics of opportunistic success. *Nat. Rev. Microbiol.* 9:749–759.

Du Fall, L.A. and Solomon, P.S. 2013. The necrotrophic effector SnToxA induces the synthesis of a novel phytoalexin in wheat. *New Phytologist.* 200:185–200.

Duffy, B., et al. 2003. Pathogen self-defense: Mechanisms to counteract microbial antagonism. *Ann. Rev. Phytopathol.* 41:501–538.

Duffy, B.K., et al. 1996. Combination of *Trichoderma koningii* with fluorescent pseudomonads for control of take-all on wheat. *Phytopathology.* https://doi.org/10.1094/phyto-86-188

El-Sharkawy, H.H.A., et al. 2015. Combined effects of *Streptomyces viridosporus* and *Trichoderma harzianum* on controlling wheat leaf rust caused by *Puccinia triticina. Plant Pathol. J.* 14 182–188.

El-Sharkawy, H.H.A., et al. 2018. Biocontrol of stem rust disease of wheat using arbuscular mycorrhizal fungi and *Trichoderma* spp. *Physiol. Mol. Plant Pathol.* 103:8491.

Etesami, H., et al. 2014. Bacterial biosynthesis of 1- aminocyclopropane-1-carboxylate (ACC) deaminase and indole-3-acetic acid (IAA) as endophytic preferential selection traits by rice plant seedlings. *J. Plant Growth Regul.* 33:654670.

Ferreira, F.V. and Musumeci, M.A. 2021. *Trichoderma* as biological control agent: scope and prospects to improve efficacy. *World J. Microbiol. & Biotechnol.* 37:17.

Fisher, M.C., et al. 2022. Tackling the emerging threat of antifungal resistance to human health. *Nat. Rev. Microbiol.* 20:557–571.

Forero-Junco, L.M., et al. 2022. Bacteriophages roam the wheat phyllosphere. *Viruses-Basel.* 14:13.

Frost, C.J., et al. 2008. Plant defense priming against herbivores: getting ready for a different battle. *Plant Physiol.* 146:818–824.

Garda-Pedrajas, M.D., et al. 2019. Mycoviruses in biological bontrol: from basic research to field implementation. *Phytopathology.* 109:1828–1839.

Gerlagh, M. 1968. Introduction of *Ophiobolus graminis* into new polders and its decline *Neth. J. Plant Pathol.* 74 Suppl. 2:97.

Gillis, A. and Mahillon, J. 2014. Phages preying on *Bacillus anthracis, Bacillus cereus,* and *Bacillus thuringiensis:* past, present and future. *Viruses-Basel.* 6:2623–2672.

Glenn, A.E., et al. 2003. Identification of intermediate and branch metabolites resulting from biotransformation of 2-benzoxazolinone by *Fusarium verticillioides. Appl. Environ. Microbiol.* 69:3165–3169.

Gong, A.-D., et al. 2015. Antagonistic mechanism of Iturin A and Plipastatin A from *Bacillus amyloliquefaciens* S76-3 from wheat spikes against *Fusarium graminearum. Plos One.* 10:e0116871–0116818.

Harrach, B.D., et al. 2013. The mutualistic fungus *Piriformospora indica* protects barley roots from a loss of antioxidant capacity caused by the necrotrophic pathogen *Fusarium culmorum. Mol Pl Micr Interac.* 26:599–605.

Hassan, M.M. 2014. Influence of protoplast fusion between two *Trichoderma* spp. on extracellular enzymes production and antagonistic activity. *Biotechnol. Biotechnol. Equip.* 28:1014–1023.

Heil, M. and Silva Bueno, J.C. 2007. Within-plant signaling by volatiles leads to induction and priming of an indirect plant defense in nature. *Proc. Natl. Acad. Sci. USA.* 104:5467–5472.

Herrero, N., et al. 2009. Mycoviruses are common among different species of endophytic fungi of grasses. *Arch. Virol.* 154:327–330.

Huang, Z.Y., et al. 2004. Transformation of *Pseudomonas fluorescens* with genes for biosynthesis of phenazine-1-carboxylic acid improves biocontrol of rhizoctonia root rot and *in situ* antibiotic production. *FEMS Microbiol. Ecol.* 49:243–251.

Hysek, J., et al. 2002. The influence of the application of mineral fertilizers with the biopreparation supresivit *(Trichoderma harzianum)* on the health and the yield of different crops. *Arch. Phytopathol Plant Prot.* 35:115–124.

Ishihara, A., et al. 2017. Induced accumulation of tyramine, serotonin, and related amines in response to *Bipolaris sorokiniana* infection in barley. *Biosci. Biotechnol. Biochem.* 81:10901098.

Jacobs, S., et al. 2011. Broad-spectrum suppression of innate immunity is required for colonization of Arabidopsis roots by the fungus *Piriformospora indica. Plant Physiol.* 156:726–740.

Jaklitsch, W.M. and Voglmayr, H. 2015. Biodiversity of *Trichoderma (Hypocreaceae)* in Southern Europe and Macaronesia. *Stud. Mycol.* 80:1–87.

Jiao, Z., et al. 2013. Characterisation of *Pseudomonas chlororaphis* subsp. *aurantiaca* strain Pa40 with the ability to control wheat sharp eyespot disease. *Ann. Appl. Biol.* 163:444–453.

Kashyap, A., et al. 2021. Blocking intruders: inducible physico-chemical barriers against plant vascular wilt pathogens. *J. Exp. Bot.* 72:184198.

Kayim, M., et al. 2022. Fungal Diseases of Wheat. In Wheat. M.-u.-R. Ansari, editor. IntechOpen, London, 85–104.

Kim, D.S., et al. 1997. *Bacillus* sp. L324-92 for biological control of three root diseases of wheat grown with reduced tillage. *Phytopathology.* 87:551–558.

Kim, S.G., et al. 2019. Isolation and characterisation of pVa-21, a giant bacteriophage with anti-biofilm potential against *Vibrio alginolyticus. Scientific Reports.* 9:10.

Kimmelshue, C., et al. 2019. The use of biological seed coatings based on bacteriophages and polymers against *Clavibacter michiganensis* subsp. *nebraskensis* in maize seeds. *Sci. Rep.* 9:11.

Kremer, R.J. and Ben-Hammouda, M. 2009. Allelopathic plants. 19. Barley *(Hordeum vulgare* L). *Allel. J.* 24:225–242.
Kubicek, C.P., et al. 2008. Fungal genus *Hypocrea/Trichoderma:* from barcodes to biodiversity. *J Zhejiang Univ-Sc B.* 9:753–763.
Kumar, N. and Pruthi, V. 2014. Potential applications of ferulic acid from natural sources. *Biotechnology reports (Amsterdam, Netherlands).* 4:86–93.
Kumar, P., et al. 2022. Aflatoxins in cereals and cereal-based products: Occurrence, toxicity, impact on human health, and their detoxification and management strategies. *Toxins.* 14:26.
Kyrychenko, A.N., et al. 2018. Hypovirulence of mycoviruses as a tool for biotechnological control of phytopathogenic fungi. *Cytol. Genet.* 52:374–384.
Larriba, E., et al. 2015. Endophytic colonization of barley *(Hordeum vulgare)* roots by the nematophagous fungus *Pochonia chlamydosporia* reveals plant growth promotion and a general defense and stress transcriptomic response. *J. Plant Res.* 128:665–678.
Lattanzio, V., et al. 1994. Antifungal activity of phenolics against fungi commonly encountered during storage. *It J Food Sci.* 1:23–30.
Lee, M.R. 2009. The history of ergot of rye (Claviceps purpurea) I: From antiquity to 1900. *J. R. Coll. Physicians Edinb.* 39:179–184.
Lewis, C.M., et al. 2018. Potential for re-emergence of wheat stem rust in the United Kingdom. *Commun. Biol.* 1:9.
Li, Q., et al. 2006. Expression profiling of rice genes in early defense responses to blast and bacterial blight pathogens using cDNA microarray. *Physiol. Mol. Plant Pathol.* 68:51–60.
Liebl, R.A. and Worsham, A.D. 1983. Inhibition of pitted morning glory *(Ipomoea lacunosa* L) and certain other weed species by phytotoxic components of wheat *(Triticum aestivum* L.) straw. *J. Chem. Ecol.* 9:1027–1043.
Liu, H., et al. 2012. Genome Sequence of the Plant Pathogen Pseudomonas syringae pv. panici LMG 2367. *J. Bacteriol.* 194:5693–5694.
Lofgren, L.A., et al. 2018. *Fusarium graminearum:* pathogen or endophyte of North American grasses? *New Phytologist.* 217:1203–1212.
Mahato, S., et al. 2018. Effect of *Trichoderma viride* as biofertilizer on growth and yield of wheat. *Malaysian Journal of Sustainable Agriculture.* 2:1–5.
Manzar, N., et al. 2021. Biocontrol potential of native *Trichoderma* spp. against anthracnose of great millet *(Sorghum bicolour* L.) from Tarai and hill regions of India. *Biol. Control.* 152:104474.
Marquez, L.M., et al. 2007. A virus in a fungus in a plant: Three-way symbiosis required for thermal tolerance. *Science.* 315:513–515.
Martin, C., et al. 2017. Role of ferulic acid in Fusarium head blight resistance of wheat spikes. In *Future IPM 3.0 towards a sustainable agriculture.* IOBC-WPRS, Riva del Garda, Italy.
Mastouri, F., et al. 2010. *Trichoderma-plant* association enhances plant antioxidant capacity and tolerance to abiotic stresses. *Hortscience.* 45:S65–S65.
Mazzola, M., et al. 1995. Variation in sensitivity of *Gaeumannomyces graminis* to antibiotics produced by fluorescent *Pseudomonas* spp. and effect on biological control of take-all of wheat. *Appl. Environ. Microbiol.* 61:2554–2559.
McKeehen, J.D., R.H. Busch, and R.G. Fulcher. 1999. Evaluation of wheat *(Triticum aestivum* L.) phenolic acids during grain development and their contribution to *Fusarium* resistance. *J. Agricult. Food Chem.* 47:1476–1482.

Meaden, S., et al. 2015. The cost of phage resistance in a plant pathogenic bacterium is context-dependent. *Evolution* 69: 1321–1328.
Meyer, J., et al. 2016. Signals that stop the rot: Regulation of secondary metabolite defences in cereals. *Physiol. Mol. Plant Pathol.* 94:156–166.
Milgroom, M.G. and Cortesi, P. 1999. Analysis of population structure of the chestnut blight fungus based on vegetative incompatibility genotypes. *Proc. Natl. Acad. Sci. USA.* 96:10518–10523.
Monteiro, R., et al. 2019. Phage therapy: Going temperate? *Trends Microbiol.* 27:368–378.
Mousa, W.K., et al. 2016. Root-hair endophyte stacking in finger millet creates a physicochemical barrier to trap the fungal pathogen *Fusarium graminearum*. *Nat. Microbiol.* 1:12.
Mukherjee, P.K., et al. 2012. Functional analysis of non-ribosomal peptide synthetases (NRPSs) in *Trichoderma virens* reveals a polyketide synthase (PKS)/NRPS hybrid enzyme involved in the induced systemic resistance response in maize. *Microbiology SGM.* 158:155–165.
Myers, J.M. and James, T.Y. 2022. Mycoviruses. *Curr. Biol.* 32:R150-R155.
Nair, M.G., et al. 1988. Ferulate esters of higher fatty alcohols and allelopathy in *Kalanchoe daigremontiana*. *J. Chem. Ecol.* 14:589–603.
Niemeyer, H.M. 1988. Hydroxamic acids (4-hydroxy-1,4-benzoxazin-3-ones), defense chemicals in the Gramineae. *Phytochem.* 27:3349–3358.
Niemeyer, H.M. 2009. Hydroxamic Acids Derived from 2-Hydroxy-2 H-1,4-Benzoxazin-3(4 H)-one: Key Defense Chemicals of Cereals. *J. Agricult. Food Chem.* 57:1677–1696.
Ongena, M. and Jacques, P. 2008. *Bacillus* lipopeptides: versatile weapons for plant disease biocontrol. *Trends Microbiol.* 16:115–125.
Ongenae, V., et al. 2021. Cell wall deficiency as an escape mechanism from phage infection. *Open Biol.* 11:7.
Ou, S.Y. and Kwok, K.C. 2004. Ferulic acid: pharmaceutical functions, preparation and applications in foods. *J. Sci. Food Agric.* 84:1261–1269.
Paudel, B., et al. 2022. Fusarium graminearum Virus-1 strain FgV1- 5D4 infection eliminates mycotoxin deoxynivalenol synthesis by *Fusarium graminearum* in FHB. *Microorganisms.* 10:10.
Peipp, H., et al. 1997. Arbuscular mycorrhizal fungus-induced changes in the accumulation of secondary compounds in barley roots. *Phytochem.* 44:581587.
Pereira, C., et al. 2022. Mycotoxins of concern in children and infant cereal food at European level: Incidence and bioaccessibility. *Toxins.* 14:21.
Perello, A., et al. 2003. Biocontrol efficacy of *Trichoderma* isolates for tan spot of wheat in Argentina. *Crop. Protect.* 22:1099–1106.
Pfordt, A., et al. 2023. Pathogenicity of *Trichoderma afroharzianum* in cereal crops. *Pathogens.* 12:11.
Pfordt, A., et al. 2020. *Trichoderma afroharzianum* ear rot - A new disease on maize in Europe. *Front. Agron.* 2:7.
Pieterse, C.M.J., et al. 2012. Hormonal modulation of plant immunity. In *Annual Review of Cell and Developmental Biology*. Vol. 28. R. Schekman, editor. 489521.
Pineda, A., et al. 2010. Helping plants to deal with insects: the role of beneficial soil-borne microbes. *Trends Plant Sci.* 15:507–514.
Planas-Marquès, M., et al. 2020. Four bottlenecks restrict colonization and invasion by the pathogen *Ralstonia solanacearum* in resistant tomato. *J. Exp. Bot.* 71: 2157–2171

Poutanen, K.S., et al. 2022. Grains - a major source of sustainable protein for health. *Nutr. Rev.* 80:16481663.

Qiang, X., et al. 2012. *Piriformospora indica* - a mutualistic basidiomycete with an exceptionally large plant host range. *Mol. Plant Pathol.* 13:508–518.

Raaijmakers, J.M., et al. 2010. Natural functions of lipopeptides from *Bacillus* and *Pseudomonas:* more than surfactants and antibiotics. *FEMS Microbiol. Rev.* 34:1037–1062.

Raaijmakers, J.M. and Weller, D.M. 1998. Natural plant protection by 2,4-diacetylphloroglucinol- producing *Pseudomonas* spp. in take-all decline soils. *Mol. Plant-Microbe Interact.* 11:144152.

Rabiey, M., et al. 2017. Potential ecological effects of *Piriformospora indica,* a possible biocontrol agent, in UK agricultural systems. *Biol. Control.* 104:1–9.

Rajer, F.U., et al. 2022. *Bacillus* spp.-mediated growth promotion of rice seedlings and suppression of bacterial blight disease under greenhouse conditions. *Pathogens.* 11:16.

Rasmussen, J.A. and Einhellig, F.A. 1977. Synergistic inhibitory effects of p-coumaric and ferulic acids on germination and growth of grain sorghum. *J. Chem. Ecol.* 3:197–205.

Reddy, M.V.B., et al. 1999. Chitosan treatment of wheat seeds induces resistance to *Fusarium graminearum* and improves seed quality. *J. Agricult. Food Chem.* 47:1208–1216.

Redman, R.S., et al. 2011. Increased fitness of rice plants to abiotic stress via habitat adapted symbiosis: a strategy for mitigating impacts of climate change. *PLoS One.* 6:e14823.

Riber, L., et al. 2023. Pheno- and genotyping of three novel bacteriophage genera that target a wheat phyllosphere *Sphingomonas* genus. *Microorganisms.* 11:17.

Roget, D.K. 1995. Decline in root rot *(Rhizoctonia solani* AG-8) in wheat in a tillage and rotation experiment at Avon, South Australia. *Aust. J. Exp. Agric.* 35:10091013.

Salmond, G.P.C. and Fineran, P.C. 2015. A century of the phage: past, present and future. *Nat. Rev. Microbiol.* 13:777–786.

Sasaki, A., et al. 2006. Infection of *Rosellinia necatrix* with purified viral particles of a member of *Partitiviridae* (RnPV1-W8). *Arch. Virol.* 151:697707.

Saunders, D.G.O., et al. 2019. Tackling the re-emergence of wheat stem rust in Western Europe. *Commun. Biol.* 2:3.

Savary, S., et al. 2019. The global burden of pathogens and pests on major food crops. *Nat. Ecol. Evol.* 3, 430–439.

Schouten, A. 2018. Natural antifungal compounds for the control of diseases in wheat and other cereals. In *Integrated disease management of wheat and barley*. R. Oliver, editor. Burleigh Dodds Science Publishing Limited, Cambridge.

Schouten, A. 2019. Saving resources: The exploitation of endophytes by plants for the biosynthesis of multifunctional defense compounds. *In* Endophyte biotechnology - Potential for agriculture and pharmacology. S. A., editor. CABI, Wallingford, UK. 122–144.

Schouten, A., et al. 2004. Defense responses of *Fusarium oxysporum* to 2,4-diacetylphloroglucinol, a broad-spectrum antibiotic produced by *Pseudomonas fluorescens. Mol. Plant-Microbe Interact.* 17:1201–1211.

Serfling, A., et al. 2007. Performance of the biocontrol fungus *Piriformospora indica* on wheat under greenhouse and field conditions. *Phytopathology.* 97:523–531.

Shakya, D.D. and Chung, H.S. 1983. Detetction of *Pseudomonas avenae* in rice seed. *SeedSci. Technol.* 11:583–587.

Shakya, D.D., et al. 1986. Transmission of *Pseudomonas avenae,* the causal agent of bacterial stripe of rice. *J. Phytopathol.-Phytopathol. Z.* 116:92–96.

Sharma, P.K., et al. 2012. Field demonstration of *Trichoderma harzianum* as a plant growth promoter in wheat *(Triticum aestivum* L). *J. Agric. Sci.* 4:65–73.

Shipton, P.J. 1972. Take-all in spring-sown cereals under continuous cultivation: disease progress and decline in relation to crop succession and nitrogen. *Ann. Appl. Biol.* 71:33–46.

Shoresh, M. and Harman, G.E. 2008. The molecular basis of shoot responses of maize seedlings to *Trichoderma harzianum* T22 inoculation of the root: a proteomic approach. *Plant Physiol.* 147:2147–2163.

Shoresh, M., et al. 2010. Induced systemic resistance and plant responses to fungal biocontrol agents. In *Annual review of phytopathology*. Vol. 48. N.K. VanAlfen, G. Bruening, and J.E. Leach, editors. 21–43.

Søltoft, M., et al. 2008. Benzoxazinoid concentrations show correlation with Fusarium Head Blight resistance in Danish wheat varieties. *Biochem. Syst. Ecol.* 36:245–259.

Song, Y.Y., et al. 2011. Induction of DIMBOA accumulation and systemic defense responses as a mechanism of enhanced resistance of mycorrhizal corn *(Zea mays* L.) to sheath blight. *Mycorrhiza*. 21:721731.

Sood, M., et al. 2020. *Trichoderma:* The "secrets" of a multitalented biocontrol agent. *Plants-Basel.* 9:25.

Stack, R.W., et al. 2003. History of fusarium head blight with emphasis on North America. In *Fusarium head blight of wheat and barley*. L.K. J. and W.R. Bushnell, editors. APS Press, St. Paul, MN, USA, 1–34.

Stone, E., et al. 2019. Understanding and exploiting phage-host interactions. *Viruses-Basel.* 11:26.

Suttle, C. A. 1994. The significance of viruses to mortality in aquatic microbial communities. *Microb. Ecol.* 28: 237–243.

Svircev, A., et al. 2018. Framing the future with bacteriophages in agriculture. *Viruses-Basel.* 10:13.

Tambong, J.T. 2022. Bacterial pathogens of wheat: symptoms, distribution, identification, and taxonomy. In Wheat. M.-u.-R. Ansari, editor. IntechOpen, London, 105–128.

Tan, R.X. and Zou, W.X. 2001. Endophytes: a rich source of functional metabolites. *Nat. Prod. Rep.* 18:448–459.

Tanwir, F., et al. 2017. Biosynthesis and chemical transformation of benzoxazinoids in rye during seed germination and the identification of a rye *Bx6-like* gene. *Phytochem.* 140:95–107.

Thomas, R. 1935. A bacteriophage in relation to Stewart's disease of corn. *Phytopathology* 25:371372.

Tian, B.N., et al. 2020. A cosmopolitan fungal pathogen of dicots adopts an endophytic lifestyle on cereal crops and protects them from major fungal diseases. *Isme J.* 14:3120–3135.

Tian, D.D., et al. 2022. Small secreted peptides encoded on the wheat *(Triticum aestivum* L.) genome and their potential roles in stress responses. *Front. Plant Sci.* 13:18.

Vargas, W.A., et al. 2008. Dimerization controls the activity of fungal elicitors that trigger systemic resistance in plants. *J. Biol. Chem.* 283:19804–19815.

Verma, S., et al. 1998. Piriformospora indica, gen. et sp. nov., a new root-colonizing fungus. *Mycologia*. 90:896–903.

Virtanen, A.I. and Hietala, P.K. 1955. 2(3)-Benzoxazolinone, an anti-fusarium factor in rye seedlings. *Acta Chem. Scand.* 9:1543–1544.

Viterbo, A., et al. 2005. Trichoderma mitogen-activated protein kinase signaling is involved in induction of plant systemic resistance. *Appl. Environ. Microbiol.* 71:6241–6246.

Vu, N.T. and Oh, C.S. 2020. Bacteriophage usage for bacterial disease management and diagnosis in plants. *Plant Pathol. J.* 36:204–217.

Wahlroos, O. and Virtanen, A.I. 1959. The precursors of 6-methoxy-benzoxazolinone in maize and wheat plants, their isolation and some of their properties. *Acta Chem. Scand.* 13:1906–1908.

Wang, S.C., P.F. Li, J.Z. Zhang, D.W. Qiu, and L.H. Guo. 2016. Generation of a high resolution map of sRNAs from *Fusarium graminearum* and analysis of responses to viral infection. *Sci. Rep.* 6:12.

Wilkes, M.A., et al. 1999. Hydroxamic acids in cereal roots inhibit the growth of take-all. *Soil Biol. Biochem.* 31:1831–1836.

Wiseman, B.M., et al. 1996. Suppression of *Rhizoctonia solani* anastomosis group 8 in Australia and its biological nature. *Soil Biol. Biochem.* 28:727–732.

Xin, A.Z. and Herburger, K. 2021. Precursor biosynthesis regulation of lignin, suberin and cutin. *Protoplasma*. 258:1171–1178.

Yang, P., et al. 2019. Fungal resistance mediated by maize wall-associated kinase ZmWAK-RLK1 correlates with reduced benzoxazinoid content. *New Phytologist*. 221:976–987.

Yin, Y.N., et al. 2023. Fungicide Resistance: Progress in Understanding Mechanism, Monitoring, and Management. *Phytopathology*. 113:707–718.

Yu, K., et al. 2019. Beneficial microbes going underground of root immunity. *Plant Cell Environ.* 42:2860–2870.

Yu, X., et al. 2010. A geminivirus-related DNA mycovirus that confers hypovirulence to a plant pathogenic fungus. *Proc. Natl. Acad. Sci. USA*. 107:8387–8392.

Yuan, Y.H. and Gao, M.Y. 2017. Jumbo bacteriophages: an overview. *Frontiers in Microbiology*. 8:9.

Zhang, H.X., et al. 2020a. A 2-kb mycovirus converts a pathogenic fungus into a beneficial endophyte for *Brassica* protection and yield enhancement. *Mol. Plant*. 13:1420–1433.

Zhang, L.H., et al. 2020b. Molecular characterization of a novel strain of Fusarium graminearum Virus 1 Infecting *Fusarium graminearum*. *Viruses-Basel*. 12:10.

Zhang, J.L., et al. 2021. *Trichoderma:* A treasure house of structurally diverse secondary metabolites with medicinal importance. *Front. Microbiol.* 12:21.

Chapter 9

Advances in the use of biocontrol agents to control diseases in potatoes

Jianjun Hao and Lijie Teng, University of Maine, USA; and Xiaoyu Zhang, Inner Mongolia Agricultural University, China

1 Introduction: rationale for using biological control

The potato, globally recognized as the third most important food crop for human consumption after rice and wheat, reached a production of over 380 million tons in 2020 (FAO, 2020, 2024; Johnson, 2008; Love et al., 2020). Being an underground tuber, the potato faces threats from a spectrum of diseases throughout its entire growth period, including foliar diseases such as late blight, which caused Irish Potato Famine, early blight, gray mold, white mold, blackleg, and soilborne diseases such as soft rot, Fusarium wilt, nematodes, and powdery scab (Akino et al., 2014; Ge et al., 2021; Hao et al., 2009; Johnson, 2008; Love et al., 2020; Perry et al., 1998; Ristaino, 2002). Most of these diseases can cause significant yield losses and impact the quality of the harvested crop. Besides direct losses, potato diseases can also lead to indirect losses due to reduced marketability, increased costs associated with disease management, and potential impacts on food security (Maas, 1975)

Conventionally, the control of potato diseases has relied on synthetic fungicides (Barkov et al., 2023; Shattock, 2002; Tsror (Lahkim) et al., 2021).

http://dx.doi.org/10.19103/AS.2024.0142.20

However, other cultural practices, such as cultivating resistant varieties and crop rotation, have been commonly implemented in potato production (Larkin & Brewer, 2020; Larkin & Halloran, 2015; Maas, 1975; Teng et al., 2024). While fungicides are effective, they pose several challenges, including the risk of pathogen resistance, environmental contamination, and adverse effects on non-target organisms (Zhang et al., 2019). Moreover, the cost of chemical fungicides and the need for specialized application equipment, like those required for soil fumigation, can be prohibitive for small-scale farmers (Maas, 1975). Public health implications of synthetic pesticides also drive the demand for safer, more eco-friendly disease control methods. Consequently, the focus is shifting toward innovative approaches such as bioprotectants and biocontrol agents for effective and environmentally responsible potato disease management (Jin et al., 2023; Lahlali et al., 2022).

Recent years have seen increasing interest in using biocontrol agents to control potato diseases, with promising results reported in field trials (Diallo et al., 2011; Lahlali et al., 2022). This chapter aims to provide an overview of these advances, including the types of biocontrol agents used for controlling soilborne and foliar diseases in potatoes, the mechanisms by which they work, and the latest research and development efforts in this area.

2 Biocontrol agents for controlling potato diseases

Biological control or biocontrol agents (BCAs) are considered microorganisms or substances derived from living organisms that can control plant diseases by suppressing the growth and spread of plant pathogens. These BCAs microorganisms include but are not limited to bacteria, fungi, oomycetes, viruses, and nematodes (Abd-El-Khair & El-Nagdi, 2013; Fravel, 2005; Hough et al., 2023; Kenneth & Baker, 1975; Lahlali et al., 2022; Narayanasamy, 2013; Shafi et al., 2017; Siddiqui & Mahmood, 1999) (Table 1). In the following context, some organisms mentioned are not specifically reported in association with potato diseases but are included in this review as illustrative examples (Table 2).

2.1 Bacterial BCAs

Bacteria including Proteobacteria and Firmicutes are among the most widely studied and utilized organisms as biocontrol agents due to their ability to suppress plant pathogens through various mechanisms (Lahlali et al., 2013) (Table 1). Prominent among these are the *Bacillus* species, particularly *B. subtilis* and *B. amyloliquefaciens* (Meng & Hao, 2017; Shafi et al., 2017). These bacteria produce a diverse array of secondary metabolites, including antibiotics, enzymes, and lipopeptides, which inhibit the growth of pathogens such as *Fusarium*, *Rhizoctonia*, and *Phytophthora* spp. *Bacillus firmus* controls nematodes like *Meloidogyne* spp. and *Globodera* spp. (Ghahremani et al.,

Table 1 Representative biological control agents (BCAs) and products on potatoes

Type	Agent	Target diseases/pathogens	Mode of action	Reference
Bacteria	*Bacillus amyloliquefaciens; B. velezensis; B. licheniformis; B. subtilis; B. thuringiensis*	*Streptomyces* spp.; *Pectobacterium carotovorum; Ralstonia solanacearum; Alternaria* spp.; *Phytophthora infestans*; *Rhizoctonia solani*; *Sclerotium rolfsii*	Antagonism (producing glycolipids, siderophores, lipopeptides, surfactin, iturin, fengycin, cell-wall degrading enzymes (CWDEs), antifungal chitinase, and antibiotics); inducing systemic resistance and enhancing plant growth	(Azaiez et al., 2018; Bahmani et al., 2021; Ben Khedher et al., 2015; Lin et al., 2018; Meng et al., 2012; Reyes-Ramírez et al., 2004)
Bacteria	*Enterobacter cloacae*	*Fusarium* spp.; *Gibberella pulicaris; Ralstonia solanacearum*	Inducing plant systemic resistance	(Mohamed et al., 2020; Schisler et al., 2000)
Bacteria	*Paenibacillus peoriae*	*Ralstonia solanacearum*	Antagonism and growth enhancement	(Bahmani et al., 2021)
Bacteria	*Pseudomonas brassicacearum; P. fluorescence; P. putida*	*Clavibacter michiganensis* subsp.*sepedonicus*; *Fusarium* spp.; *Pectobacterium* spp.; *Pythium aphanidermatum; Ralstonia solanacearum; Rhizoctonia solani; Streptomyces* spp.; *Phytophthora capsici*; *Phytophthora infestans; Potato Virus Y*	Antibiosis (enzymes: chitinase and β-1,3-glucanase; hydrogen cyanide HCN; secondary metabolites interferer; piericidin A; phenazines (PCN) and cyclic lipopeptides (sessilins and orfamides); nonribosomal peptides (nunamycin and nunapeptin); induced systemic resistance	(de la Cruz et al., 1992; Naveen Kumar, 2007)
Bacteria	*Streptomyces hygroscopicus*	*Phytophthora infestans*	Antimicrobiosis (enzymatic activities)	(Jin et al., 2023)
Fungi	*Clonostachys* spp.	*Alternaria solani; A. grandis*	Antagonism (enzymes); mycoparasitism	(da Silva et al., 2021)
Fungi	*Coniothyrium minitans*	*Sclerotinia sclerotiorum*	Mycoparasitism; sclerotial degradation; antagonistic metabolites and CWDEs; niche competition; systemic resistance	(Budge & Whipps, 2001; Ojaghian, 2010; Zeng, 2012b)
Fungi	*Epicoccum nigrum* *Mixture of arbuscular mycorrhizal fungi*	*Phytophthora infestans;* *Pectobacterium atrosepticum*	Antagonisis; systemic plant resistance	(Bagy et al., 2019; Li et al., 2013)

(*Continued*)

Table 1 (*Continued*)

Type	Agent	Target diseases/pathogens	Mode of action	Reference
Fungi	*Paecilomyces lilacinus*	Root-knot nematode	Parasitism; antagonistic metabolites; CWDEs; nutrient competition; plant defense enhancement	(Antas et al., 2012; Khan et al., 2003; Musa et al., 2023)
Fungi	*Rhizoctonia zeae*	*Rhizoctonia solani*	Mycoparasitism; antagonism (antifungal metabolites); CWDEs; nutrient and space competition; plant defense induction	(Escande & Echandi, 1991; Larkin & Brewer, 2020)
Fungi	*Rhodotorula* sp.	*Alternaria alternata*	Inducing systemic plant resistance; antagonism (fatty acids, carotenoids, and other secondary metabolites); competition for nutrients and space; hydrolytic enzymes;	(Yan et al., 2014)
Fungi	*Talaromyces flavum; T. asperellum; T. harzianum; atroviride*	*Sclerotinia sclerotiorum; Ralstonia solanacearum*; *Rhizoctonia solani; Fusarium* spp.; *Globodera pallida; Phytophthora infestans*	Antagonism (antifungal metabolites; enzymes); induce systemic resistance; biofilm; chitinases; mycoparasitism; competition; volatile organic compounds	(Abd-El-Khair & El-Nagdi, 2013; Ojaghian, 2011)
Fungi	*Verticillium bigutattum*	*Rhizoctonia solani*	Mycoparasitism; antagonism (secondary metabolites); CWDEs; competition; inducing plant defense	(Demirci et al., 2009)
Nematodes	*Steinernema glaseri; Heterorhabditis indica*	*Meloidogyne incognita*	Parasitism; infection and toxin; reducing the invasion of nematodes	(Javed et al., 2012)
Viruses	Bacteriophages	*Streptomyces scabies*	Lytic infection, phage adsorption, toxins, and enzymes	(Abdelrhim et al., 2021; Zhang et al., 2024)

Table 2 Examples of potato diseases managed using biological control products

Diseases	Biocontrol Agents	Disease control	Scales	References
Early blight (*Alternaria grandis*)	*Clonostachys* spp.	83.1% to 92.9%	Greenhouse	(da Silva et al., 2021)
Late blight (*Phytophthora infestans*)	6 *Streptomyces strains* & biochar *Bacillus subtilis* *Streptomyces* sp., *Trichoderma atroviride*	10% to 59.25%	Greenhouse, field, lab	(Fu et al., 2022; Jin et al., 2023)
Common scab (*Streptomyces* spp.)	*Bacillus altitudinis* *Bacillus amyloliquefaciens* Bacteriophages SscP1EGY *Enterobacter cloacae* *Pseudomonas fluorescens* *Streptomyces* sp.	30.9% to 92.4%	Greenhouse, field	(Abdelrhim et al., 2021; Al-Mughrabi, 2010; Salem & Abd El-Shafea, 2018; =Zhang et al., 2024; Zhang et al., 2020)
Soft rot (*Pectobacterium carotovorum*, formally *Erwinia carotovora* subsp. *carotovora*)	*Bacillus amyloliquefaciens* *Bacillus subtilis* *Pseudomonas aeruginosa* *Pseudomonas fluorescence* *Streptomyces* spp.	60% to 100%	Greenhouse, Lab	(Azaiez et al., 2018; Salem & Abd El-Shafea, 2018)
Bacterial wilt (*Ralstonia solanacearum*)	*Bacillus licheniformis* *Bacillus pumilus* *Chryseobacterium indologenes* *Chryseobacterium lathyri* *Enterobacter cloacae* *Microbacterium phyllosphaerae* *Paenibacillus peoriae* *Pseudomonas brassicacearum* *Pseudomonas putida* *Trichoderma asperellum*	10.7% to 55.2%	Greenhouse, field, lab	(Bahmani et al., 2021; Mohamed et al., 2020)

(*Continued*)

Table 2 (*Continued*)

Diseases	Biocontrol Agents	Disease control	Scales	References
Stem canker and black scurf (*Rhizoctonia solani*)	*Bacillus subtilis* *Cladorrhinum foecundissimum* *Laetisaria arvalis* *Paenibacillus polymyxa* *Penicillium* sp. *Pseudomonas fluorescens* *Rhizoctonia zeae* *Stilbella aciculosa* *Trichoderma harzianum* *Trichoderma virens* *Verticillium bigutattum*	14.7% to 94.3%	Greenhouse, field, lab	(Ben Khedher et al., 2015; Brewer & Larkin, 2005; Demirci et al., 2009; Mohamed et al., 2020)
Fusarium dry rot (*Fusarium* spp.)	*Enterobacter cloacae* *Pseudomonas fluorescens* *Pseudomonas fluorescens*	8.3% to 46.1%	Storage, field	(Al-Mughrabi, 2010; Schisler et al., 2000)
Root rot (*Fusarium solani*)	*Trichoderma hamatum*	Up to 25%	Greenhouse	(Abd-El-Khair & El-Nagdi, 2013)
Ring rot (*Clavibacter michiganensis* subsp. *sepedonicus*)	*Fluorescent pseudomonads*	23.4% to 26.7%	Greenhouse	(de la Cruz et al., 1992)
Root-knot nematode	*Paecilomyces lilacinus*	83.7% egg reduction and 87.5% egg reduction	field	(Abd-El-Khair & El-Nagdi, 2013)
potato stem rot (*Sclerotinia sclerotiorum*)	*Talaromyces flavum* *Trichoderma* spp.	16.4% to 70.6%	Greenhouse, lab	(Ojaghian, 2011)
silver scurf (*Helminthosporium solani*)	*Enterobacter cloacae* *Pseudomonas fluorescens*	43.3% to 88.5%	Field	(Al-Mughrabi, 2010)

2020). Another significant genus is *Pseudomonas*, notably *Pseudomonas fluorescens* (Arseneault et al., 2013, 2015; Haas & Défago, 2005), recognized for its production of siderophores and antibiotics that facilitate iron sequestration and the suppression of soilborne pathogens. Additionally, *Streptomyces* species have been effectively employed in controlling diseases like common scab in potatoes caused by *Streptomyces scabies* (Brader et al., 2017; Jin et al., 2023; Sarwar et al., 2018). Other frequently reported biocontrol bacteria include fluorescent pseudomonads, *Lysobacter*, *Paenibacillus*, *Pectobacterium*, and *Myxococcus*. For instance, *Lysobacter* is known for its antimicrobial activities, with biocontrol attributed to the bioactive metabolites it produces. Preventive applications of *Lysobacter* culture filtrate have shown a reduction in infections caused by *Phytophthora infestans* (Siddiqui & Mahmood, 1999). Bacteria such as *Cronobacter muytjensii*, *Saccharothrix texasensis*, *Methylobacterium*, and *Streptomyces melanosporofaciens* are also recognized for producing volatiles and other compounds with biocontrol activities. Besides direct pathogen inhibition, many of these bacteria induce systemic resistance and promote host plant growth (Mishra et al., 2014; Pieterse et al., 2014; Suresh et al., 2022; Van Loon et al., 1998), making them invaluable in integrated pest management strategies aimed at reducing reliance on chemical pesticides in agriculture. Meta-analysis shows that bacilli reduce disease by 60% (Serrão et al., 2024).

2.2 Fungal and oomycetal BCAs

Fungi are also widely recognized for their role as biocontrol agents, offering an eco-friendly alternative to chemical pesticides in managing plant diseases (Table 1). Mycorrhizal fungi further contribute by enhancing the overall health of potato plants, making them more resilient to root pathogens. Among the most notable fungal BCAs are species of *Trichoderma*, particularly *Trichoderma harzianum* and *T. viride* (Coventry et al., 2006; Harman, 2006; Harman et al., 2004; Ordentlich et al., 1990). These fungi are effective against a wide range of soilborne pathogens, such as *Rhizoctonia solani*, *Sclerotinia sclerotiorum*, through mechanisms like mycoparasitism, where the fungi directly attack and degrade the cell walls of pathogens, and competition for nutrients and space (Chet et al., 1981; Harman, 2006; Harman et al., 2004). *Gliocladium* spp. are also effective in controlling diseases like damping-off, caused by *Pythium* spp. (Harris & Lumsden, 1997), through their antagonistic activities. Additionally, *Coniothyrium minitans* is specifically used to combat *Sclerotinia* diseases by parasitizing the sclerotia of *Sclerotinia sclerotiorum* (Ojaghian, 2010; Zeng et al., 2012b). In addition, *Talaromyces flavum, Cladosporium rosea, Epicoccum,* binucleate *Rhizoctonia,* and *Verticillium bigutattum* also have been reported as BCAs (Bagy et al., 2019; Brewer & Larkin, 2005; Escande & Echandi, 1991;

Fravel, 2005; Ojaghian, 2011; Silva-Valderrama et al., 2021; Vanajothi et al., 2022).

Predatory fungi, also known as nematophagous fungi, have developed specialized mechanisms to trap and consume nematodes, particularly plant-pathogenic nematodes (PPNs), offering a natural form of biological control against nematode pests in the soil (Bagy et al., 2019; Druzhinina et al., 2011; Li et al., 2022; Zhang et al., 2021). The genus *Arthrobotrys* employs adhesive networks or constricting rings to capture PPNs (Tkalenko et al., 2021). Once the nematode is trapped, the fungal hyphae penetrate and digest the nematode, deriving nutrients from it.

While some oomycetes are notorious for causing devastating plant diseases, others have been explored for their potential as biocontrol agents. For example, *Pythium oligandrum* (Mulligan et al., 1995) is antagonistic against other pathogenic species of *Phytophthora* and *Pythium*. *Aphanomyces cochlioides* (Islam, 2008) is effective in controlling damping-off diseases caused by *Pythium* spp.

2.3 Viruses for biocontrol

Viruses are increasingly being explored as biocontrol agents. These biological entities, known as bacteriophages or phages, specifically target and infect bacterial pathogens, making them a highly selective and effective tool in biocontrol (Fujiwara et al., 2011; McKenna et al., 2001). For instance, bacteriophages have been successfully employed to control *Ralstonia solanacearum* in potatoes and tomatoes (Fujiwara et al., 2011). *Dickeya* spp., *Pectobacterium* spp., *Xanthomonas* spp., and *Streptomyces* spp. are also controlled by phages (Buttimer et al., 2017; Czajkowski, 2015; McKenna et al., 2001).

Mycoviruses can cause hypovirulence in their fungal hosts, meaning they reduce the virulence of the fungus, thereby lessening its impact on crops. Depending on the species, beneficial viruses target pathogens including fungi, bacteria, nematodes, oomycetes, and viruses (Hough et al., 2023; Nuss, 2005). Mild strains of *Potato Virus Y* (PVY) have been investigated for their potential to cross-protect potatoes against more severe strains or other viral diseases (Valkonen, 2007).

2.4 Entomopathogenic nematodes (EPNs) for biocontrol

Beneficial EPNs such as *Steinernema* and *Heterorhabditis* species have shown promise in suppressing root-knot nematodes (*Meloidogyne* spp.) and potato cyst nematodes (*Globodera* spp.) (Grewal et al., 2005; Perry et al., 1998). These nematodes not only reduce the nematode population but also improve soil health, making them valuable tools in sustainable agriculture, although none of the products are on the market. Nematodes like *Caenorhabditis elegans* consume various bacteria in large amounts, which has the potential to control bacterial diseases (Samuel et al., 2016).

2.5 Broader scope of biocontrol plant-based extracts and cultural practices

To a broader extent, materials derived from living organisms can be used for biocontrol, with botanical fungicides and plant extracts playing a significant role. These extracts, derived from various plant parts such as leaves, seeds, bark, and roots, possess antimicrobial, antifungal, and insecticidal properties that make them valuable tools in integrated pest management (IPM) strategies (Hostettmann, 1998; Li et al., 2023; Ouanas et al., 2017; Uppal et al., 2008). For instance, compounds like chitosan, neem oil, garlic extract, essential oils, and tannins exhibit notable antifungal activity (Bautista-Baños et al., 2006; Davis et al., 2007; Hussain et al., 2017). Beyond their direct antimicrobial effects, plant-derived compounds can induce systemic acquired resistance (SAR) in plants, providing long-lasting protection against pathogens (Abass & Fayed, 2021).

In addition, living plants themselves can be employed in disease control strategies, such as intercropping, which is effective in reducing disease incidence, improving soil health, and enhancing biodiversity (Zhu et al., 2017). Intercropping involves growing two or more crops in proximity. For example, onions and garlic release sulfur compounds into the soil, which have antifungal properties that help suppress potato pathogens (El-Anany, 2021). Intercropping marigolds with potatoes can reduce nematode populations in the soil, indirectly reducing nematode-related diseases as marigolds produce compounds like alpha-terthienyl, which are toxic to nematodes (Bairwa et al., 2017). Neem extracts have been shown to be effective against a range of soilborne pathogens, including *Phytophthora infestans*. Field trials have demonstrated that the use of neem-based biocontrol products can reduce the incidence of soilborne diseases in potatoes and increase potato yield (Adhikari et al., 2023; Mahmood et al., 2022).

3 Mechanisms of biological control: direct actions

BCAs suppress plant diseases through various mechanisms of action, which can be grouped into direct and indirect actions (Lahlali et al., 2022) (Table 1).

3.1 Competition for nutrients and space

Some BCAs outcompete plant pathogens for essential nutrients and spaces on the plant surface such as rhizosphere, endosphere, and phyllosphere, effectively suppressing pathogen growth and colonization (Brader et al., 2017; Compant et al., 2005; Köhl et al., 2019). One key mechanism involves the production of siderophores, which have a high affinity for iron (Fe^{3-}). Siderophores bind and solubilize iron from the environment, which is crucial because iron is typically present in insoluble forms that are not readily available for uptake by organisms (Weller, 1988). *Pseudomonas fluorescens* produces siderophores such as pyoverdine, which competes with pathogens like *Pythium* spp. for iron (Cornelis & Andrews, 2010; Kloepper et al., 1980). By binding and transporting iron, siderophores deprive pathogens of this critical nutrient, thus inhibiting their growth. Pseudomonads, some bacilli (Krewulak & Vogel, 2008), and *Azospirillum brasilense* (Bashan et al., 2013) produce siderophores (Compant et al., 2005; Köhl et al., 2019). This competition for nutrients and space effectively suppresses the ability of pathogens to thrive and cause diseases (Abo-Zaid et al., 2023; Haas & Défago, 2005; Krewulak & Vogel, 2008). Some non-pathogenic *Streptomyces* spp. can suppress and compete with pathogenic strains to occupy infection sites and inhibit pathogen colonization (Neeno-Eckwall et al., 2001; Ryan & Kinkel, 1997).

3.2 Antagonism

The production of antimicrobial compounds is a critical mechanism by which BCAs suppress plant pathogens through antagonism. The antagonistic compounds include antibiotics or antimicrobial substances, enzymes, siderophores, peptides, and secondary plant metabolites, which directly inhibit or kill pathogens, preventing their growth and colonization (Czajkowski et al., 2012; Haas & Défago, 2005; Millard & Taylor, 1927) (Table 1).

Biocontrol agents produce various types of enzymes, such as lytic enzymes chitinases and glucanases. β-1,3-glucanase breaks down glucans present in the cell walls of many pathogenic fungi, and chitinases break down chitin, which is a major component of fungal cell walls, thereby directly inhibiting their growth. The *Pseudomonas* genus produces chitinases and glucanases that inhibit *Phytophthora capsici* and *Rhizoctonia solani* (Naveen Kumar, 2007).

There are many non-enzymatic substances and secondary metabolites that inhibit or suppress plant pathogens. For example, *Pseudomonas* produces metabolites like phenazines that have antimicrobial properties and inhibit pathogen growth through redox activity and interference with cellular processes, pyoluteorin that disrupt pathogens' cellular functions, 2,4-diacetylphloroglucinol (DAPG) that has antifungal properties and inhibits pathogen growth through various mechanisms including interference with microbial metabolism, and cyclic lipopeptides (surfactin, iturin, and fengycin, etc.) which exhibit antimicrobial properties and disrupt pathogen cell membranes (Arseneault et al., 2015; Clough et al., 2024; Ma et al., 2016;

Michelsen et al., 2015; Oni et al., 2019; Raaijmakers et al., 2006). Hydrogen cyanide (HCN) is a volatile organic compound that acts as a toxicant to pathogens and can inhibit their growth.

Pseudomonas chlororaphis subsp. *aureofaciens* can produce PCA, a phenazine compound that effectively inhibits some fungal pathogens, like some species of *Phytophthora, Fusarium, Pythium,* and *sclerotinia* (Delaney et al., 2001; Oni et al., 2019). *Bacillus subtilis* produces a variety of antifungal compounds, such as lipopeptides (surfactin and fengycin), and cyclic peptides. These compounds can inhibit the growth of fungal pathogens in the soil. Surfactin disrupts the cell membranes of pathogens and inhibits their growth (Lin et al., 2018). *Bacillus velezensis* produces iturin A being a key functional metabolite against *P. infestans*. This compound disrupts the pathogen's cell membrane, causing oxidative stress and mitochondrial damage. *Bacillus velezensis* strain BAC03 produces a polypeptide that inhibits many pathogens including *Streptomyces scabies* (Meng et al., 2012). The *Lysobacter* genus is recognized for its antimicrobial activities, with biocontrol attributed to the bioactive metabolites they produce. These include compounds such as pyrazines and pyrrole, which have been found to be effective against *P. infestans* (Islam, 2008; Kim et al., 2019). Many bacteria and fungi produce volatile organic compounds (VOCs) with biocontrol efficacy, including well-known molecules like hydrogen cyanide and novel compounds such as dimethyl disulfide and various alkenes (Meng et al., 2012).

3.3 Signaling interference

BCAs can interfere with pathogen signaling, a key mechanism of their antagonistic activity (Anand et al., 2023; Dong et al., 2007). Pathogens rely on signaling systems for their virulence, such as quorum sensing to coordinate activities, the production of toxins, biofilm formation, and host colonization. BCAs produce compounds that can mimic these signaling molecules, effectively disrupting the pathogen's communication network. These mimic compounds bind to the pathogen's receptors but do not activate them, thereby blocking the real signals. This competitive inhibition prevents the pathogen from detecting and responding to its own signals, leading to a disruption in coordinated behaviors necessary for pathogenicity, such as toxin production and surface attachment. By interrupting these crucial signaling pathways, BCAs can effectively suppress pathogen growth and reduce their ability to cause disease.

Pathogens use quorum sensing to regulate gene expression in response to population density (Dong et al., 2007). BCAs can produce quorum quenching enzymes that degrade these signaling molecules. For example, certain *Bacillus* and *Pseudomonas* species produce lactonases and acylase enzymes that hydrolyze N-acyl-homoserine lactones (AHLs), common quorum sensing molecules in Gram-negative bacteria (Dong et al., 2007; Hosseinzadeh et al., 2017; Teplitski et al., 2003). By degrading AHLs, BCAs disrupt the communication network of pathogens, preventing the expression of genes responsible for virulence and biofilm formation. Hydrogen cyanide (HCN) produced by some *Pseudomonas* species interferes with the signaling pathways of competing microbes. This disruption prevents the pathogens from coordinating activities essential for their survival and virulence (Anand et al., 2023). Biosurfactants and other metabolites produced by BCAs disrupt pathogen biofilms, which are structured communities of microorganisms protected by an extracellular matrix. By breaking down biofilms, BCAs expose pathogens to environmental stress and host immune responses, reducing their survival and virulence (Davey et al., 2003).

3.4 Predation and parasitism

Parasitism entails a parasite living on or in a host organism, ultimately leading to its demise. Some can penetrate the cell walls of fungi or consume bacterial cells (Druzhinina et al., 2011). For example, *Coniothyrium minitans* colonizes on the sclerotia of *Sclerotinia sclerotiorum* and degrade the entire structure (Zeng et al., 2012a, 2012b). *Pochonia chlamydosporia* and *Pasteuria penetrans* parasitizes the eggs of root-knot nematodes, thereby controlling their spread in the soil (Tobin et al., 2008). *Pythium oligandrum* operates through several mechanisms, including mycoparasitism, where it parasitizes the hyphae of pathogenic fungi and oomycetes, effectively reducing their populations in the soil (Mulligan et al., 1995).

Certain fungi have evolved fascinating strategies to trap and consume PPNs, providing a natural form of biological control against these soil-dwelling pests. These fungi, often referred to as nematophagous fungi, employ various trapping mechanisms to capture nematodes. For example, *Arthrobotrys oligospora* forms adhesive networks or loops that ensnare nematodes, such as *Globodera* (potato cyst nematodes) and *Meloidogyne* species (root-knot nematodes), as they move through the soil (Neuenschwander, 2003; Wang et al., 2018). Once a nematode is trapped, the fungus penetrates its cuticle with specialized hyphae, eventually digesting and absorbing the nematode's nutrients. *Monacrosporium* species, also known as orbilia, are another group of nematophagous fungi. These fungi utilize constricting rings to capture their nematode prey. *Dactylellina* (formerly *Dactylaria*) produces sticky knobs or constricting rings that actively trap nematodes (Lei et al., 2023). Once the nematode is captured, the fungal hyphae invade its body, digesting it from the inside out. Upon contact, the rings swell rapidly, tightening around the nematode and holding it in place while the fungus digests it.

Mononchus spp. are free-living, predatory nematodes that prey on other nematodes, including plant-parasitic nematodes. The nematode *Caenorhabditis elegans* helps maintain a healthy soil microbiome by feeding on bacteria, potentially limiting the growth of some harmful bacterial species, although it is not commercially used as a biocontrol agent (Akyazi et al., 2012).

4 Mechanisms of biological control: indirect actions

4.1 Induction of plant resistance

BCAs play a crucial role in enhancing plant defense mechanisms through induced systemic resistance (ISR) or systemic acquired resistance (SAR), enabling plants to better withstand a wide range of pathogens (Kloepper et al., 2004; Pieterse et al., 2014; Van Loon et al., 1998). Various factors can elicit or enhance a plant's systemic resistance, including signal molecules, volatile organic compounds (VOCs), extreme temperatures, pH levels, and pathogen infections (Akyazi et al., 2012).

BCAs stimulate the production of plant-derived compounds like phytoalexins and other defensive chemicals that interfere with pathogen signaling and inhibit infections. For instance, VOCs such as acetoin and 2,3-butanediol produced by *Bacillus* species inhibit pathogen growth and induce systemic resistance. Additionally, *Bacillus subtilis* and *B. amyloliquefaciens* have been reported to stimulate ISR by surfactin, a nonvolatile elicitor (Cawoy et al., 2014).

There are some examples. *Enterobacter cloacae* PS14 has demonstrated ISR induction in potatoes against *Ralstonia solanacearum* (Mohamed et al., 2020). *Pseudomonas fluorescens* and *Rhodotorula* sp. have shown efficacy in inducing resistance against *Potato Virus Y (*PVY*)* in field trials. *Trichoderma harzianum* enhances plant resistance by colonizing root systems and producing elicitors that activate the plant's defense pathways (Harman, 2006; Harman et al., 2004). *Trichoderma asperellum* T34 induces systemic resistance in potatoes against *Ralstonia solanacearum* (Mohamed et al., 2020).

EPNs such as *Steinernema* spp. contribute to plant resistance by releasing metabolites absorbed by plant roots, indirectly boosting the plant's defenses (Kerry, 2000). The production of siderophores (e.g. pyoverdin, pyochelin, and salicylic acid), antibiotics/ surfactins (e.g. lipopeptides), and enzymes (e.g. chitinases, proteases, and lipases) by BCAs, which stimulate the plant's immune response (Cawoy et al., 2014; Van Loon et al., 1998). Leucinostatins A and B produced by the fungus *Purpureocillium lilacinum* effectively inhibit the growth of *P. infestans*.

Key molecules and enzymes involved in plant defense mechanisms and stress tolerance include salicylic acid (SA), jasmonic acid (JA), ethylene (ET), indole-3-acetic acid (IAA), abscisic acid (ABA), cytokinin, chitinase, glucanase, brassinosteroids (BRs), polyamines, proline, gibberellins (GAs), nitric oxide

(NO), hydrogen peroxide (H_2O_2), glutathione (GSH), trehalose, jasmonates (JAs), and betaines.

Salicylic acid (SA) plays a central role in SAR, inducing the production of pathogenesis-related (PR) proteins like chitinases and glucanases, which degrade pathogen cell walls, enhancing the plant's ability to fend off infections. Additionally, supplementing with SA when using *Bacillus subtilis* strains 10-4 and 26D to control *Phytophthora infestans* and *Fusarium oxysporum* has been shown to enhance penetration and colonization on potato tubers during storage (Lastochkina et al., 2020).

Jasmonic acid (JA) is involved in defense responses against necrotrophic pathogens and herbivorous insects, triggering the production of defensive compounds such as protease inhibitors and secondary metabolites. Ethylene modulates plant responses to both biotic and abiotic stresses, often working in conjunction with SA and JA to regulate stress-related gene expression and promote the senescence of infected tissues.

Indole-3-acetic acid (IAA), a form of auxin, regulates growth and development in response to environmental cues, modulating root architecture to enhance water and nutrient uptake, particularly under stress conditions. Auxin indirectly enhances potato resistance to *Streptomyces scabiei* by inhibiting the toxicity of Thaxtomin A (Tegg et al., 2008).

4.2 Plant growth promotion

BCAs are instrumental in both suppressing plant diseases and enhancing plant growth through various mechanisms. Initially discovered in rhizosphere bacteria, these plant growth-promoting rhizobacteria (PGPR) improve plant health by enhancing nutrient uptake and producing growth-stimulating hormones. This concept extends to other microorganisms, including fungi, collectively termed plant growth-promoting microorganisms (PGPM).

For example, *Azospirillum brasilense* enhances root growth and nutrient absorption by producing auxins and other phytohormones (Bashan & de-Bashan, 2010). *Bacillus subtilis* produces a range of compounds, including antibiotics, enzymes, and phytohormones, that stimulate root growth, enhance nutrient uptake, and improve plant resilience to stress (Kloepper et al., 2004). Furthermore, siderophores produced by BCAs improve iron availability to plants, promoting better plant health and resistance to diseases. Siderophores produced by BCAs can indirectly promote plant growth by enhancing the availability of iron to plants, especially in iron-limited soils. Plants can take up the iron-siderophore complex through specific receptors, utilizing the iron for their metabolic needs, which leads to improved plant health and increased resistance to diseases. Additionally, Indole-3-Acetic Acid (IAA), a phytohormone produced by many BCAs, promotes plant growth by altering root architecture, thus enhancing nutrient uptake.

Fungi, like arbuscular mycorrhizal fungi (AMF), form mutualistic associations with plant roots, increasing surface area for water and nutrient absorption, particularly phosphorus, which is crucial for plant growth (Smith & Read, 2008). *Trichoderma* spp., beyond their biocontrol role, secrete enzymes and secondary metabolites that enhance root growth, improve nutrient uptake, and induce systemic resistance in plants (Harman et al., 2004).

4.3 Stress alleviation

BCAs help plants adapt to various abiotic stresses, including extreme temperatures, salinity, and drought, by promoting growth through mechanisms such as increasing proline levels, enhancing phytohormone production, and upregulating stress-responsive genes. For instance, *Bacillus amyloliquefaciens* has been shown to improve salt stress tolerance in plants like rice and *Arabidopsis thaliana* by promoting root growth and nutrient uptake (Chauhan et al., 2019; Liu et al., 2017; Nautiyal et al., 2013; Vacheron et al., 2015). BCAs can also assist plants in coping with abiotic stress, making them more resistant to diseases (Yang et al., 2009). *Pseudomonas fluorescens* and *Azospirillum brasilense* enhance plant tolerance to drought by improving water uptake and increasing the production of stress-related hormones such as abscisic acid (ABA) (Vardharajula et al., 2011). *Bacillus subtilis* and *Trichoderma harzianum* help plants cope with salinity by enhancing potassium uptake and modulating the ionic balance within plant cells (Ruzzi & Aroca, 2015).

Certain BCAs can mitigate heavy metal stress by reducing the uptake of toxic metals and enhancing the availability of essential nutrients. For example, arbuscular mycorrhizal fungi (AMF) form symbiotic relationships with plant roots, sequestering heavy metals in their hyphae and preventing translocation to the plant (Göhre & Paszkowski, 2006). This symbiosis also promotes the uptake of essential nutrients like phosphorus and zinc, often limited in contaminated soils.

In response to temperature stress, BCAs such as *Pseudomonas putida* and *Bacillus amyloliquefaciens* produce heat shock proteins and other stress-related proteins that protect cellular structures and enhance antioxidant enzyme expression, mitigating oxidative stress caused by temperature fluctuations (Zhang et al., 2008). BCAs are thus critical in helping plants alleviate various abiotic stresses, thereby enhancing growth and productivity. By improving root development, nutrient uptake, and stress-related hormonal balance, BCAs are indispensable in sustainable agriculture practices.

4.4 Symbiosis

Mycorrhizae refer to the symbiotic associations between certain fungi and plant roots, playing a crucial role in plant health and soil ecology. These fungi, particularly arbuscular mycorrhizal fungi (AMF) like *Glomus* spp., penetrate the plant root cells, forming structures known as arbuscules, which facilitate the exchange of nutrients (Hashem et al., 2018; Smith & Read, 2008). Through this symbiosis, the fungi receive carbohydrates produced by the plant through photosynthesis, while the plant benefits from enhanced nutrient uptake, particularly phosphorus and other essential minerals. Mycorrhizae also improve soil structure by binding soil particles together, thereby increasing soil porosity and water retention. Moreover, these fungi contribute to plant disease resistance by outcompeting root pathogens for space and resources and by triggering the plant's systemic resistance mechanisms. This relationship is especially beneficial in nutrient-poor soils, where mycorrhizae can significantly boost plant growth and resilience. As a result, mycorrhizal fungi are increasingly being utilized as biocontrol agents in sustainable agriculture, reducing the need for chemical fertilizers and pesticides, and promoting healthier, more resilient crops. Their role in enhancing soil health and plant productivity makes them a vital component of integrated pest management (IPM) strategies aimed at achieving sustainable agricultural systems.

4.5 Environmental modification

BCAs can modify the environment in ways that enhance plant growth and health while creating less favorable conditions for pathogen survival. These modifications include improving soil structure, altering soil pH, and enhancing nutrient availability, which collectively create a more conducive environment for plant development.

Arbuscular mycorrhizal fungi (AMF) produce glomalin, a glycoprotein that binds soil particles together, improving soil porosity and water retention (Rillig, 2004). Certain mycorrhizal fungi can help stabilize soil pH in acidic soils, creating a more neutral environment that is conducive to plant growth. *Bacillus* species produce organic acids that help lower soil pH in alkaline environments, thereby making nutrients more available to plants (Hayat et al., 2010). Phosphate-solubilizing bacteria (PSB) such as *Pseudomonas* spp. and *Bacillus* spp. produce organic acids that solubilize bound phosphates in the soil, making them accessible to plants (Rodríguez & Fraga, 1999). Similarly, nitrogen-fixing bacteria like *Rhizobium* and *Azospirillum* convert atmospheric nitrogen into forms that plants can readily use, thus enhancing nitrogen availability and reducing the need for synthetic fertilizers (Rillig, 2004).

Trichoderma spp. and *Bacillus subtilis* play a crucial role in the decomposition of organic matter (Zhai et al., 2020; Zin & Badaluddin, 2020). These microorganisms produce enzymes such as cellulases and proteases that break down organic residues into simpler compounds, enriching the soil with humus and essential nutrients. This decomposition process not only improves soil fertility but also promotes the growth of beneficial microbial communities, creating a healthier rhizosphere (Harman et al., 2004).

By modifying the environment, BCAs can effectively suppress soilborne diseases. The introduction of beneficial microorganisms can outcompete pathogenic microbes for space and resources, reducing the incidence of diseases such as root rot and wilt (Whipps, 2001). Additionally, the production of antimicrobial compounds by BCAs can directly inhibit pathogen growth, further enhancing plant health. BCAs also employ other strategies such as producing enzymes that degrade pathogen virulence factors or altering the pH to create inhospitable conditions for pathogens (Budge & Whipps, 2001; Chet et al., 1981; Whipps, 2001). For instance, organic acids produced by BCAs can lower the pH of the environment, which is often detrimental to pathogen survival. Siderophore production by BCAs allows them to thrive in iron-scarce environments, giving them a competitive advantage in diverse ecological niches.

5 Commercial applications of biocontrol

5.1 Advanced biotechnological approaches

Advances in technology and knowledge have led to the development of new and improved biocontrol agents that are more effective, cost-efficient, and sustainable (De Corato, 2021; Hills et al., 2020; Li et al., 2022; Wang & Long, 2023) (Table 3). The realm of biocontrol agents is experiencing exciting progress, thanks to cutting-edge research that has unveiled new microorganisms with powerful anti-pathogen properties, genetic enhancements to existing agents, refined formulations, and the strategic use of bioinformatics and genomics to fine-tune microbial consortia (Serrão et al., 2024). Breakthroughs in bioinformatics and genomics have greatly improved our ability to identify and understand the microbial populations in soil and on plant surfaces. This rich repository of data allows for the careful selection or genetic engineering of microbial consortia that offer optimal defense against plant pathogens. For instance, the discovery of specific microbes that bolster plant growth and resilience has given rise to microbial inoculants that not only boost crop yields but also lower disease rates.

5.2 Challenges and opportunities in the development and commercialization of biocontrol products

5.2.1 Commercialization

There are many challenges in the application of BCAs. A major obstacle to moving from laboratory research to the market commercialization of microbial pesticides is complexity and long-time processing of the required studies for product registration.

Table 3 Selected commercial biocontrol products used for controlling potato diseases

Trade names	Active ingredients	Application methods	Target pathogens	Companies/ distributors
Actinovate® SP	*Streptomyces lydicus* WYEC 108	Chemigation, in-furrow application, and soil drench	*Botrytis*, *Sclerotinia*, *Pythium*, *Phytophthora* and *Alternaria*, and others	Novozymes
Asperello T34 Biocontrol	*Trichoderma asperellum strain T34*	Foliar spray, seed treatment, in-furrow application, and soil drench	*Botrytis cinerea*, *Fusarium*, *Pythium*, *Ralstonia solanacearum*, *Rhizoctonia*, and others	Biocontrol Technologies S.L.
BioSave® 10 LP	*Pseudomonas syringae* Strain ESC-10	Postharvest seed treatment	*Fusarium*, *Botrytis*, and others	Lallemand Plant Care
Brandt Organics Aleo®	Garlic oil	Soil and foliar spray	*Ralstonia*, *Phytophthora infestans*, *Sclerotinia sclerotiorum*, and others	Brandt Consolidated, Inc.
Cease®	*Bacillus subtilis* strain QST 713	Foliar spray and soil drench	*Alternaria, Xanthomonas campestris, Phytophthora infestans*, and others	BioWorks
Cinnerate®	Cinnamon oil	Foliar and soil spray	A broad spectrum of diseases	Sym-Agro
Companion®	*Bacillus subtilis* GB03	Soil drench, foliar spray, and seed treatment	*Pythium*, *Fusarium solani*, *Rhizoctonia solani*, *Botrytis cinerea*, and others	Growth Products
Double Nickel 55®	*Bacillus amyloliquefaciens* strain D747	Ground and aerial application, and chemigation	*Phytophthora infestans*, *Rhizoctonia solani*, *Sclerotinia sclerotiorum*, *Alternaria, Botrytis*, and others	Certis
GreenFurrow BacStop	Clove, rosemary, peppermint, cottonseed oil, Thyme, garlic, cinnamon	Foliar spray	*Streptomyces*, *Phytophthora infestans*, and others	Guaranteed by Mar Vista Resource
LifeGard® WG	*Bacillus mycoides* strain J	Ground and aerial application, and chemigation	*Phytophthora infestans*, *Alternaria dauci*, *Sclerotinia sclerotiorum*, and others	Certis
MeloCon® WG.	*Paecilomyces lilacinus* strain 251	Soil drench and soil application	Nematodes	Certis

Mycostop®	*Streptomyces* sp. strain K61	Drip irrigation, soil drench, foliar spray, and seed treatment	*Alternaria*, *Rhizoctonia solani*, *Fusarium*, *Phytophthora*, and *Pythium*	Verdera Oy
Polyversum®	*Pythium oligandrum* strain M1	Foliar spray	*Rhizoctonia solani*, *Phytophthora*, *Alternaria*, *Botrytis cinerea*, and others	Gowan
RootShield® PLUS+ WP	*Trichoderma harzianum* strain T-22 and *T. virens* strain G-41	Soil application, seed treatment, foliar application, and chemigation	*Phytophthora*, *Pythium*, *Fusarium*, *Rhizoctonia*, and others	BioWorks
Serenade® ASO	*Bacillus subtilis* strain QST 713	Soil drench, in-furrow application	*Pythium*, *Rhizoctonia*, *Fusarium*, *Phytophthora*, and others	Bayer
Serifel®	*Bacillus amyloliquefaciens* strain MBI 600	Foliar spray, in-furrow application, and soil drench	*Alternaria solani*, *Botrytis*, *Fusarium*, *Phytophthora infestans*, and others	BASF
Timorex Act ®	Tea tree oil (*Melaleuca alternifolia*)	Soil spray and soil drench	*Alternaria*, *Botrytis*, *Pythium*, *Verticillium*, *Rhizoctonia solani*, and others	Stockton (Israel) Ltd.
Trilogy®	Neem oil (*Azadirachta indica*)	Foliar spray	A broad spectrum of diseases	Certis
Regalia® Biofungicide Concentrate	Extract of *Reynoutria sachalinensis*	Foliar and soil application	*Fusarium*, *Pythium*, *Rhizoctonia*, *Verticillium*, *Xanthomonas campestris*, *Alternaria solani*	Marrone Bio Innovations

5.2.2 Factor affecting BCAs

The effectiveness of biocontrol agents (BCAs) can be influenced by several factors, including pathogen population density, environmental conditions, and the formulation and application method of the BCA (Fravel, 2005; Harman, 2006; Meng & Hao, 2017). Environmental conditions such as microclimate, soil type, and the presence of other microorganisms significantly affect BCA's efficacy, with certain BCAs performing better in specific conditions. Tailored approaches are therefore necessary for different regions (Harman et al., 2004).

Key factors for the success of bacterial BCAs in the field include their survival and colonization abilities. For example, *Pseudomonas* strains can persist on potato leaves for extended periods in greenhouse conditions, but their populations decline more rapidly in the field. In contrast, bacteria like *Bacillus* spp., which produce endospores, have a survival advantage in extreme conditions, making them common active ingredients in many BCA products. The adaptation and efficacy are dependent on the species and strains of microbial BCAs. To overcome this, multiple species and strains of microorganisms have been packed into one product (Schisler et al., 2004).

Unfortunately, not all BCAs remain effective indefinitely, as plant pathogens can develop resistance over time. For instance, *Ralstonia solanacearum* has evolved tolerance to *Pseudomonas* species after multiple generations of exposure, potentially reducing the efficacy of antimicrobial compounds like allyl-isothiocyanate produced by Brassica plants (Alderley et al., 2022; Clough et al., 2024). This resistance development needs further study to ensure the long-term efficacy of BCAs.

6 Case studies: managing potato diseases using various biocontrol technologies

6.1 Late blight

Multiple BCAs have been studied for the control of late blight, such as fungal endophytes studied in planta (de Vries et al., 2018), extracts of *Myxococcus xanthus* suppressing late blight in vitro (Wu et al., 2021), *Pseudomonas putida* in vitro and in vivo (Shafi et al., 2017), and *Streptomyces* spp. (Jin et al., 2023) and *Trichoderma harzianum* (Harman, 2006) in fields.

Trichoderma harzianum is a well-known BCA that acts against *Phytophthora infestans* by producing antifungal enzymes and secondary metabolites (Chet et al., 1981; Harman, 2006; Harman et al., 2004). It also induces systemic resistance in the host plant. *Trichoderma virens* produces antibiotics and enzymes that degrade the cell walls of *P. infestans* and enhances the plant's natural defense mechanisms.

Bisvertinolone is an active compound from *Trichoderma longibrachiatum* SFC100166. Volatile organic compounds (VOCs) from *Trichoderma atroviride* strains have strong inhibitory effects against *P. infestans* (Elsherbiny et al., 2020). These VOCs cause significant ultrastructural damages to *P. infestans*, including cell deformation, collapse, and degradation of cytoplasmic organelles.

Other fungi and oomycetes have also been studied for biological control of late blight. *Saccharomyces cerevisiae* can suppress late blight by competing with *P. infestans* for nutrients and space and by producing antifungal compounds. *Chaetomium cochliodes, Chaetomium aureum, Saccharothrix texasensis, Aureobasidium pullulans, Pythium oligandrum,* and *Purpureocillium lilacinum* displayed remarkably good inhibitory effects on *P. infestans* (Akino et al., 2014; Wang & Long, 2023).

Not surprisingly, bacteria are a primary source of BCAs. *Bacillus amyloliquefaciens* produces antifungal metabolites and enzymes that inhibit the growth of *P. infestans* and protect potato plants from late blight. *Streptomyces griseoviridis* produces antifungal antibiotics that can suppress the growth of *P. infestans* and reduce the severity of late blight in potatoes. *Pseudomonas fluorescens* suppresses *P. infestans* through the production of siderophores, antibiotics, and other secondary metabolites that inhibit pathogen growth. *Pseudomonas putida* produces antifungal compounds and induces systemic resistance in potatoes, leading to reduced late blight symptoms (Van Loon et al., 1998). *Bacillus subtilis* is effective against late blight by producing lipopeptides, such as iturins and surfactins, which have strong antifungal properties against *P. infestans* (Shafi et al., 2017).

A mixture of six *Streptomyces* strains effectively controlled potato late blight both in vitro and in fields, and the effect was enhanced by co-application of biochar (Jin et al., 2023). In this study, different strains may share or have distinctive antibacterial compounds, enhancing the adaptation and efficacies of BCAs. These include enzymes for tyrosine hydrolysis, starch hydrolysis, cellulase, and lipase. Secondary metabolites from alkaloid, phenylpropanoid, and terpenoid classes, along with lipids, proteins, and peptides, can induce plant resistance against late blight (Dufková et al., 2023; Li et al., 2023).

Other bacteria, such as *Lysobacter capsica, Myxococcus xanthus, Cronobacter muytjensii,* and *Methylobacterium* sp. have been considered potential BCAs. In addition, extracts from compost have been shown to suppress *P. infestans* by enhancing microbial activity in the soil, leading to natural biocontrol of the pathogen.

6.2 Common scab

Common scab can result in significant reductions in potato yield and tuber quality, which is caused by multiple species of *Streptomyces*, such as *S. scabies,*

S. bottropensis, and *S. stelliscabiei* (Braun et al., 2017; Hao et al., 2009; Keinath & Loria, 1991; Wharton et al., 2007). There are no fully effective methods to control this disease. As such, biological control has shown potential to manage common scab. Products of *Bacillus* spp. are available, including *Bacillus amyloliquefaciens, B. velezensis, B. altitudinis,* and *B. subtilis.* These products have shown promising potential for suppressing common scab, while also exhibiting plant growth promotion in potatoes (Arseneault et al., 2015; Hao et al., 2009; Li et al., 2021; Lin et al., 2018; Meng et al., 2012; Meng & Hao, 2017; Tao et al., 2023; Zhang et al., 2024).

Non-pathogenic *Streptomyces* spp., such as *Streptomyces lydicus* (Yuan & Crawford, 1995), can effectively control potato common scab, while promoting potato growth, increasing yield, enhancing potato sprouting as well as shoot and root development (Sarwar et al., 2018; Zhang et al., 2020).

Pseudomonads have been applied to control common scab because they produce various antimicrobial substances with strong efficacy. For example, *Pseudomonas fluorescens, P. chlororaphis,* have been reported to control common scab (Duffy & Défago, 1999; Kobayashi et al., 2015; Sarwar et al., 2019). *Pseudomonas fluorescens* LBUM223 reduces potato common scab through the biosynthesis of phenazine-1-carboxylic acid (PCA), which directly alter the transcriptional activity of the key gene, *txtA* gene, a major pathogenicity factor (Arseneault et al., 2013, 2015, 2016).

Brevibacillus laterosporus BL12 suppresses potato scab by decreasing the abundance of the pathogenic *S. bottropensis* and affecting the soil microbial community composition and enriching some potentially beneficial *Pseudomonas* (Li et al., 2021). In addition, *Micromonospora* spp. can be a candidate for controlling common scab (Duffy & Défago, 1999; Kobayashi et al., 2015; Sarwar et al., 2019).

Bacteriophages have been shown to reduce the symptoms of common scab on potatoes, which can lyse *Streptomyces stelliscabiei* (Zhang et al., 2024). In another study, phage SscP1EGY significantly reduced the severity of common scab on potato tubers inoculated with only *S. scabies* (Abdelrhim et al., 2021).

Extracts from microbes containing isatropolone C, such as *Streptomyces* A1RT have been studied for their potential to suppress *S. scabies* (Sarwar et al., 2018). Due to its long shelf life, these microbial extracts hold promise for commercial applications. Additionally, various plant extracts, including myrrh, garlic, cayenne, barberry, frankincense, wild indigo root, lavender, and Canadian goldenrod, have been shown to reduce the growth of *S. scabies* (Gutierrez et al., 2022; Paré et al., 2018). Essential oils from thyme and suico effectively inhibit *S. scabies,* making them useful for controlling potato common scab (Prieto et al., 2020). Moreover, the medicinal plant *Geranium pratense* has demonstrated a promising effect against common scab caused by *S. scabies,* with geraniin identified as the primary antimicrobial compound (Ushiki et al., 1998).

7 Future trends and conclusions

Biological control methods for managing potato diseases hold a promising future, driven by relentless research and technological breakthroughs. These sustainable alternatives to chemical pesticides offer a greener approach to agriculture. The development of stronger and more durable biocontrol agents, along with advanced application techniques, presents the potential to significantly lessen our dependence on traditional pesticides. Progress in areas such as genetic engineering, the identification of new biocontrol agents, and their integration with other pest management strategies, promises a transformative shift in potato agriculture. Sustained research and innovation are vital for enhancing the efficiency, eco-friendliness, and cost-effectiveness of bioprotectants and biocontrol agents, thereby reinforcing their importance in potato farming.

Naturally isolated microorganisms may have some disadvantages, such as lower adaptation, survival, and efficacy of disease control. Genetic modification can be used to enhance the production of antimicrobial compounds, improve the ability of biocontrol agents to colonize plant surfaces or increase their effectiveness against specific pathogens. For example, the introduction of genes encoding chitinases, lytic enzymes that can degrade fungal cell walls, into the genome of *Bacillus subtilis* has been shown to increase its efficacy against fungal pathogens.

Using mixtures of microorganisms and other products can enhance the efficacy and adaptation for BCA products (Serrão et al., 2024). The SoilGard product uses a combination of *Gliocladium virens* and other microorganisms to control soil-borne plant pathogens. BioWorks' RootShield Plus contains a mixture of *Trichoderma harzianum* and *Trichoderma virens* to protect plant roots from various soil pathogens. HortiKleen® employs a blend of beneficial bacteria and fungi to improve plant health and suppress disease.

8 Conclusions

Biocontrol products registered take account for around 5% to 15% of total pesticide sales globally, and it is a growing area (Brader et al., 2017; Fravel, 2005; ISAAA, 2019; Montesinos, 2003). The biopesticide market is estimated to value at USD 7.54 billion in 2023 and potentially grow to USD 28.61 billion by 2023 (Fortune Business Insights, 2023). The increasing interest in bioprotectants or BCAs as sustainable alternatives for managing potato diseases reflects a shift toward more environmentally friendly agricultural practices (Amons, 2022; Egamberdieva et al., 2023; Fravel, 2005; Kenneth & Baker, 1975; Montesinos,

2003). BCAs, derived from natural sources, present a safer option compared to synthetic chemical controls. They offer a reduced risk of environmental contamination and minimal impact on non-target organisms, thus supporting overall ecosystem health (De Corato, 2021; Fravel, 2005; Kinkel, 2008).

One of the key advantages of BCAs is their ability to avoid harmful residues in the food chain, making them a healthier alternative for human consumption. Moreover, the use of biocontrol agents helps mitigate the issue of pathogen resistance. Unlike chemical fungicides, which typically employ a single mode of action, BCAs utilize multiple mechanisms to suppress pathogens. This multifaceted approach makes it more challenging for pathogens to develop resistance, ensuring long-term efficacy and sustainability in disease control strategies.

In addition to controlling plant diseases, BCAs offer broader agricultural benefits. They enhance the growth of beneficial microorganisms in the soil, improve nutrient cycling, and contribute to a healthier soil ecosystem, promoting the overall health of the crop. These benefits align with the principles of sustainable agriculture and are particularly valuable in integrated disease management strategies, which aim to reduce chemical inputs while maintaining crop productivity and environmental health. This growing emphasis on biocontrol underscores the potential for BCAs to play a critical role in the future of sustainable disease management in potato production.

9 References

Abass, J. & Fayed, A. (2021). Induced resistance to Potato Virus Y Potyvirus by plant extracts and salicylic acid. *Journal of Plant Protection and Pathology*, *12*(9). https://doi.org/10.21608/jppp.2021.90564.1029

Abd-El-Khair, H. & El-Nagdi, W. M. A. (2013). Field application of bio-control agents for controlling fungal root rot and root-knot nematode in potato. *Archives of Phytopathology and Plant Protection*, *47*(10), 1218–1230. https://doi.org/10.1080/03235408.2013.837632

Abdelrhim, A. S., et al. (2021). A new Streptomyces scabies-infecting bacteriophage from Egypt with promising biocontrol traits. *Archives of Microbiology*, *203*(7), 4233–4242. https://doi.org/10.1007/s00203-021-02415-2

Abo-Zaid, G. A., et al. (2023). Evaluation of bio-friendly formulations from siderophore-producing fluorescent pseudomonas as biocontrol agents for the management of soil-borne fungi, Fusarium oxysporum and Rhizoctonia solani. *Agriculture*, *13*(7), 1418. https://doi.org/10.3390/agriculture13071418

Adhikari, M., et al. (2023). Integrated management of late blight of potato in Pokhara, Kaski, Nepal. *Nepal Agriculture Research Journal*, *15*(1). https://doi.org/10.3126/narj.v15i1.51508

Akino, S., et al. (2014). Phytophthora infestans: a review of past and current studies on potato late blight. *Journal of General Plant Pathology*, *80*(1), 24–37. https://doi.org/10.1007/s10327-013-0495-x

Akyazi, F., et al. (2012). Biodiversity of nematodes in potato growing areas of Ordu, Turkey. *Journal of Animal and Veterinary Advances*, *11*(15). https://doi.org/10.3923/javaa.2012.2660.2664

Alderley, C. L., et al. (2022). Plant pathogenic bacterium can rapidly evolve tolerance to an antimicrobial plant allelochemical. *Evolutionary Applications*, *15*(5). https://doi.org/10.1111/eva.13363

Al-Mughrabi, K. I. (2010). Biological control of Fusarium dry rot and other potato tuber diseases using Pseudomonas fluorescens and Enterobacter cloacae. *Biological Control*, *53*(3), 280–284. https://doi.org/10.1016/j.biocontrol.2010.01.010

Amons, S. (2022). Biological protection of plants in the system of organic farming. *Agriculture and Forestry*. https://doi.org/10.37128/2707-5826-2022-2-13

Anand, A., et al. (2023). Biological hydrogen cyanide emission globally impacts the physiology of both HCN-emitting and HCN-perceiving Pseudomonas. *MBio*, *14*(5). https://doi.org/10.1128/mbio.00857-23

Antas, P. R. Z., et al. (2012). Neglected and emerging fungal infections: Review of hyalohyphomycosis by Paecilomyces lilacinus focusing in disease burden, in vitro antifungal susceptibility and management. In *Microbes and Infection* (Vol. 14, Issue 1). https://doi.org/10.1016/j.micinf.2011.08.004

Arseneault, T., et al. (2013). Phenazine production by Pseudomonas sp. LBUM223 contributes to the biological control of potato common scab. *Phytopathology*, *103*(10), 995–1000. https://doi.org/10.1094/PHYTO-01-13-0022-R

Arseneault, T., et al. (2015). Pseudomonas fluorescens LBUM223 increases potato yield and reduces common scab symptoms in the field. *Phytopathology*, *105*(10), 1311–1317. https://doi.org/10.1094/PHYTO-12-14-0358-R

Arseneault, T., et al. (2016). Biocontrol of potato common scab is associated with high Pseudomonas fluorescens LBUM223 populations and phenazine-1-carboxylic acid biosynthetic transcript accumulation in the potato geocaulosphere. *Phytopathology*, *106*(9), 963–970. https://doi.org/10.1094/PHYTO-01-16-0019-R

Azaiez, S., et al. (2018). Biological control of the soft rot bacterium Pectobacterium carotovorum by Bacillus amyloliquefaciens strain Ar10 producing glycolipid-like compounds. *Microbiological Research*, *217*, 23–33. https://doi.org/10.1016/j.micres.2018.08.013

Bagy, H. M. M. K., et al. (2019). Efficacy of arbuscular mycorrhizal fungi and endophytic strain Epicoccum nigrum ASU11 as biocontrol agents against blackleg disease of potato caused by bacterial strain Pectobacterium carotovora subsp. atrosepticum PHY7. *Biological Control*, *134*. https://doi.org/10.1016/j.biocontrol.2019.03.005

Bahmani, K., et al. (2021). Isolation and identification of endophytic bacteria from potato tissues and their effects as biological control agents against bacterial wilt. *Physiological and Molecular Plant Pathology*, *116*, 101692. https://doi.org/10.1016/j.pmpp.2021.101692

Bairwa, A., et al. (2017). Effect of non-solanaceous crops on potato cyst nematodes. *Indian Journal of Nematology*, *47*(1). https://doi.org/10.1163/156854102765216696

Barkov, V. A., et al. (2023). Chemical protection of potatoes from fungal diseases, taking into account the stability of the variety. *Agricultural Science Euro-North-East*, *24*(3), 389–398. https://doi.org/10.30766/2072-9081.2023.24.3.389-398

Bashan, Y. & de-Bashan, L. E. (2010). How the plant growth-promoting bacterium azospirillum promotes plant growth-a critical assessment. In *Advances in Agronomy* (Vol. 108, Issue C, pp. 77–136). https://doi.org/10.1016/S0065-2113(10)08002-8

Bashan, Y., et al. (2013). Advances in plant growth-promoting bacterial inoculant technology: formulations and practical perspectives (1998–2013). *Plant and Soil, 378*(1–2), 1–33. https://doi.org/10.1007/s11104-013-1956-x

Bautista-Baños, S., et al. (2006). Chitosan as a potential natural compound to control pre and postharvest diseases of horticultural commodities. *Crop Protection*, *25*(2), 108–118. https://doi.org/10.1016/j.cropro.2005.03.010

Ben Khedher, S., et al. (2015). Efficacy of Bacillus subtilis V26 as a biological control agent against Rhizoctonia solani on potato. *Comptes Rendus. Biologies*, *338*(12), 784–792. https://doi.org/10.1016/j.crvi.2015.09.005

Brader, G., et al. (2017). Ecology and genomic insights into plant-pathogenic and plant-nonpathogenic endophytes. *Annual Review of Phytopathology*, *55*, 61–83. https://doi.org/10.1146/annurev-phyto-080516-035641

Braun, S., et al. (2017). Potato common scab: a review of the causal pathogens, management practices, varietal resistance screening methods, and host resistance. *American Journal of Potato Research*, *94*(4), 283–296. https://doi.org/10.1007/s12230-017-9575-3

Brewer, M. T. & Larkin, R. P. (2005). Efficacy of several potential biocontrol organisms against Rhizoctonia solani on potato. *Crop Protection*, *24*(11), 939–950. https://doi.org/10.1016/j.cropro.2005.01.012

Budge, S. P. & Whipps, J. M. (2001). Potential for integrated control of Sclerotinia sclerotiorum in glasshouse lettuce using Coniothyrium minitans and reduced fungicide application. *Phytopathology*, *91*(2), 221–227. https://doi.org/10.1094/PHYTO.2001.91.2.221

Buttimer, C., et al. (2017). Bacteriophages and bacterial plant diseases. *Frontiers in Microbiology*, *8*(1), 1–15. https://doi.org/10.3389/fmicb.2017.00034

Cawoy, H., et al. (2014). Plant defense stimulation by natural isolates of Bacillus depends on efficient surfactin production. *Molecular Plant-Microbe Interactions*, *27*(2), 87–100. https://doi.org/10.1094/MPMI-09-13-0262-R

Chauhan, P. S., et al. (2019). Transcriptional alterations reveal Bacillus amyloliquefaciens-rice cooperation under salt stress. *Scientific Reports*, *9*(1), 11912. https://doi.org/10.1038/s41598-019-48309-8

Chet, I., et al. (1981). Trichoderma hamatum: Its hyphal interactions with Rhizoctonia solani and Pythium spp. *Microbial Ecology*, *7*(1), 29–38. https://doi.org/10.1007/BF02010476

Clough, S. E., et al. (2024). Plant pathogenic bacterium Ralstonia solanacearum can rapidly evolve tolerance to antimicrobials produced by Pseudomonas biocontrol bacteria. *Journal of Evolutionary Biology*, *37*(2). https://doi.org/10.1093/jeb/voae002

Compant, S., et al. (2005). Use of plant growth-promoting bacteria for biocontrol of plant diseases: principles, mechanisms of action, and future prospects. *Applied and Environmental Microbiology*, *71*(9), 4951–4959. https://doi.org/10.1128/AEM.71.9.4951-4959.2005

Cornelis, P. & Andrews, S. C. (2010). Iron uptake and homeostasis in microorganisms. In *Analysis* (Issue June). Caister Academic Press. https://doi.org/10.21775/9781910190159

Coventry, E., Noble, R., Mead, A., Marin, F. R., Perez, J. A., & Whipps, J. M. (2006). Allium white rot suppression with composts and Trichoderma viride in relation to sclerotia viability. *Phytopathology*, *96*(9), 1009–1020. https://doi.org/10.1094/PHYTO-96-1009

Czajkowski, R. (2015). Bacteriophages of soft rot Enterobacteriaceae: a minireview. *FEMS Microbiology Letters*, *363*(2), fnv230. https://doi.org/10.1093/femsle/fnv230

Czajkowski, R., et al. (2012). Characterization of bacterial isolates from rotting potato tuber tissue showing antagonism to Dickeya sp. biovar 3 in vitro and in planta. *Plant Pathology*, *61*(1), 169–182. https://doi.org/10.1111/j.1365-3059.2011.02486.x

da Silva, H. A. O., et al. (2021). Biocontrol of potato early blight and suppression of *Alternaria grandis* sporulation by *Clonostachys* spp. *Plant Pathology*, *70*(7), 1677–1685. https://doi.org/10.1111/ppa.13402

Davey, M. E., et al. (2003). Rhamnolipid surfactant production affects biofilm architecture in Pseudomonas aeruginosa PAO1. *Journal of Bacteriology*, *185*(3), 1027–1036. https://doi.org/10.1128/JB.185.3.1027-1036.2003

Davis, R. M., et al. (2007). Efficacy of germination stimulants of sclerotia of Sclerotium cepivorum for management of white rot of garlic. *Plant Disease*, *91*(2), 204–208. https://doi.org/10.1094/PDIS-91-2-0204

De Corato, U. (2021). Soil microbiome manipulation gives new insights in plant disease-suppressive soils from the perspective of a circular economy: a critical review. In *Sustainability (Switzerland)* (Vol. 13, Issue 1). https://doi.org/10.3390/su13010010

de la Cruz, A. R., et al. (1992). Biological suppression of potato ring rot by fluorescent pseudomonads. *Applied and Environmental Microbiology*, *58*(6), 1986–1991. https://doi.org/10.1128/aem.58.6.1986-1991.1992

de Vries, S., et al. (2018). Broad-spectrum inhibition of Phytophthora infestans by fungal endophytes. *FEMS Microbiology Ecology*, *94*(4), fiy037. https://doi.org/10.1093/FEMSEC/FIY037

Delaney, S. M., et al. (2001). phzO, a gene for biosynthesis of 2-hydroxylated phenazine compounds in Pseudomonas aureofaciens 30–84. *Journal of Bacteriology*, *183*(1), 318–327. https://doi.org/10.1128/JB.183.1.318-327.2001

Demirci, E., et al. (2009). Biological control of Rhizoctonia solani on potato by Verticillium biguttatum. *African Journal of Biotechnology*, *8*(11), 2503–2507.

Diallo, S., et al. (2011). Mechanisms and recent advances in biological control mediated through the potato rhizosphere. *FEMS Microbiology Ecology*, *75*(3), 351–364. https://doi.org/10.1111/j.1574-6941.2010.01023.x

Dong, Y. H., et al. (2007). Quorum-quenching microbial infections: mechanisms and implications. *Philosophical Transactions of the Royal Society B: Biological Sciences*, *362*(1483), 1483. https://doi.org/10.1098/rstb.2007.2045

Druzhinina, I. S., et al. (2011). Trichoderma: The genomics of opportunistic success. *Nature Reviews Microbiology*, *9*(10), 749–759. https://doi.org/10.1038/nrmicro2637

Duffy, B. K. & Défago, G. (1999). Environmental factors modulating antibiotic and siderophore biosynthesis by Pseudomonas fluorescens biocontrol strains. *Applied and Environmental Microbiology*, *65*(6), 2429–2438. https://doi.org/10.1128/aem.65.6.2429-2438.1999

Dufková, H., et al. (2023). Secondary metabolites, other prospective substances, and alternative approaches that could promote resistance against Phytophthora infestans. *Agronomy*, *13*(7), 1822. https://doi.org/10.3390/agronomy13071822

Egamberdieva, D., et al. (2023). Bacterial bioprotectants: biocontrol traits and induced resistance to phytopathogens. In *Microbiology Research* (Vol. 14, Issue 2). https://doi.org/10.3390/microbiolres14020049

El-Anany, A. M. A. (2021). Studies on intercropping systems of garlic and green onion to potatoes and impact that on growth, yield, and resistance late blight disease.

Annals of Agricultural Science, Moshtohor, *59*(1). https://doi.org/10.21608/assjm.2021.176631

Elsherbiny, E. A., et al. (2020). Trichoderma volatile organic compounds as a biofumigation tool against late blight pathogen Phytophthora infestans in postharvest potato tubers. *Journal of Agricultural and Food Chemistry*, *68*(31). https://doi.org/10.1021/acs.jafc.0c03150

Escande, A. R. & Echandi, E. (1991). Protection of potato from Rhizoctonia canker with binucleate Rhizoctonia fungi. *Plant Pathology*, *40*(2), 197–202. https://doi.org/10.1111/j.1365-3059.1991.tb02367.x

FAO. (2020). *Food and Agricultural Organization of the United Nations FAOSTAT Statistical Database*. FAO. https://www.fao.org/faostat/en/#home

FAO. (2024). *International Day of Potato: At Inaugural Celebration, FAO Highlights Crop's Significance and Further Potential*. Food and Agriculture Organization of the United Nations. https://www.fao.org/newsroom/detail/international-day-of-potato--at-inaugural-celebration--fao-highlights-crop-s-significance-and-further-potential/en

Fortune Business Insights. (2023). *Biopesticides Market Size, Share & Industry Analysis, By Type (Bioinsecticide, Biofungicide, Bionematicide, and Others), By Source (Microbials and Biochemicals), By Mode of Application (Foliar Application, Seed Treatment, Soil Application, and Others), By Crop (Cereals, Oilseeds, Fruits & Vegetables, and Others), and Regional Forecast, 2024-2032 Source. https://www.fortunebusinessinsights.com/industry-reports/biopesticides-market-100073*

Fravel, D. R. (2005). Commercialization and implementation of biocontrol. *Annual Review of Phytopathology*, *43*, 337–359. https://doi.org/10.1146/annurev.phyto.43.032904.092924

Fu, X., et al. (2022). Biological control of potato late blight by Streptomyces sp. FXP04 and potential role of secondary metabolites. *Biological Control*, *169*, 104891. https://doi.org/10.1016/j.biocontrol.2022.104891

Fujiwara, A., et al. (2011). Biocontrol of Ralstonia solanacearum by treatment with lytic bacteriophages. *Applied and Environmental Microbiology*, *77*(12). https://doi.org/10.1128/AEM.02847-10

Ge, T., et al. (2021). Genotyping Dickeya dianthicola causing potato blackleg and soft rot outbreak associated with inoculum geography in the United States. *Plant Disease*, *105*, 1976–1983. https://doi.org/10.1094/pdis-10-20-2138-re

Ghahremani, Z., et al. (2020). Bacillus firmus strain I-1582, a nematode antagonist by itself and through the plant. *Frontiers in Plant Science*, *11*. https://doi.org/10.3389/fpls.2020.00796

Göhre, V. & Paszkowski, U. (2006). Contribution of the arbuscular mycorrhizal symbiosis to heavy metal phytoremediation. *Planta*, *223*(6), 1115–1122. https://doi.org/10.1007/s00425-006-0225-0

Grewal, P. S., et al. (2005). Nematodes as biocontrol agent. In *Nematodes as Biocontrol Agent*. https://doi.org/10.21273/hortsci.46.2.333a

Gutierrez, J., et al. (2022). Plant natural products as antimicrobials for control of Streptomyces scabies: a causative agent of the common scab disease. *Frontiers in Microbiology*, *12*. https://doi.org/10.3389/fmicb.2021.833233

Haas, D. & Défago, G. (2005). Biological control of soil-borne pathogens by fluorescent pseudomonads. *Nature Reviews Microbiology*, *3*(4), 307–319. https://doi.org/10.1038/nrmicro1129

Hao, J. J., et al. (2009). Characterization of a new Streptomyces strain, DS3024, that causes potato common scab. *Plant Disease*, *93*(12), 1329–1334. https://doi.org/10.1094/PDIS-93-12-1329

Harman, G. E. (2006). Overview of mechanisms and uses of Trichoderma spp. *Phytopathology*, *96*(2). https://doi.org/10.1094/PHYTO-96-0190

Harman, G. E., et al. (2004). Trichoderma species - Opportunistic, avirulent plant symbionts. *Nature Reviews Microbiology*, *2*(1), 43–56. https://doi.org/10.1038/nrmicro797

Harris, A. R. & Lumsden, R. D. (1997). Interactions of Gliocladium virens with Rhizoctonia solani and Pythium ultimum in non-sterile potting medium. *Biocontrol Science and Technology*, *7*(1). https://doi.org/10.1080/09583159731027

Hashem, A., et al. (2018). Arbuscular mycorrhizal fungi regulate the oxidative system, hormones and ionic equilibrium to trigger salt stress tolerance in Cucumis sativus L. *Saudi Journal of Biological Sciences*, *25*(6). https://doi.org/10.1016/j.sjbs.2018.03.009

Hayat, R., et al. (2010). Soil beneficial bacteria and their role in plant growth promotion: a review. *Annals of Microbiology*, *60*(4), 579–598. https://doi.org/10.1007/s13213-010-0117-1

Hills, K., et al. (2020). Improving soil health in Pacific Northwest potato production: a review. *American Journal of Potato Research*, *97*(1), 1–22. https://doi.org/10.1007/s12230-019-09742-7

Hosseinzadeh, S., et al. (2017). Attenuation and quantitation of virulence gene expression in quorum-quenched Dickeya chrysanthemi. *Archives of Microbiology*, *199*(1), 51–61. https://doi.org/10.1007/s00203-016-1276-7

Hostettmann, K. (1998). Strategy for the biological and chemical evaluation of plant extracts. *Pure and Applied Chemistry*, *70*(11), 1–9. http://www.iupac.org/symposia/proceedings/phuket97/hostettmann.html

Hough, B., et al. (2023). Fungal viruses unveiled: a comprehensive review of mycoviruses. *Viruses*, *15*(5), 1202. https://doi.org/10.3390/v15051202

Hussain, A., et al. (2017). Management of aphid borne Potato Virus Y (PVY) through chemical and bio-chemical methods. *World Journal of Biology and Biotechnology*, *2*(2), 159–163. https://doi.org/10.33865/wjb.002.02.0108

ISAAA. (2019). Brief 55: global status of commercialized biotech/GM crops: 2019. In *ISAAA Brief 55*. https://www.isaaa.org/resources/publications/briefs/55/

Islam, M. T. (2008). Disruption of ultrastructure and cytoskeletal network is involved with biocontrol of damping-off pathogen Aphanomyces cochlioides by Lysobacter sp. strain SB-K88. *Biological Control*, *46*(3). https://doi.org/10.1016/j.biocontrol.2008.02.006

Javed, N., et al. (2012). Effect of Steinernema glaseri and Heterorhabditis indica on the plant vigour and root knot nematodes in tomato roots at different densities and time of applications. *Pakistan Journal of Zoology*, *44*(4), 1165–1170. https://www.cabidigitallibrary.org/doi/pdf/10.5555/20123377553

Jin, L., et al. (2023). Biological control of potato late blight with a combination of Streptomyces strains and biochar. *Biological Control*, *183*, 105248. https://doi.org/10.1016/j.biocontrol.2023.105248

Johnson, D. A. (2008). *Potato Health Management* (2nd ed.). The American Phytopathological Society. ISBN 978-0-89054-353-5

Keinath, A. P. & Loria, R. (1991). Effects of inoculum density and cultivar resistance on common scab of potato and population dynamics of Streptomyces scabies. *American Potato Journal*, *68*(8), 515–524. https://doi.org/10.1007/BF02853768

Kenneth, F. & Baker, R. J. C. (1975). Biological control of plant pathogens. In K. F. Baker and R. J. Cook (Eds.), *Experimental Agriculture* (Vol. 11, Issue 2, p. 433). Freeman. https://doi.org/10.1017/s001447970000661x

Kerry, B. R. (2000). Rhizosphere interactions and the exploitation of microbial agents for the biological control of plant-parasitic nematodes. *Annual Review of Phytopathology*, *38*(1), 423–441. https://doi.org/10.1146/annurev.phyto.38.1.423

Khan, A., et al. (2003). Testing the nematophagous biological control strain Paecilomyces lilacinus 251 for paecilotoxin production. *FEMS Microbiology Letters*, *227*(1). https://doi.org/10.1016/S0378-1097(03)00654-2

Kim, Y. T., et al. (2019). Effects of Lysobacter antibioticus hs124, an effective biocontrol agent against Fusarium graminearum, on crown rot disease and growth promotion of wheat. *Canadian Journal of Microbiology*, *65*(12). https://doi.org/10.1139/cjm-2019-0285

Kinkel, L. (2008). Soil health: managing the soil microflora to enhance potato health. In D. A. Johnson (Ed.), *Potato Health Management2* (2nd ed., pp. 11–14). APS Press. ISBN 978-0-89054-353-5

Kloepper, J. W., et al. (1980). Enhanced plant growth by siderophores produced by plant growth-promoting rhizobacteria. *Nature*, *286*, 885–886. https://doi.org/10.1038/286885a0

Kloepper, J. W., et al. (2004). The nature and application of biocontrol microbes: Bacillus spp. induced systemic resistance and promotion of plant growth by Bacillus spp. *Phytopathology*, *94*(11), 1259–1266. https://doi.org/10.1094/PHYTO.2004.94.11.1259

Kobayashi, Y. O., et al. (2015). Biological control of potato scab and antibiosis by antagonistic Streptomyces sp. WoRs-501. *Journal of General Plant Pathology*, *81*(6), 439–448. https://doi.org/10.1007/s10327-015-0614-y

Köhl, J., et al. (2019). Mode of action of microbial biological control agents against plant diseases: relevance beyond efficacy. *Frontiers in Plant Science*, *10*, 845. https://doi.org/10.3389/fpls.2019.00845

Krewulak, K. D. & Vogel, H. J. (2008). Structural biology of bacterial iron uptake. *Biochimica et Biophysica Acta (BBA) - Biomembranes*, *1778*(9), 1781–1804. https://doi.org/10.1016/j.bbamem.2007.07.026

Lahlali, R., and Hijri, M. (2010). Screening, identification and evaluation of potential biocontrol fungal endophytes against Rhizoctonia solani AG3 on potato plants. *FEMS Microbiology Letters*, *311*(2), 152–159. https://doi.org/10.1111/j.1574-6968.2010.02084.x

Lahlali, R., et al. (2013). Evidence that the biofungicide Serenade (Bacillus subtilis) suppresses clubroot on canola via antibiosis and induced host resistance. *Phytopathology*, *103*(3), 245. https://doi.org/10.1094/PHYTO-06-12-0123-R

Lahlali, R., et al. (2022). Biological control of plant pathogens: a global perspective. *Microorganisms*, *10*(3), 596. https://doi.org/10.3390/microorganisms10030596

Larkin, R. P. & Brewer, M. T. (2020). Effects of crop rotation and biocontrol amendments on rhizoctonia disease of potato and soil microbial communities. *Agriculture (Switzerland)*, *10*(4), 1–20. https://doi.org/10.3390/agriculture10040128

Larkin, R. P. & Halloran, J. M. (2015). Management effects of disease-suppressive rotation crops on potato yield and soilborne disease and their economic implications in potato production. *American Journal of Potato Research*, *91*(5), 429–439. https://doi.org/10.1007/s12230-014-9366-z

Lastochkina, O., et al. (2020). Effects of endophytic Bacillus subtilis and salicylic acid on postharvest diseases (Phytophthora infestans, Fusarium oxysporum) development in stored potato tubers. *Plants*, *9*(1). https://doi.org/10.3390/plants9010076

Lei, H., et al. (2023). Secondary metabolites from the nematode-trapping fungus Dactylellina haptotyla YMF1.03409. *Microorganisms*, *11*(11). https://doi.org/10.3390/microorganisms11112693

Li, C., et al. (2021). Biocontrol of potato common scab by Brevibacillus laterosporus BL12 is related to the reduction of pathogen and changes in soil bacterial community. *Biological Control*, *153*, 104496. https://doi.org/10.1016/j.biocontrol.2020.104496

Li, F., et al. (2023). Crude extracts and secondary metabolites of Epichloë bromicola against Phytophthora infestans. *Chemistry and Biodiversity*, *20*(1), e202200841. https://doi.org/10.1002/cbdv.202200841

Li, Y., et al. (2013). The inhibitory effect of Epicoccum nigrum strain XF1 against Phytophthora infestans. *Biological Control*, *67*(3). https://doi.org/10.1016/j.biocontrol.2013.09.007

Li, Y., et al. (2022). The biocontrol of potato dry rot by microorganisms and bioactive substances: a review. *Physiological and Molecular Plant Pathology*, *122*, 101919. https://doi.org/10.1016/j.pmpp.2022.101919

Lin, C., et al. (2018). Biological control of potato common scab by Bacillus amyloliquefaciens Ba01. *PLoS One*, *13*(4), 0196520. https://doi.org/10.1371/journal.pone.0196520

Liu, S., et al. (2017). Transcriptome profiling of genes involved in induced systemic salt tolerance conferred by Bacillus amyloliquefaciens FZB42 in Arabidopsis thaliana. *Scientific Reports*, *7*(1), 10795. https://doi.org/10.1038/s41598-017-11308-8

Love, S. L., et al. (2020). A short history of potato production systems. In J. Stark, M. Thornton, & M. Nolte (Eds.), *Potato Production Systems* (pp. 1–17). Springer. https://doi.org/10.1007/978-3-030-39157-7_1

Ma, Z., et al. (2016). Role of phenazines and cyclic lipopeptides produced by Pseudomonas sp. CMR12a in induced systemic resistance on rice and bean. *Environmental Microbiology Reports*, *8*(5). https://doi.org/10.1111/1758-2229.12454

Maas, P. W. T. (1975). Soil fumigation and crop rotation to control spraing disease in potatoes. *Netherlands Journal of Plant Pathology*, *81*(4), 138–143. https://doi.org/10.1007/BF01976805

Mahmood, B., et al. (2022). Management of late blight of potato caused by Phytophthora infestans through botanical aqueous extracts. *International Journal of Phytopathology*, *11*(1). https://doi.org/10.33687/phytopath.011.01.4131

McKenna, F., et al. (2001). Novel in vivo use of a polyvalent Streptomyces phage to disinfest Streptomyces scabies-infected seed potatoes. *Plant Pathology*, *50*(6). https://doi.org/10.1046/j.1365-3059.2001.00648.x

Meng, Q. & Hao, J. J. (2017). Optimizing the application of Bacillus velezensis BAC03 in controlling the disease caused by Streptomyces scabies. *BioControl*, *62*(4), 535–544. https://doi.org/10.1007/s10526-017-9799-7

Meng, Q., et al. (2012). Characterizing a novel strain of Bacillus amyloliquefaciens BAC03 for potential biological control application. *Journal of Applied Microbiology*, *113*(5), 1165–1175. https://doi.org/10.1111/j.1365-2672.2012.05420.x

Michelsen, C. F., et al. (2015). Nonribosomal peptides, key biocontrol components for Pseudomonas fluorescens in5, isolated from a Greenlandic suppressive soil. *MBio*, *6*(2). https://doi.org/10.1128/mBio.00079-15

Millard, W. A. & Taylor, C. B. (1927). Antagonism of micro-organisms as the controlling factor in the: inhibition of scab by green-manuring. *Annals of Applied Biology*, *14*(2), 202–216. https://doi.org/10.1111/j.1744-7348.1927.tb07076.x

Mishra, A. K., et al. (2014). Plant growth-promoting rhizobacterial strain-mediated induced systemic resistance in tea (Camellia sinensis (L.) O. Kuntze) through defense-related enzymes against brown root rot and charcoal stump rot. *Applied Biochemistry and Biotechnology*, *174*(2), 506–521. https://doi.org/10.1007/s12010-014-1090-0

Mohamed, B. F. F., et al. (2020). Approving the biocontrol method of potato wilt caused by Ralstonia solanacearum (Smith) using Enterobacter cloacae PS14 and Trichoderma asperellum T34. *Egyptian Journal of Biological Pest Control*, *30*(1), 60. https://doi.org/10.1186/s41938-020-00262-9

Montesinos, E. (2003). Development, registration and commercialization of microbial pesticides for plant protection. *International Microbiology*, *6*(4), 245–252. https://doi.org/10.1007/s10123-003-0144-x

Mulligan, D. F. C., et al. (1995). Monitoring and manipulation of populations of Pythium oligandrum, Pythium mycoparasiticum and a Papulaspora species in soil. *Soil Biology and Biochemistry*, *27*(10), 1333–1343. https://doi.org/10.1016/0038-0717(95)00049-K

Musa, M., et al. (2023). An endophytic fungal isolate Paecilomyces lilacinus produces bioactive secondary metabolites and promotes growth of Solanum lycopersicum under heavy metal stress. *Agronomy*, *13*(3). https://doi.org/10.3390/agronomy13030883

Narayanasamy, P. (2013). Abiotic biological control agents for crop disease management. In P. Narayanasamy (Ed.), *Biological Management of Diseases of Crops* (pp. 511–632). Springer Netherlands. https://doi.org/10.1007/978-94-007-6380-7_8

Nautiyal, C. S., et al. (2013). Plant growth-promoting bacteria Bacillus amyloliquefaciens NBRISN13 modulates gene expression profile of leaf and rhizosphere community in rice during salt stress. *Plant Physiology and Biochemistry*, *66*, 1–9. https://doi.org/10.1016/j.plaphy.2013.01.020

Naveen Kumar, A., et al. (2007). Role of chitinase and β-1,3-glucanase activities produced by a fluorescent pseudomonad and in vitro inhibition of Phytophthora capsici and Rhizoctonia solani. *Canadian Journal of Microbiology*, *53*(2), 207–212. https://doi.org/10.1139/w06-119

Neeno-Eckwall, E. C., et al. (2001). Competition and antibiosis in the biological control of potato scab. *Canadian Journal of Microbiology*, *47*(4), 332–340. https://doi.org/10.1139/w01-010

Neuenschwander, P. (2003). Biological control in IPM systems in Africa. In P. Neuenschwander, C. Borgemeister, & J. Langewald (Eds.), *Biological Control in IPM Systems in Africa*. CABI. https://doi.org/10.1079/9780851996394.0000

Nuss, D. L. (2005). Hypovirulence: mycoviruses at the fungal-plant interface. *Nature Reviews Microbiology*, *3*(8), 632–642. https://doi.org/10.1038/nrmicro1206

Ojaghian, M. R. (2010). Biocontrol of potato white mold using Coniothyrium minitans and resistance of potato cultivars to Sclerotinia sclerotiorum. *Plant Pathology Journal*, *26*(4). https://doi.org/10.5423/PPJ.2010.26.4.346

Ojaghian, M. R. (2011). Potential of Trichoderma spp. and Talaromyces flavus for biological control of potato stem rot caused by Sclerotinia sclerotiorum. *Phytoparasitica*, *39*(2), 185–193. https://doi.org/10.1007/s12600-011-0153-9

Oni, F. E., et al. (2019). Phenazines and cyclic lipopeptides produced by Pseudomonas sp. CMR12a are involved in the biological control of Pythium myriotylum on cocoyam (Xanthosoma sagittifolium). *Biological Control*, *129*. https://doi.org/10.1016/j.biocontrol.2018.10.005

Ordentlich, A., et al. (1990). Integrated control of Verticillium dahliae in potato by Trichoderma harzianum and captan. *Crop Protection*, *9*(5), 363–366. https://doi.org/10.1016/0261-2194(90)90008-U

Ouanas, S., et al. (2017). Protection against bacterial soft rot by olive extracts is related to general defence induction in potato tubers. *Plant Pathology*, *66*(3), 404–411. https://doi.org/10.1111/ppa.12583

Paré, M. C., et al. (2018). Canadian goldenrod residues and extracts inhibit the growth of Streptomyces scabiei, the causal agent of potato common scab. *Canadian Journal of Plant Pathology*, *40*(1). https://doi.org/10.1080/07060661.2017.1406992

Perry, R. N., et al. (1998). Effect of the entomopathogenic nematodes, Steinernema feltiae and S. carpocapsae, on the potato cyst nematode, Globodera rostochiensis, in pot trials. *Biocontrol Science and Technology*, *8*(1), 175–0. https://doi.org/10.1080/09583159830531

Pieterse, C. M. J., et al. (2014). Induced systemic resistance by beneficial microbes. *Annual Review of Phytopathology*, *52*. https://doi.org/10.1146/annurev-phyto-082712-102340

Prieto, M. C., et al. (2020). Thyme and suico essential oils: promising natural tools for potato common scab control. *Plant Biology*, *22*(1). https://doi.org/10.1111/plb.13048

Raaijmakers, J. M., et al. (2006). Cyclic lipopeptide production by plant-associated *Pseudomonas* spp.: diversity, activity, biosynthesis, and regulation. *Molecular Plant-Microbe Interactions®*, *19*(7), 699–710. https://doi.org/10.1094/mpmi-19-0699

Reyes-Ramírez, A., et al. (2004). Antifungal activity of Bacillus thuringiensis chitinase and its potential for the biocontrol of phytopathogenic fungi in soybean seeds. *Journal of Food Science*, *69*(5), 131–134. https://doi.org/10.1111/j.1365-2621.2004.tb10721.x

Rillig, M. C. (2004). Arbuscular mycorrhizae and terrestrial ecosystem processes. *Ecology Letters*, *7*(8), 740–754. https://doi.org/10.1111/j.1461-0248.2004.00620.x

Ristaino, J. B. (2002). Tracking historic migrations of the Irish potato famine pathogen, Phytophthora infestans. *Microbes and Infection*, *4*(13), 1369–1377. https://doi.org/10.1016/S1286-4579(02)00010-2

Rodŕiguez, H. & Fraga, R. (1999). Phosphate solubilizing bacteria and their role in plant growth promotion. *Biotechnology Advances*, *17*(4–5), 319–339. https://doi.org/10.1016/s0734-9750(99)00014-2

Ruzzi, M. & Aroca, R. (2015). Plant growth-promoting rhizobacteria act as biostimulants in horticulture. *Scientia Horticulturae*, *196*, 124–134. https://doi.org/10.1016/j.scienta.2015.08.042

Ryan, A. D. & Kinkel, L. L. (1997). Inoculum density and population dynamics of suppressive and pathogenic Streptomyces strains and their relationship to biological control of potato scab. *Biological Control*, *10*(3), 180–186. https://doi.org/10.1006/bcon.1997.0564

Salem, E. A. & Abd El-Shafea, Y. M. (2018). Biological control of potato soft rot caused by Erwinia carotovora subsp. carotovora. *Egyptian Journal of Biological Pest Control*, *28*(1). https://doi.org/10.1186/s41938-018-0100-x

Samuel, B. S., et al. (2016). Caenorhabditis elegans responses to bacteria from its natural habitats. *Proceedings of the National Academy of Sciences of the United States of America*, *113*(27). https://doi.org/10.1073/pnas.1607183113

Sarwar, A., et al. (2019). A potential biocontrol agent Streptomyces violaceusniger AC12AB for managing potato common scab. *Frontiers in Microbiology*, *10*(FEB), 222. https://doi.org/10.3389/fmicb.2019.00202

Sarwar, A., et al. (2018). Biological control of potato common scab with rare isatropolone C compound produced by plant growth promoting Streptomyces A1RT. *Frontiers in Microbiology*, *9*, 1126. https://doi.org/10.3389/fmicb.2018.01126

Schisler, D. A., et al. (2004). Formulation of Bacillus spp. for bicontrol of plant diseases *Phytopathology*, *94*(11), 1267. http://www.ncbi.nlm.nih.gov/pubmed/18944465

Schisler, D. A., et al. (2000). Biological control of fusarium dry rot of potato tubers under commercial storage conditions. *American Journal of Potato Research*, *77*(1), 29–40. https://doi.org/10.1007/bf02853659

Serrão, C. P., et al. (2024). Bacillus species as tools for biocontrol of plant diseases: a meta-analysis of twenty-two years of research, 2000–2021. *World Journal of Microbiology and Biotechnology*, *40*(4), 110. https://doi.org/10.1007/s11274-024-03935-x

Shafi, J., et al. (2017). Bacillus species as versatile weapons for plant pathogens: a review. *Biotechnology and Biotechnological Equipment*, *31*(3). https://doi.org/10.1080/13102818.2017.1286950

Shattock, R. (2002). Compendium of Potato Diseases, Second Edition. W.R. Stevenson. *Plant Pathology*, *51*(4), 581–521. https://doi.org/10.1046/j.1365-3059.2002.06934.x

Siddiqui, Z. A. & Mahmood, I. (1999). Role of bacteria in the management of plant parasitic nematodes: a review. *Bioresource Technology*, *69*(2), 167–179. https://doi.org/10.1016/S0960-8524(98)00122-9

Silva-Valderrama, I., et al. (2021). Biocontrol potential of grapevine endophytic and rhizospheric fungi against trunk pathogens. *Frontiers in Microbiology*, *11*. https://doi.org/10.3389/fmicb.2020.614620

Smith, S. E. & Read, D. J. (2008). *Mycorrhizal Symbiosis* (3rd ed.). Academic Press.

Suresh, P., et al. (2022). Pseudomonas fluorescens VSMKU3054 mediated induced systemic resistance in tomato against Ralstonia solanacearum. *Physiological and Molecular Plant Pathology*, *119*, 101836. https://doi.org/10.1016/j.pmpp.2022.101836

Tao, H., et al. (2023). Biological control of potato common scab and growth promotion of potato by Bacillus velezensis Y6. *Frontiers in Microbiology*, *14*, 1295107. https://doi.org/10.3389/fmicb.2023.1295107

Tegg, R. S., et al. (2008). Auxin-induced resistance to common scab disease of potato linked to inhibition of thaxtomin A toxicity. *Plant Disease*, *92*(9), 1321–1328. https://doi.org/10.1094/PDIS-92-9-1321

Teng, L., et al. (2024). Examining resistance of potato variety and clones for pink rot in 2023. *Plant Disease Management Reports*, *18*, V049. https://www.plantmanagementnetwork.org/pub/trial/pdmr/volume18/abstracts/V049.asp

Teplitski, M., et al. (2003). Chemical identification of N-acyl homoserine lactone quorum-sensing signals produced by Sinorhizobium meliloti strains in defined medium.

Archives of Microbiology, *180*(6), 494–497. https://doi.org/10.1007/s00203-003-0612-x
Tkalenko, H., et al. (2021). In vitro screening of new strains of predacious nematophagous fungi for biocontrol suitability when produced in liquid culture. *Agricultural Science and Practice*, *8*(2). https://doi.org/10.15407/agrisp8.02.003
Tobin, J. D., et al. (2008). Effect of the fungus Pochonia chlamydosporia and fosthiazate on the multiplication rate of potato cyst nematodes (Globodera pallida and G. rostochiensis) in potato crops grown under UK field conditions. *Biological Control*, *46*(2). https://doi.org/10.1016/j.biocontrol.2008.03.014
Tsror (Lahkim), L., et al. (2021). Fungicide field treatments to control potato leak caused by Pythium ultimum. *American Journal of Potato Research*, *98*(2), 115–121. https://doi.org/10.1007/s12230-021-09822-7
Uppal, A. K., et al. (2008). Biological control of potato Verticillium wilt under controlled and field conditions using selected bacterial antagonists and plant extracts. *Biological Control*, *44*(1), 90–100. https://doi.org/10.1016/j.biocontrol.2007.10.020
Ushiki, J., et al. (1998). Medicinal plants for suppressing soil-borne plant diseases: II. Suppressive effect of geranium pratense l. On common scab of potato and identification of the active compound. *Soil Science and Plant Nutrition*, *44*(2), 157–165. https://doi.org/10.1080/00380768.1998.10414436
Vacheron, J., et al. (2015). Alleviation of abiotic and biotic stresses in plants by azospirillum. In F. D. Cassan, Y. Okon, & C. Cassán (Ed.), *Handbook for Azospirillum: Technical Issues and Protocols* (pp. 333–365). Springer. https://doi.org/10.1007/978-3-319-06542-7_19
Valkonen, J. P. T. (2007). Viruses: economical losses and biotechnological potential. In *Potato Biology and Biotechnology: Advances and Perspectives* (pp. 619–641). https://doi.org/10.1016/B978-044451018-1/50070-1
Van Loon, L. C., et al. (1998). Systemic resistance induced by rhizosphere bacteria. *Annual Review of Phytopathology*, *36*, 453–483. https://doi.org/10.1146/annurev.phyto.36.1.453
Vanajothi, R., et al. (2022). Exploring the biostimulants in plant science. In *New and Future Developments in Microbial Biotechnology and Bioengineering: Sustainable Agriculture: Revitalization through Organic Products*. https://doi.org/10.1016/B978-0-323-85579-2.00008-3
Vardharajula, S., et al. (2011). Drought-tolerant plant growth promoting Bacillus spp.: effect on growth, osmolytes, and antioxidant status of maize under drought stress. *Journal of Plant Interactions*, *6*(1), 1–14. https://doi.org/10.1080/17429145.2010.535178
Wang, B., et al. (2018). Integrated metabolomics and morphogenesis reveal volatile signaling of the nematode-trapping fungus Arthrobotrys oligospora. *Applied and Environmental Microbiology*, *84*(9). https://doi.org/10.1128/AEM.02749-17
Wang, W. & Long, Y. (2023). A review of biocontrol agents in controlling late blight of potatoes and tomatoes caused by Phytophthora infestans and the underlying mechanisms. In *Pest Management Science*. https://doi.org/10.1002/ps.7706
Weller, D. (1988). Biological control of soilborne plant pathogens in the rhizosphere with bacteria. *Annual Review of Phytopathology*, *26*(1), 379–407. https://doi.org/10.1146/annurev.phyto.26.1.379
Wharton, P., et al. (2007). Common Scab of Potato. *Michigan Potato Diseases*, *April*, 1–4. https://findingaids.lib.msu.edu/repositories/2/archival_objects/325044

Whipps, J. M. (2001). Microbial interactions and biocontrol in the rhizosphere. *Journal of Experimental Botany*, *52*(SPEC. ISS. MAR.), 487–511. https://doi.org/10.1093/jxb/52.suppl_1.487

Wu, Z., et al. (2021). Biocontrol mechanism of Myxococcus xanthus B25-I-1 against Phytophthora infestans. *Pesticide Biochemistry and Physiology*, *175*. https://doi.org/10.1016/j.pestbp.2021.104832

Yan, F., et al. (2014). Effect of rhamnolipids on Rhodotorula glutinis biocontrol of Alternaria alternata infection in cherry tomato fruit. *Postharvest Biology and Technology*, *97*. https://doi.org/10.1016/j.postharvbio.2014.05.017

Yang, J., et al. (2009). Rhizosphere bacteria help plants tolerate abiotic stress. In *Trends in Plant Science* 14 (1): 1–4. https://doi.org/10.1016/j.tplants.2008.10.004.

Yuan, W. M. & Crawford, D. L. (1995). Characterization of Streptomyces lydicus WYEC108 as a potential biocontrol agent against fungal root and seed rots. *Applied and Environmental Microbiology*, *61*(8), 3119–3128. https://doi.org/10.1128/aem.61.8.3119-3128.1995

Zeng, W., et al. (2012a). Field management of Sclerotinia stem rot of soybean using biological control agents. *Biological Control*, *60*(2), 141–147. https://doi.org/10.1016/j.biocontrol.2011.09.012

Zeng, W., et al. (2012b). Use of Coniothyrium minitans and other microorganisms for reducing Sclerotinia sclerotiorum. *Biological Control*, *60*(2), 225–232. https://doi.org/10.1016/j.biocontrol.2011.10.009

Zhai, J., et al. (2020). Reed decomposition under Bacillus subtilis addition conditions and the influence on water quality. *Ecohydrology and Hydrobiology*, *20*(4). https://doi.org/10.1016/j.ecohyd.2019.11.003

Zhang, H., et al. (2008). Soil bacteria confer plant salt tolerance by tissue-specific regulation of the sodium transporter HKT1. *Molecular Plant-Microbe Interactions®*, *21*(6), 737–744. https://doi.org/10.1094/mpmi-21-6-0737

Zhang, S., et al. (2024). Isolation, characterization and application of noble bacteriophages targeting potato common scab pathogen Streptomyces stelliscabiei. *Microbiological Research*, *283*, 127699. https://doi.org/10.1016/j.micres.2024.127699

Zhang, X., et al. (2019). Evaluation of the risk of development of fluopicolide resistance in Phytophthora erythroseptica. *Plant Disease*, *103*, 284–288. https://doi.org/10.1094/PDIS-02-18-0366-RE

Zhang, X. Y., et al. (2020). A novel Streptomyces sp. strain PBSH9 for controlling potato common scab caused by Streptomyces galilaeus. *Plant Disease*, *104*(7), 1986–1993. https://doi.org/10.1094/PDIS-07-19-1469-RE

Zhang, X., et al. (2021). A new strain of Volutella citrinella with nematode predation and nematicidal activity, isolated from the cysts of potato cyst nematodes in China. *BMC Microbiology*, *21*(1). https://doi.org/10.1186/s12866-021-02385-x

Zhu, J. H., et al. (2017). Advances in the mechanism of crop disease control by intercropping. *Chinese Journal of Ecology*, *36*(4), 1117–1126. https://doi.org/10.13292/j.1000-4890.201704.016

Zin, N. A. & Badaluddin, N. A. (2020). Biological functions of Trichoderma spp. for agriculture applications. In *Annals of Agricultural Sciences* (Vol. 65, Issue 2). https://doi.org/10.1016/j.aoas.2020.09.003

Chapter 10

Advances in bioherbicides/ biological control of weeds

Te Ming Tseng, Varsha Singh, Worlanyo Segbefia, Ncomiwe Maphalala, Tabata D. Oliveira, and Ziming Yue, Mississippi State University, USA

1 Introduction

Weed control plays a critical role in mitigating competition between weed species and crops, thereby preventing yield decreases (Zimdahl, 2018). Various techniques exist to minimize weed–crop competition, including chemical, cultural, biocontrol, and mechanical control methods (Zimdahl, 2018; Van Antwerpen et al., 2022). With the scarcity of labor in the agricultural industry, there is a global trend toward increased herbicide use for weed management in crop fields (Aktar et al., 2009; Singh et al., 2022). Unfortunately, improper herbicide use, such as targeting specific weed species, incorrect timing of application during plant growth stages, and unfavorable weather conditions, has led to the accumulation of chemical residues in the soil, proliferation of

http://dx.doi.org/10.19103/AS.2024.0142.21

weed species, and the development of herbicide-resistant weed biotypes (Motmainna et al., 2021).

The contemporary agricultural landscape heavily relies on chemical herbicides to control weed infestations, designed for targeting specific weed species to promote efficient crop growth and prevent yield losses (Ehrlich and Harte, 2015). However, this overreliance has given rise to a myriad of issues, necessitating a critical examination of their impact on soil health, ecological balance, and human health (Varah et al., 2020; Kughur, 2012). Despite their efficacy, current herbicides create major challenges, including residue accumulation due to improper use such as targeting specific weed species, incorrect timing of application, and unfavorable weather conditions, resulting in the accumulation of chemical residues in the soil and posing risks to subsequent crops and the overall ecosystem (Thirkell et al., 2017). The continuous evolution of weed species has led to the development of herbicide-resistant biotypes, posing a serious threat to effective weed control and requiring alternative strategies to manage evolving weed populations. Additionally, synthetic herbicides are known to persist in the environment, which has prompted discussions about potential effects on soil properties, crop productivity, and broader ecological and human health considerations. This has led to increased interest in exploring alternative weed management approaches that may reduce long-term herbicide use (Lin et al., 2015; Li et al., 2016).

Addressing the emergence of herbicide-resistant weeds is essential due to their continuous evolution and the potential economic impact surpassing that caused by other pests (Soltys et al., 2013; Yue et al., 2022). It is important to explore new approaches for integrated weed management that incorporate alternative techniques, aiming to reduce reliance on chemical control methods while promoting agricultural productivity and innovation (Ram and Pathak, 2016; EPA, 2023). Among the potential solutions for sustainable agriculture, the use of bioherbicides, which are products derived from living organisms, particularly microorganisms or their byproducts, shows potential (Pacanoski, 2015).

The term "bioherbicides" refers to the use of microorganisms (fungi, bacteria, viruses, algae, etc.), insects, or plant extracts to promote sustainable weed control (Crump et al., 1999). Recent studies have demonstrated the weed-suppressive ability of bioherbicides in various plant species, including both monocotyledons and dicotyledons (Snyder et al., 2009; Kato-Noguchi et al., 2016; Algandaby et al., 2016; Scavo et al., 2023). Another sustainable strategy for weed control is the use of allelopathic compounds released from cover crops or crop residues, which offer multiple benefits such as inhibiting weed growth, reducing soil compaction, and increasing moisture and nutrient content (Farooq et al., 2011; Mushtaq et al., 2020). Several studies have investigated the effectiveness of specific plant-based formulations in controlling weed species. For example, Petrova et al. (2015) found that lavender and spearmint formulations significantly reduced the germination of Johnson grass, white pigweed, twitch, and curly dock in wheat. Similarly, scientists at the Western Research Center discovered that the bottlebrush plant effectively prevented the growth of many weeds, with the herbicide compound leptospermone playing a key role (Mitchell et al., 2001). However, some weed species exhibit adverse allelopathic effects on economically significant crops in agriculture. Santa Maria feverfew, for instance, is an aggressive weed species that suppresses important crops like canola and wheat (Khaliq et al., 2016). Nevertheless, recent studies have shown that the use of microorganisms, specifically *Pseudomonas putidas*, can help mitigate the negative effects of Santa Maria feverfew while also promoting beneficial allelopathic compounds that support soil microbial communities and eco-friendly approaches for weed control (Mishra et al., 2013).

The use of bioherbicides in agriculture provides an alternative approach to weed control, offering potential benefits such as lower toxicity and reduced persistence in the environment compared to synthetic herbicides (Pyon et al., 2017; Aneja et al., 2017). This highlights the potential of bioherbicides as effective tools within weed management strategies.

2 Bioherbicides available in the market

Bioherbicide concept has experienced the change with time since its entering into the market in the 1960s; the early concept of bioherbicide was limited to living microbes (Zheng, 2020), mostly plant pathogens and/or their (secondary) metabolites as their active ingredients, the recent concept has included plant secondary metabolites (Hasan et al., 2021). Compared to research on bioherbicides (El-Mergawi and Al-Humaid, 2019; Zheng, 2020; Hasan et al., 2021), bioherbicides in the market are much less common. Table 1 lists some of the bioherbicides that occurred on the market.

From the chronological list, we can see the trend from living microbial organisms or their secondary metabolites to plant secondary metabolites. This is interesting as the history of human utilization of plant secondary metabolites (natural products) and the number (more than 45 000) of the known compounds are much longer and larger than those of the microbial secondary metabolites. One reason that plant secondary metabolites were not easy to be marketed as bioherbicides might be due to plant hormones. The easiest way to get plant secondary metabolites is plant extracts which are usually mixtures of many compounds. Their effects are also the cumulative effects of many compounds. The US EPA is in charge of pesticide registration, and active ingredient characterizations are usually required, with the remaining as non-active ingredients. While the main active ingredients are easy to characterize,

Table 1 Bioherbicides that are/were on the market

Brand name and company/country	Source organism/ active ingredients	Target weed	Reference	Marketed year
(brand unknown)(USA)	*Acremonium diospyri*	*Diospyros virginiana* L. trees in rangelands	Aneja et al. (2013); Uludag et al. (2018)	1960
Lubao(China)	*Colletrichum gloeosporioides* f. sp. *cuscutae*	*Cuscuta* spp. in soybean	Templeton (1992); Uludag et al. (2018)	1963
DeVineAbbott Laboratories(USA)	facultative fungus, *Phytophthora palmivora* Butl.	*Morrenia odorata*	Zheng (2020); Uludag et al. (2018)	1981
Casst™(USA)	*Austroplebeia cassia* (a stingless bee)	*Cassia obtusifolia* L., *Celits occidentalis* L., *Crotalaria spectabilis* DC. in soybean and peanut	Charudattan et al. (1986); Uludag et al. (2018)	1983
Dr. BiosedgeTifton Innovation Corporation (USA)	Uredispore preparation of *Pucinia canaliculata*	*Cyperus esculentus* in rice fields	Phatak (1992); Uludag et al. (2018); Zheng (2020)	1987
CollegoUpjohn (USA)	*Collectotrichum gloeosporoids* f. sp. *Aeschenomene* (Cga)	*Aschenomene virginica*	Zheng (2020)	1992
BiomalPhilombios (Canada)	*Collectotrichum gloeosporoids* f. sp. *malvae* (Cgm)	In wheat and other crop fields in grassland region of Canada and northern plain region of the US	Zheng (2020)	1992
Stumpout™South Africa	*Cylindrobasidium laeve*	*Poa annua* L. in golf courses; *Acacia mearnsii* (De Wild) and *Acacia pycnantha* (Benth.) in native vegetations	Shamoun and Hintz (1998); Uludag et al. (2018)	1997
CampericoBy Mycogen (USA) and Japan Tobacco ("JT")	*Xanthamonas campstris pv. Poae* (JT-P482)	Bluegrass on golf courses	Zheng (2020)	1997
BiochonKoppert (the Netherlands)	*Chondrostereum purpureum*	Woody weeds	Zheng (2020)	1997
SolviNix™(USA)	Tobacco mild green mosaic tobamovirus (TMGMV)	*Solanum viarum Dunal* in rangelands	Aneja et al. (2013); Uludag et al. (2018)	2009

Di-Bak ParkisoniaBioherbicides Australia Pty Ltd. (Australia)	Natural fungus	*Parkisonia aculeate*	Zheng (2020)	2013
EMERION™ 7005Emery Oleochemicals (Malaysia)	Naturally derived fats and oils	Weeds and grasses for food crops, field crops, pastures, landscapes, and lawns	Zheng (2020)	2014
Verdesian Life Sciences (USA)	*Pseudomonas fluorescens strain D7*	Cheatgrass	Zheng (2020)	2015
Commercialized-PhomaBy Premier Tech (Canada)	Indiginous fungus *Phoma macrostoma*	Broadleaf weeds such as dandelion	Zheng (2020)	2017
MBI014Marrone Bio Innovations Inc. (USA)	*Burkholderia rinojensis* (strain A396) heat-killed	Pigweed family	Zheng (2020)	2018 submitted registration package
HomeplateCertis(USA)	Mixture of fatty acids	Burndown, harvest aid, and desiccant	Zheng (2020)	2019
HerbiScan project by Plant Advanced Technologies (France)	Plant origin	New herbicide	Zheng (2020)	No product yet

the plant hormones in plant extract, which usually exist in trace amounts and are active, are not easy to characterize. Moreover, some plant hormones are not known, for example, during the past 60 years, when bioherbicides entered the market, several plant hormones were newly identified, such as jasmonates, salicylic acid (Farhangi-Abriz and Ghassemi-Golezani, 2018), and strigolactones (Mishra et al., 2016). Homeplate sets up a new way to commercialize plant secondary metabolites as bioherbicides: all the fatty acids are plant secondary metabolites. The formulation is a mixture of different kinds of fatty acids without restraining their sources. In this way, it easily passed US EPA registration and entered into the market. Now Homeplate has registered in 45 states in the US. As suggested by Hasan et al. (2021), the chemical industries might be able to elaborate appropriate formulations of the bioherbicides to improve their activity and synthesize the relevant allelochemical compounds whenever this is more convenient than obtaining them from nature. On the other hand, microbial or animal hormones don't need to face this problem as those hormones might not work on plants. That is why living microbes and their secondary metabolites are easy to get registered. An animal product that got EPA registered is putrescent egg solids as an animal repellent; its active ingredients were not even clearly characterized but still registered even if it might contain animal hormone in trace amount. Animal hormones do not work on plants. Hence, plant extract usually is a way to discover herbicides rather than to commercialize directly.

3 Plant extracts as bioherbicides

The key attributes of a bioherbicide involve their host specificity, efficacy, crop tolerance, mode of action, environmental fate, optimum temperature and moisture, and toxicology (Bailey, 2014). Keeping these points in mind, developing plant-based bioherbicides involves identifying and isolating bioactive compounds from plants, followed by formulation and testing for efficacy. Extracts from various parts of plants, such as leaves, stems, roots or even the whole plant, have been found to have weed-suppressive properties (Yue et al., 2022). These extracts can interfere with the weed's photosynthesis process, disrupting the plant's ability to produce energy, inhibit seed germination, or disrupt the weed's cell membranes or growth hormones (Hasan et al., 2021).

The most prominent instances of bioherbicides from plant extract include sorgaab and sunfaag, the phytotoxic water extracts derived from sorghum and sunflower (*Helianthus annuus* L.) leaves, respectively (Soltys et al., 2013). These extracts can be utilized effectively for weed control in crops like wheat, rice, cotton, and soybean without causing any loss in crop yield (Soltys et al., 2013). Extracts from cover crops such as alfalfa (*Medicago sativa*) and rye (*Secale cereale*) have been used as an allelochemical for weed suppression in some cash crops (Shilling et al., 1986; Lemessa and Wakjira, 2015). Whole plant extract from cover crops has also been used for weed suppression separately and with other weedicides (Scavo and Mauromicale, 2021). Root extracts from cotton (*Gossypium spp.*) have been shown to contain phenolic compounds such as ferulic acid, gallic acid, p-hydroxybenzoic acid, and vanillin, which can inhibit the growth of nearby weeds (Gui-Ying et al., 2015). Similar allelopathic effects from cover crop residues, including those of sorghum and rye, have also been reported to suppress weed development (Hoffman et al., 1996).

By utilizing plant extracts as pre-emergence or early post-emergence herbicides, it has been possible to reduce the required doses of synthetic herbicides due to the synergistic effects they exhibit (Soltys et al., 2013). These extracts display high allelopathic potential in mulch, intercropping systems, or crop rotation. However, it should be noted that not all allelopathic plant extracts demonstrate sufficient effectiveness in inhibiting weed growth or germination when applied as sprays. This limitation may arise from factors such as the masking of one compound's activity by another within the water solution or the inability of the extract to penetrate the plant's cuticle (Dayan et al., 2009b).

4 Natural byproducts as bioherbicides

During the past decades, there has been increased interest in agricultural practices that do not depend so much on chemical means of controlling weeds, thereby calling on farmers and stakeholders in the agriculture industry to support alternative measures to prevent the development of herbicide-resistant weeds. The call for adding biological agents to sustainable integrated weed management practices has increased (Cordeau et al., 2016; Radhakrishnan et al., 2018). Farmers desire to increase yield at a reduced cost. There have been numerous studies to buttress the point that in order to control weed seed germination and growth, bacteria, fungi, plant extracts, and other products are very instrumental (Harding & Raizada, 2015; Priya Dharsini et al., 2017; Radhakrishnan et al., 2018). However, there are only a few microbes and plant products (one plant extract, three bacteria, and nine fungi) that have been commercialized currently, even though a lot of these microbes and plant products have been laboratory-tested and field-trialed successfully (Cordeau et al., 2016).

The application of synthetic fertilizers in crop production incurs inherent costs. The pioneering efforts of American, Canadian, European, and Ukrainian farmers in the commercialization of bioherbicides began in 1980 (Charudattan, 1986; Bailey, 2014; Cordeau et al., 2016). Bioherbicides are derived from natural sources, including living organisms or their metabolites, and are utilized as a safe method for controlling weed populations (Hoagland et al., 2007; Bailey, 2014). Being of natural origin (microbes, minerals, and plants), bioherbicides are generally considered to pose less risk to humans and the surrounding environment (Roberts, 2022). Furthermore, they exhibit host- or non-host-specificity in weed control (Radhakrishnan et al., 2018). With their biodegradable nature, low toxicity, and reduced persistence in the soil, bioherbicides present a favorable alternative to chemical weed control (Hasan et al., 2021). By leveraging these characteristics, bioherbicides offer a promising option for weed management while aiming to minimize potential risks and support productive agricultural practices.

Currently, over 22 differently formulated bioherbicides are registered on the market for controlling weeds (Roberts, 2022). Besides, bioherbicides encourage compatibility with organic farming. Organic farmers produce crops devoid of synthetic herbicide use. This tag has made organic products the more preferred crops than those crops to which synthetic herbicides have been applied (Hasan et al., 2021). According to Zimdhal (2022), organic farmers can use bioherbicides to control weeds without sacrificing their ethical standards because they adhere to the standards for organic certification. Organic farming plays an important role as it conserves the environment, prioritizes crop rotation, boots biodiversity, improves soil fertility, and tends to minimize the presence of pesticide residue in food (Barański, 2014).

By preventing many of the plant's physiological processes and causing cell death by destabilizing the cell membrane, bioherbicides destroy the target weed species (Travaini et al., 2016; Motmainna, 2021; Hasan et al., 2021). The phytotoxic potential of essential oils (EOs) involves mitotic suppression, membrane depolarization, a decrease in chlorophyll content, cellular respiration, oxidative damage, chlorosis, leaf burning, and reduced plant development (Raveau et al., 2020).

Bailey et al. (2010) attempted to determine the model pathway of microbial infection to control weed growth. They suggested that *Phoma macrostoma* colonizes and enters the root hairs of a host and grows in the intercellular spaces to reach the root core. The proliferation of mycelium infects the vascular trachea to prevent the supply of food and nutrition (Aloni, 2006).

In their study, Daba et al. (2022) reported that EOs derived from aromatic plants were considered a practical and safe herbicide option for reducing the impact of weeds in coffee farms. Therefore, they selected five aromatic plants based on their efficacy as herbicides: *Eucalyptus citriodora*, *Cymbopogon martini*, *Cymbopogon citratus*, *Thymus schimperi*, and *Cymbopogon winterianus*. Their findings proved that the oil of *C. citratus* included 45 different compounds, with trans-Geranyl acetate (16.82%), geraniol (55.27%), (6R,7R)-Bisabolone (7.23%), and trans-beta-Ocimene (6.18%) being the predominant ones. They concluded that EO chemical composition and the biological activity of plants are closely connected. According to Pacanoski (2015), bioherbicides have the advantage of target weed selectivity, minimizing resistance problems, reduced effects on non-target crops, and fewer to no residue problems. The allelopathy phenomenon is widely used as an eco-friendly strategy for weed management (Mushtaq et al., 2020). According to Dayan and Zaccaro (2012), herbicide formulations specifically target plant biological processes, which inhibit plant growth and ultimately lead to plant death. They also believed that these biological mechanisms inhibit plant growth enzymes by interfering with photosynthesis and disrupting plant hormone balance. Besides, they are a major cause of soil degradation, leaching, and runoff.

5 Allelochemicals as bioherbicides

The biochemical compounds, called allelochemicals, are secondary metabolites released into the environment through root, stem, leaves, or decomposing plant material (Weston and Duke, 2003). Allelochemicals, sometimes referred to as nature's own herbicides, are generally water-soluble and have a shorter half-life, which may reduce their persistence in the environment (Gavazzi et al., 2010). Allelopathic crops release allelochemical compounds into the environment through various processes, including leaching, decomposition, root exudation, and volatilization, contributing to the effectiveness of allelopathic crops in controlling weeds and promoting sustainable agriculture (Bonanomi et al., 2006; Rice, 2012; Fuller et al., 2021).

Some allelochemicals can inhibit seed germination, root growth, or photosynthesis, while others can induce plant stress responses or activate defense mechanisms (Gechev et al., 2006; Shah and Smith, 2020). Allelochemicals belong to different classes, such as phenolics (e.g. flavonoids, and tannins), terpenoids (e.g. menthol, limonene, and eucalyptol), alkaloids (e.g. caffeine and nicotine), and fatty acids (e.g. linoleic and palmitic acid) (Mastelić and Jerković, 2002; Wu et al., 2007; Iqbal et al., 2019). The allelopathic effect of allelochemicals extracted from several crops/cover crops may also suppress the main cash crop, such as corn (*Zea mays*) (Liebman and Davis, 2000; Reddy and Koger, 2004; Lemessa and Wakjira, 2015). Studies have been conducted on crops like oat, wheat, barley, rye, sweet potato, and sorghum (Liu and Lovett, 1993; Pérez et al., 1991; Pérez et al., 1993; Nimbal et al., 1996; Singh et al., 2022) to use their allelopathic effect for weed suppression. Allelochemical production increases when the allelopathic crop varieties grow in close proximity to the competing weed plants (Chen et al., 2010; Kato-Noguchi, 2011). Major weed suppressive allelochemicals from different plants are listed in Table 2.

The use of allelopathic crops is considered a feasible alternative for reducing reliance on synthetic herbicides, offering potential agronomic and management benefits (Perry et al., 2009; Ferguson et al., 2013). However, further research is needed to develop effective strategies for integrating allelopathic compounds into herbicide management programs.

Table 2 Major allelochemicals from different plants reported to reduce weed growth and development

Allelopathic plant	Allelochemical	Target weed	Reference
Brassica spp., *Raphanus sativus*	Glucosinolates, Isothiocynates	*Amaranthus hybridus* L., *Echinochloa cruss-galli*, *Daucus carota* L., *Capsella bursa-pastoris* L., *Cuscuta* spp., *Convolvulus arvensis* L.	Soltys et al. (2013)
Secale cereale	PLA, BHA, AZOB, DIBOA, and BOA	*Avena fatua*, *Lepidum sativum*, *E. crus-galli*, *Bilderdykia convolvulus*, *Panicum miliaceum*, *Polygonum aviculare*, *Lactuca sativa*, *Veronica persica*, *Lycopersicon esculentum*, *Lamium amplexicaule*, and *Chenopodium album*	Kruse et al. (2000)
Sorghum bicolor	Sorgoleone	*Cyperus rotundus* L., *Phalaris minor*, *Cassia obtusifolia*, *Ambrosia atrtemisiflora*, *Solanum nigrum*, *Amaranthus retroflexus*	Soltys et al. (2013); Thi et al., (2015)
Artemisia annua L.	Artemisinin	*L. sativa* L., *Euphorbia heterophylla* L., *Amaranthus returoflexus*, *Portulaca oleracea* L., *Ipomoea lacunosa* L., *Limna minor*	Magiero et al. (2009); Lydon et al. (1997)
Ipomoea batatas	Chlorogenic acid, caffeic acid, trans-cinnamic acid	*Amaranthus palmeri*, *Mikania micrantha*, *Sesbanea herbacea*, *Echinochloa colona*	Werle et al. (2022); Soni et al. (2019); Shen et al. (2018)
Oryza sativa	Momilactone	*A. lividus*, *E. colona* L., *Digitaria sanguinalis*, *Poa annua* L.	Motmainna et al. (2021)
Piper longum L.	Sarmentine	*D. sanguinalis*, *A. retroflexus*, *Echinochloa crus-galli*, *Leptochloa filiformis*, *P. annua*, *Taraxacum* sp., *C. album*, *I. purpurea*, *Raphanus crispus*, *Sherardia arvensis*	Dayan et al. (2015); Soltys et al. (2013)
Juglans nigra L.	Juglone	*Hordeum vulgare* L., *Lamium amplexicaule* L., *Sinapis arvensis*, *Papaver rhoeas* L., *C. arvense*, *Triticum vulgare*	Hasan et al. (2021)
Eucalyptus citriodora	Essential oils	*Parthenium hysterophorus*, *S. arvensis*, *Sonchus oleraceus*, *Xanthium strumarium*, and *Avena fatua*	Singh et al. (2005)
Helianthus annuus L.	Sunfaag	*C. rotundus*, *Phoenix canariensis*, *Coronopus didymus*	Motmainna et al. (2021)

6 Advantages of bioherbicides

Weed management is a significant challenge for the farmers, and it has become increasingly difficult due to the development of herbicide resistance in the weed species (Weerarathne et al., 2017). Also, the synthetic herbicides are prohibited for use under the organic production systems (Nwosisi et al., 2019). The demand for a wider range of options for weed control is increasing in response to growing interest in alternative products (Cai and Gu, 2016). An effective strategy in weed control is the use of bioherbicides, which are derived from natural sources such as microorganisms, plant extracts, microbial phytotoxins, or phytopathogenic (Hasan et al., 2021). Bioherbicides are often used as part of an integrated weed management approach, which combines different weed control strategies for optimal results.

A herbicide's ideal or anticipated inhibitory impact on a specific weed produces herbicide effectiveness (Mushtaq et al., 2020). The species present in a weed community must be identified for farmers to choose an appropriate herbicide for weed management. One of the key considerations in choosing the right herbicides for weed control is their efficacy (Jursík, 2011; Hasan, 2021). The bioactive compound/allelochemical concentration, plant growth stage, formulation type, spray preparation, application method, type of soil, and environmental parameters (e.g. light, CO_2, temperature, humidity) are only a few of the variables that might affect how effective bioherbicides are (Hasan et al., 2021). The efficacy of these bioherbicides is significantly influenced by specific local environmental factors, including temperature, moisture levels, and soil type (Bailey, 2014). Therefore, in selecting bioherbicides for weed control, caution should be exercised.

Bioherbicides typically do not persist in the environment for extended periods, thereby reducing the risk of soil and water contamination and minimizing harm to non-target organisms (CE et al., 2019; Bailey et al., 2010). Additionally, the water-soluble nature of some allelochemicals makes their application simple and low-cost, and they do not require surfactants (Soltys et al., 2013; Dayan et al., 2009a). Furthermore, the chemical structures of bioherbicides are considered to present fewer persistence concerns compared to those of synthetic herbicides (Hasan et al., 2021). They have versatile applications, being suitable for use in various environments, including natural settings like roadsides, pastures, and forests, as well as cultivated areas such as row crops, orchards, and turfgrass (Bailey, 2014). Allelochemical-based bioherbicides are low in toxicity, have shorter persistence in the environment, and often employ multiple modes of action, thus decreasing the likelihood of development of herbicide resistance (Bailey, 2014).

7 Disadvantages of bioherbicides

Despite their potential benefits, plant-based bioherbicides may have some limitations. The key limiting factor for the use of bioherbicides is their effectiveness. They can be less potent or have slower action compared to synthetic herbicides, requiring multiple applications. Additionally, the effectiveness of bioherbicides may vary depending on environmental conditions and target weed species (Cordeau et al., 2016). In addition, obtaining a sufficient quantity of an allelopathic compound becomes a challenge when its chemical synthesis is not feasible (Soltys et al., 2013). In addition, the structural complexity of numerous allelochemicals often renders them too costly to be extensively utilized as agrochemicals (Hasan et al., 2021).

Bioherbicides possess a relatively shorter environmental half-life, which can help reduce the likelihood of off-target effects. However, to be an effective herbicide, it is necessary for the product to persist for a sufficient duration to achieve the desired impact on weed species (Grayson et al., 2017).

The structural and chemical properties and mode of action of numerous allelochemicals in plants have been extensively studied. However, it is unfortunate that only a small fraction of these compounds have been effectively incorporated into agricultural practices. This limited adoption can be attributed to the need for extensive field experiments to validate their efficacy. Also, information about the *in vivo* synthesis process is often necessary to isolate a particular compound of interest from the numerous compounds synthesized by plants.

8 Future trends

Recent achievements and developments in the field of bioherbicides have resulted in over 22 different formulations currently being registered on the market for weed control. Although many more have previously been developed or are currently being formulated, low customer demand and the high costs currently associated with the production of those formulations appear to limit their long-term success. In 2016, it was reported that the global bioherbicide market was valued at US$ 1.28 billion, and with continued development in this field, it is expected to further increase its market share to US$ 4.14 billion by 2024. Although advancements in this field are clear, there are still several challenges associated with the commercialization process that may hinder their widespread success, and these barriers need to be addressed to ensure that all developed bioherbicides can be economically and commercially viable in the long term (Roberts et al., 2022).

9 Future directions

Despite the recent developments in the research of plant-based bioherbicides, there is still much room for scientists to explore new solutions and improve existing ones. Take an example in Malaysia, a new bioherbicide product, WeedLock, which was developed from a plant extract and marketed locally by EntoGenex Industries Sdn. Bhd. since 2017, is a contact non-selective herbicide that controls a wide range of weed species. Nonetheless, WeedLock is currently only available as a ready-to-use formulation, making it economically prohibitive for the product to be used in agricultural settings. In addition, studies on the crop selectivity, appropriate formulation, and mechanism of action of this promising bioherbicide are still lacking. Hence, this opens more opportunities for researchers and industries to venture into the advancement of this product as an alternative to the current synthetic herbicides in both the ornamental and agricultural fields (Hasan et al., 2021).

One promising direction is the discovery and development of novel herbicides based on natural products that are by-products of microorganisms or extracts of plants. Only a small fraction of the world's microbial and plant biodiversity has been screened for herbicidal activity. It is known that interesting herbicidal compounds with novel mechanisms of action have been discovered, and some compounds have been very successful commercial herbicides (e.g. glufosinate based on phosphinothricin, a breakdown product of bialaphos discovered from *Streptomyces viridochromogenes* and *S. hygroscopicus* by researchers in Japan). One company, Marrone Bio Innovations, has screened water extracts from approximately 15 000 microorganisms (e.g. bacteria, fungi, and actinomycetes) and 350 plant extracts and found several novel herbicidal compounds with novel modes of action. Some microbial strains and plant extracts are in development at Marrone Bio Innovations and show promise as cost-effective commercial herbicides. Still, based on success rates, developing a cost-effective and broad-spectrum bioherbicide is more technically challenging than a bio-fungicide, bioinsecticide, or bionematicide (Westwood et al., 2018). Research with natural phytotoxins has demonstrated several additional novel potential herbicide target sites. But, in most of these cases, the natural herbicide is too expensive, too toxic, or lacks appropriate physicochemical properties (e.g. uptake and translocation) to be a good herbicide. Nevertheless, these compounds can point the way to the discovery of new target sites (i.e. new MOAs) that can be the focus of in vitro screening of less expensive and/or toxic compounds with better physicochemical properties (Westwood et al., 2018).

One barrier to the success of bioherbicides is the formulation and commercialization process. Because they contain a living biotic agent, the viability and stability of many bioherbicides need to be steadily maintained, requiring consideration of how these biological agents can remain active over the extended time frame needed from the development to the application stage. This would require the identification of the optimal storage conditions for a particular bioherbicide, noting that each case could vastly differ. In addition, the high costs associated with the formulation and commercialization of many bioherbicides need to be addressed for their use to be sustainable over large areas. This could be partly achieved with increased educational awareness of the benefits and use of bioherbicides and improvements to technology, such as smart spray systems (Roberts et al., 2022).

10 Future research

1 *Investigate how climatic scenarios may influence the performance of selected bioherbicides, allowing them to be used in future weed management programs with greater confidence.*

One of the many barriers that restrict the use and success of bioherbicides is the influence of environmental conditions. Factors that have been reported to strongly influence the success of a bioherbicide include humidity, soil type, temperature, and UV light (which can degrade the bioherbicide), together with adequate water availability and quality. These factors influence the formulation process and the performance of a bioherbicide, which can lower its efficiency when applied directly to a plant. It has also been reported that increased temperatures and carbon dioxide (CO_2) levels in the atmosphere, resulting from natural variability and broader environmental changes, are likely to significantly change weed population dynamics in the future. In this regard, current weed management strategies may not provide adequate control of many weeds, with evidence showing that increased CO_2 concentrations and temperature have resulted in more occurrences of herbicide resistance. This strongly supports the use and need for improved weed management, a direction that includes bioherbicide applications. It is anticipated that a plethora of questions will continue to emerge relating to the overall impacts and modes of action of bioherbicides due to structural and physiological variations and evolutionary adaptations that will inevitably occur in weeds in response to ongoing environmental shifts (Roberts et al., 2022).

2 *Advancing the formulation of bioherbicide active substances.*

A major factor that affects how well a bioherbicide works is its formulation. Techniques employing multiple surfactants or nano-formulation could be considered to enhance the penetration and absorption of active herbicidal compounds in plants (Hasan et al., 2021). The formulation of bioherbicide-active substances is one of the most critical factors affecting the efficacy of weed control in the field. For bioherbicides based on microbial metabolites and natural products, formulations for protecting the chemical nature and enhancing the entry of the active substance into the plant must be developed to optimize deployment and efficacy in the field. Formulation development has traditionally emphasized the adaptation of ingredients used in the chemical pesticide industry with little consideration to compatibility with bioherbicidal products; therefore, collaboration with food-processing, pharmaceutical, and similar industries where nanotechnology is used should be pursued in developing formulations compatible with bioherbicide substances. Nano-formulations of biochemical bioherbicides have the potential to increase biocontrol efficiency because large surface areas of nanoparticles result in a lower volume of bioherbicide required, thereby increasing concentration in a smaller package with reduced costs. The inherent small particle size greatly limits the susceptibility of the bioherbicide to environmental conditions of desiccation, heat, and UV radiation that are encountered during storage, delivery, and application (Kremer, 2019).

3 *Investigate isolating and identifying new allelopathic compounds from invasive weeds.*

Upon validating these compounds' bioactivity under laboratory and field conditions, these can be recommended for sustainable agriculture herbicide development. Due to their easy degradability and high cost, the direct use of allelochemical extracts as bioherbicides in the field is not feasible in many cases. This could be resolved if chemical industries elaborated appropriate formulations of the bioherbicides to improve their activity and synthesized the relevant allelochemical compounds whenever this is more convenient than obtaining them from nature (Hasan et al., 2021).

4 *Identify new and potential plant species containing a wide range of phytotoxic compounds acting at the molecular or cellular level of target plants.*

The antioxidant enzymatic activity related to the mechanism of inhibition and tissue injury of these compounds can be investigated (Hasan et al., 2021). Some natural products with demonstrated bioherbicidal activity include extracts of dried distiller grains, corn gluten meal, and various mustard seed-based meals. Extracts of whole or parts of plants that are allelopathic toward weeds are considered potential bioherbicides and include allelochemicals that have been identified, such as sorgoleone from sorghum (*Sorghum bicolor*) and others that are complex mixtures and have not been fully characterized such as extracts of rice hulls and Eucalyptus. Essential oils that are highly volatile liquids from specific herbs and commonly extracted for healing and pharmaceutical properties also possess allelopathic activity toward many weeds and are considered important potential bioherbicides, especially for organic agriculture systems (Kremer, 2019).

11 Conclusion

The bioherbicides market has witnessed remarkable growth, as evidenced by data from IMARC Group, revealing a substantial increase from US$2.4 billion in 2022 to a projected US$5.0 billion by 2028. This surge is indicative of the escalating interest and investment in the field of bioherbicides. The convergence of biological science with market demands is evident in the momentum gained by employing natural compounds, particularly in weed control. Leveraging ancient knowledge of allelopathic plants has provided a robust foundation for a groundbreaking approach to weed control.

Despite significant strides in understanding the chemical composition of natural compounds and their phytotoxic potential, there remains a substantial research gap. While controlled conditions in studies offer valuable insights into direct effects, field experiments are essential for validation, considering varying environmental conditions. Addressing this disparity, the integration of omics techniques, advanced chemistry and biochemistry methods, and other innovative tools emerges as a promising avenue. These approaches hold potential for developing selective herbicides while also aiming to minimize associated risks.

The methodologies discussed in this chapter contribute significantly to unraveling the intricacies of bioherbicides and phytotoxic interactions. This heightened comprehension of the complex interplay between allelopathic and natural compounds propels us closer to pioneering strategies for weed management. Additionally, it lays the groundwork for unearthing novel herbicides and identifying herbicidal target sites. Continuous research in this field remains essential for advancing agricultural practices and developing effective weed control methodologies. As the market continues to burgeon, this research will play a pivotal role in shaping the trajectory of the bioherbicides industry, fostering innovation, and meeting the evolving needs of agriculture on a global scale.

12 Where to look for further information section

- Charudattan, R. (2001). Biological control of weeds by means of plant pathogens: Significance for integrated weed management in modern agro-ecology. Biological Control, 46(2), 229–260. A foundational review of the role of microbial agents in weed control, with emphasis on application strategies and challenges.
- Harding, D. P., & Raizada, M. N. (2015). Controlling weeds with fungi, bacteria and viruses: A review. Frontiers in Plant Science, 6, 659. Provides a comprehensive overview of microbial bioherbicides, current developments, and prospects for commercialization.
- CABI Invasive Species Compendium (https://www.cabi.org/isc) – a database offering detailed profiles on invasive weeds and biological control agents used against them worldwide.

- International Organization for Biological Control (IOBC) (https://www.iobc-global.org) – a key platform for international collaboration and information on biological control practices and standards.
- Weed Science Society of America (WSSA) (https://wssa.net) – provides access to journals, symposia, and updates on the latest in weed science research, including bioherbicide development.

13 References

Aktar, M. W., et al. (2009). Impact of pesticides use in agriculture: their benefits and hazards. *Interdiscip. Toxicol.* 2, 1–12. https://doi.org/10.2478/v10102-009-0001-7.

Algandaby, M. M. and El-Darier, S. M. (2016). Management of the noxious weed; Medicago polymorpha L. via allelopathy of some medicinal plants from Taif region, Saudi Arabia. *Saudi J. Biol. Sci.* https://doi.org/10.1016/j.sjbs.2016.02.013.

Aloni, R., et al. (2006). Role of cytokinin and auxin in shaping root architecture: regulating vascular differentiation, lateral root initiation, root apical dominance and root gravitropism. *Ann. Bot.* 97, 883–893. https://doi.org/10.1093/aob/mcl027.

Aneja, K. R., et al. (2017). "Bioherbicides: strategies, challenges and prospects," in *Developments in fungal biology and applied mycology*, eds. T. Satyanarayana, et al. Springer Singapore, 449–470. https://doi.org/10.1007/978-981-10-4768-8_23.

Aneja, K. R., et al. (2013). "Potential bioherbicides: indian perspectives," in *Biotechnology: prospects and applications*, eds. R. K. Salar, et al. Springer India, 197–215. https://doi.org/10.1007/978-81-322-1683-4_15.

Bailey, J. and Jeger, M. eds. (1992). *Colletotrichum: biology, pathology and control*. GB: CABI. https://doi.org/10.1079/9780851987569.0000.

Bailey, K. L. (2014). "The bioherbicide approach to weed control using plant pathogens," in *Integrated pest management*. Elsevier, 245–266. https://doi.org/10.1016/B978-0-12-398529-3.00014-2.

Bailey, K. L., et al. (2010). Social and economic drivers shaping the future of biological control: a Canadian perspective on the factors affecting the development and use of microbial biopesticides. *Bio. Control.* 52, 221–229. https://doi.org/10.1016/j.biocontrol.2009.05.003.

Barański, M., et al. (2014). Higher antioxidant and lower cadmium concentrations and lower incidence of pesticide residues in organically grown crops: a systematic literature review and meta-analyses. *Br. J. Nutr.* 112, 794–811. https://doi.org/10.1017/S0007114514001366.

Bonanomi, G., et al. (2006). Phytotoxicity dynamics of decaying plant materials. *New Phytol.* 169, 571–578. https://doi.org/10.1111/j.1469-8137.2005.01611.x.

Cai, X. and Gu, M. (2016). Bioherbicides in organic horticulture. *Horticulturae* 2, 3. https://doi.org/10.3390/horticulturae2020003.

Ce, O., et al. (2019). Natural pesticides (biopesticides) and uses in pest management-a critical review. *Asian J. Biotechnol. Genet. Eng.* 2, 1–18.

Charudattan, R. (1986). Evaluation of Alternaria cassiae as a mycoherbicide for sicklepod (Cassia obtusifolia) in regional field tests. *Evaluation of Alternaria cassiae as a mycoherbicide for sicklepod (Cassia obtusifolia) in regional field tests.*

Chen, K.-J., et al. (2010). 2,4-Dihydroxy-7-methoxy-1,4-benzoxazin-3-one (DIMBOA) and 6-methoxy-benzoxazolin-2-one (MBOA) levels in the wheat rhizosphere and their effect on the soil microbial community structure. *J. Agric. Food Chem.* 58, 12710–12716. https://doi.org/10.1021/jf1032608.

Cordeau, S., et al. (2016). Bioherbicides: dead in the water? A review of the existing products for integrated weed management. *Crop Prot.* 87, 44–49. https://doi.org/10.1016/j.cropro.2016.04.016.

Crump, N. S., et al. (1999). Clarifying the nomenclature in microbial weed control. *Biocontrol Sci. Technol.* 9, 89–97. https://doi.org/10.1080/09583159929947.

Daba, A., et al. (2022). Phytochemical composition of essential oils from aromatic plants inherited with bioherbicidal activity in arabica coffee production system of Ethiopia. *J. Agric. Food Res.* 10, 100368. https://doi.org/10.1016/j.jafr.2022.100368.

Dayan, F. E. and Zaccaro, M. L. de M. (2012). Chlorophyll fluorescence as a marker for herbicide mechanisms of action. *Pestic. Biochem. Physiol.* 102, 189–197. https://doi.org/10.1016/j.pestbp.2012.01.005.

Dayan, F. E., et al. (2009a). Natural products in crop protection. *Bioorg. Med. Chem.* 17, 4022–4034. https://doi.org/10.1016/j.bmc.2009.01.046.

Dayan, F. E., et al. (2009b). Dynamic root exudation of sorgoleone and its in planta mechanism of action. *J. Exp. Bot.* 60, 2107–2117. https://doi.org/10.1093/jxb/erp082.

Dayan, F. E., et al. (2015). Sarmentine, a natural herbicide from Piper species with multiple herbicide mechanisms of action. *Front. Plant Sci.* 6, 222. https://doi.org/10.3389/fpls.2015.00222.

Ehrlich, P. R. and Harte, J. (2015). To feed the world in 2050 will require a global revolution. *PNAS* 112(48), 14743–14744. https://doi.org/10.1073/pnas.1519841112.

El-mergawi, R. A., and Al-humaid, A. I. (2019). Searching for natural herbicides in methanol extracts of eight plant species. *Bulletin of the National Research Centre*, 43(22).

Epa, U. (2023). "United States environmental protection agency." Herbicides: Overview. Available at: https://www.epa.gov/caddis-vol2/herbicides Accessed: April, 2023.

Farhangi-Abriz, S. and Ghassemi-Golezani, K. (2018). How can salicylic acid and jasmonic acid mitigate salt toxicity in soybean plants? *Ecotoxicol. Environ. Saf.* 147, 1010–1016. https://doi.org/10.1016/j.ecoenv.2017.09.070.

Farooq, M., et al. (2011). The role of allelopathy in agricultural pest management. *Pest management science*, 67(5), 493–506.

Ferguson, J. J., et al. (2013). Allelopathy: how plants suppress other plants: HS944/hs186, 3/2013. *Edis* 2013.

Fuller, M. G., et al. (2021). Assessing the weed-suppressing potential of cotton chromosome substitution lines using the stair-step assay. *Plants* 10. https://doi.org/10.3390/plants10112450.

Gavazzi, C., et al. (2010). Sustainable weed control by allelochemicals from rye cover crops: from the greenhouse to field evidence. *Allelopathy J.* 25, 259–274.

Gechev, T. S., et al. (2006). Reactive oxygen species as signals that modulate plant stress responses and programmed cell death. *Bioessays* 28, 1091–1101. https://doi.org/10.1002/bies.20493.

Grayson, M. L., et al. (2017). *Kucers' the Use of antibiotics: a clinical review of antibacterial, antifungal, antiparasitic, and antiviral drugs*. Boca Raton, FL: CRC Press.

Gui-Ying, J., et al. (2015). Allelochemicals from cotton (Gossypium hirsutum) rhizosphere soil: Inhibitory effects on cotton seedlings. *Allelopathy Journal*, 35(2).

Harding, D. P. and Raizada, M. N. (2015). Controlling weeds with fungi, bacteria and viruses: a review. *Frontiers in plant science*, 6, p.659.

Hasan, M., et al. (2021). Bioherbicides: an eco-friendly tool for sustainable weed management. *Plants* 10. https://doi.org/10.3390/plants10061212.

Hoagland, R. E., et al. (2007). Bioherbicides: research and risks. *Toxin Rev.* 26, 313–342. https://doi.org/10.1080/15569540701603991.

Hoffman, et al 1996. Separating the effects of sorghum (Sorghum bicolor) and rye (Secale cereale) root and shoot residues on weed development. *Weed Science*, 44(2), 402–407.

Iqbal, A., et al. (2019). Plants are the possible source of allelochemicals that can be useful in promoting sustainable agriculture. *Fresenius Environ. Bull.* 28, 1052–1061.

Jursík, M., et al. (2011). Important aspects of chemical weed control: environmental factors affecting herbicide efficacy. *Listy Cukrovarnické a Reparské* 127, 348–351.

Kato-Noguchi, H. (2011). Barnyard grass-induced rice allelopathy and momilactone B. *J. Plant Physiol.* 168, 1016–1020. https://doi.org/10.1016/j.jplph.2010.12.021.

Kato-Noguchi, H., et al. (2016). A potent phytotoxic substance in Aglaia odorata Lour. *Chem. Biodivers.* 13, 549–554. https://doi.org/10.1002/cbdv.201500175.

Khaliq, A., et al. (2016). Phytotoxic activity of parthenium against wheat and canola differ with plant parts and bioassays techniques. *Planta daninha* 34, 11–24. https://doi.org/10.1590/S0100-83582016340100002.

Kremer, R. J. (2019). "Bioherbicides and nanotechnology: current status and future trends," in *Nano-biopesticides today and future perspectives.* Elsevier, 353–366. https://doi.org/10.1016/B978-0-12-815829-6.00015-2.

Kruse, M., et al. (2000). Ecological effects of allelopathic plants-a review. NERI technical report 315.

Kughur, P. G. (2012). The effects of herbicides on crop production and environment in Makurdi Local Government Area of Benue State, Nigeria. *J. Sustain. Dev. Africa* 14(4), 23–29.

Lemessa, F. and Wakjira, M. (2015). Cover crops as a means of ecological weed management in agroecosystems. *J. Crop Sci. Biotechnol.* 18, 123–135. https://doi.org/10.1007/s12892-014-0085-2.

Li, X., et al. (2016). Improving crop nutrient efficiency through root architecture modifications. *J. Integr. Plant Biol.* 58(3), 193–202. https://doi.org/10.1111/jipb.12381.

Liebman, M. and Davis, A. S. (2000). Integration of soil, crop and weed management in low-external-input farming systems. *Weed Res.* 40, 27–47. https://doi.org/10.1046/j.1365-3180.2000.00164.x.

Lin, G., et al. (2015). Arbuscular mycorrhizal fungal effects on plant competition and community structure. *J. Ecol.* 103(5), 1224–1232. https://doi.org/10.1111/1365-2745.12436.

Liu, D. L. and Lovett, J. V. (1993). Biologically active secondary metabolites of barley. II. Phytotoxicity of barley allelochemicals. *J. Chem. Ecol.* 19, 2231–2244. https://doi.org/10.1007/BF00979660.

Lydon, J., et al. (1997). Allelopathic activity of annual wormwood (*Artemisia annua*) and the role of artemisinin. *Weed Sci.* 45, 807–811. https://doi.org/10.1017/S0043174500089001.

Magiero, E. C., et al. (2009). Efeito alelopático de Artemisia annua L. na germinação e desenvolvimento inicial de plântulas de alface (Lactuca sativa L.) e leiteiro

(Euphorbia heterophylla L.). *Rev. Bras. Plantas Med.* 11, 317–324. https://doi.org/10.1590/S1516-05722009000300014.

Mastelić, J. and Jerković, I. (2002). Volatile constituents from the leaves of young and old Ailanthus altissima (Mili.) Swingle tree. *Croatica Chemica Acta* 75, 189–197.

Mishra, S., et al. (2013). Unravelling the beneficial role of microbial contributors in reducing the allelopathic effects of weeds. *Appl. Microbiol. Biotechnol.* 97, 5659–5668. https://doi.org/10.1007/s00253-013-4885-y.

Mishra, S., et al. (2016). The role of strigolactones and their potential cross-talk under hostile ecological conditions in plants. *Front. Physiol.* 7, 691. https://doi.org/10.3389/fphys.2016.00691.

Mitchell, G., et al. (2001). Mesotrione: a new selective herbicide for use in maize. *Pest Manag. Sci.* 57, 120–128. https://doi.org/10.1002/1526-4998(200102)57:2<120::AID-PS254>3.0.CO;2-E.

Motmainna, M., et al. (2021). Assessment of allelopathic compounds to develop new natural herbicides: a review. *AJ* 52, 21–40. https://doi.org/10.26651/allelo.j/2021-52-1-1305.

Mushtaq, W., et al. (2020). "Future prospective," in *Allelopathy: potential for green agriculture*. SpringerBriefs in Agriculture. Springer International Publishing, 67–69. https://doi.org/10.1007/978-3-030-40807-7_8.

Nimbal, C. I., et al. (1996). Phytotoxicity and distribution of sorgoleone in grain sorghum germplasm. *J. Agric. Food Chem.* 44, 1343–1347. https://doi.org/10.1021/jf950561n.

Nwosisi, S., et al. (2019). Mulch treatment effect on weed biomass and yields of organic sweetpotato cultivars. *Agronomy*, 9(4), p.190.

Pacanoski, Z. (2015). Bioherbicides. Herbicides, physiology of action, and safety. *Intech Open Sci.* https://doi.org/10.5772/61528.

Pérez, F. J. and Ormenoñuñez, J. (1991). Difference in hydroxamic acid content in roots and root exudates of wheat (Triticum aestivum L.) and rye (Secale cereale L.): possible role in allelopathy. *J. Chem. Ecol.* 17, 1037–1043. https://doi.org/10.1007/BF01402932.

Pérez, F. J. and Ormeno-Nunez, J. (1993). Weed growth interference from temperate cereals: the effect of a hydroxamic-acids-exuding rye (Secale cereale L.) cultivar. *Weed Research*, 33(2), 115–119.

Perry, L. G., et al. (2009). Native cover crops suppress exotic annuals and favor native perennials in a greenhouse competition experiment. *Plant Ecol.* 204, 247–259. https://doi.org/10.1007/s11258-009-9588-1.

Petrova, S. T., et al. (2015). A case study of allelopathic effect on weeds in wheat. *Ecologia Balkanica* 7.

Phatak, S. C. (1992). Development and commercialization of rust (Puccinia canaliculata) for biological control of yellow nutsedge (Cyperus esculentus L.). *Proc. 1st Inter. Weed Control Congress* 2, 388.

Priya Dharsini, P., et al. (2017). Spectroscopic Identification and Molecular Modeling of Diethyl 7-hydroxytrideca-2, 5, 8, 11-tetraenedioate: a Herbicidal Compound from Streptomyces sp. *Arabian Journal for Science and Engineering*, 42, 2217–2227.

Pyon, J. Y., et al. (2017). Status and perspective of bioherbicde development for organic weed management. *Weed Turfgrass Sci.* 6, 1–10.

Radhakrishnan, R., et al. (2018). Bioherbicides: current knowledge on weed control mechanism. *Ecotoxicol. Environ. Saf.* 158, 131–138. https://doi.org/10.1016/j.ecoenv.2018.04.018.

Ram, R. A. and Pathak, R. K. (2016). Organic aapproaches for sustainable production of horticultural crops: a review. *Prog. Hort.* 48, 1. https://doi.org/10.5958/2249-5258.2016.00001.4.

Raveau, R., et al. (2020). Essential oils as potential alternative biocontrol products against plant pathogens and weeds: a review. *Foods* 9. https://doi.org/10.3390/foods9030365.

Reddy, K. N. and Koger, C. H. (2004). Live and killed hairy vetch cover crop effects on weeds and yield in glyphosate-resistant corn. *Weed Technol.* 18, 835–840. https://doi.org/10.1614/WT-03-228R.

Rice, E. L. (2012) Allelopathy (Second edition). Academic Press. https://doi.org/10.1016/C2009-0-03024-4

Roberts, J., et al. (2022). Achievements, developments and future challenges in the field of bioherbicides for weed control: ā global review. *Plants* 11. https://doi.org/10.3390/plants11172242.

Scavo, A. and Mauromicale, G., (2021). Crop allelopathy for sustainable weed management in agroecosystems: Knowing the present with a view to the future. *Agronomy*, 11(11), p.2104.

Scavo, A., et al. (2023). Wheat response and weed-suppressive ability in the field application of a Nanoencapsulated Disulfide (DiS-NH2) bioherbicide mimic. *Agronomy* 13, 1132. https://doi.org/10.3390/agronomy13041132.

Shah, A. and Smith, D. L. (2020). Flavonoids in agriculture: chemistry and roles in, biotic and abiotic stress responses, and microbial associations. *Agronomy* 10, 1209. https://doi.org/10.3390/agronomy10081209.

Shamoun, S. F. and Hintz, W. E. (1998). *Development and registration of Chondrostereum purpureum as a mycoherbicide for hardwood weeds in conifer reforestation sites and utility rights-of-way*. Proceedings of the IVth International Bioherbicides Workshop Programme and Abstracts, Glasgow, Scotland.

Shen, S., et al. (2018). Allelopathic effects of three sweet potato cultivars (Ipomoea batatas) on the invasive plant Mikania micrantha. *Pak. J. Biol. Sci.* 21, 8–15. https://doi.org/10.3923/pjbs.2018.8.15.

Shilling, D. G., et al. (1986). Isolation and identification of some phytotoxic compounds from aqueous extracts of rye (Secale cereale L.). *J. Agric. Food Chem.* 34, 633–638. https://doi.org/10.1021/jf00070a011.

Singh, V., et al. (2022). Allelopathy: an eco-friendly approach to control palmer amaranth using allelopathic sweetpotato. *Front. Agron.* 4. https://doi.org/10.3389/fagro.2022.930378.

Snyder, A., et al. (2009). Seed meals from brassicaceae oilseed crops as soil amendments: influence on carrot growth, microbial biomass nitrogen, and nitrogen mineralization. *Horts.* 44, 354–361. https://doi.org/10.21273/HORTSCI.44.2.354.

Soni, B., et al. (2019). Identification and quantification of allelochemicals from selected sweet potato (Ipomoea batatas (L.) lam.) cultivars. *American Journal of Plant Sciences*, 10(12), p.2354.

Soltys, D., et al. (2013). "Allelochemicals as bioherbicides – present and perspectives," in *Herbicides - Current research and case studies in use*, ed. A. Price. InTech. https://doi.org/10.5772/56185.

Thi, H. L., et al. (2015). Allelopathy in Sorghum bicolor L. *Moenoa a review on environmentally friendly solution for weed control. Res. Crops* 16, 657–662.

Thirkell, T. J., et al. (2017). Are mycorrhizal fungi our sustainable saviours? Considerations for achieving food security. *J. Ecol.* 105(4), 921–929. https://doi.org/10.1111/1365-2745.12738.

Travaini, M. L., et al. (2016). Khellin and Visnagin, Furanochromones from Ammi visnaga (L.) Lam., as Potential Bioherbicides. *J. Agric. Food Chem.* 64, 9475–9487. https://doi.org/10.1021/acs.jafc.6b02462.

Uludag, A., et al. (2018). "Biological weed control," in *Non-chemical weed control*. Elsevier, 115–132. https://doi.org/10.1016/B978-0-12-809881-3.00007-3.

Van Antwerpen, R., et al. (2022). A review of field management practices impacting root health in sugarcane. *Advances in Agronomy*, 173, pp.79–162.

Varah, A., et al. (2020). The costs of human-induced evolution in an agricultural system. *Nat. Sustain*. 3(1), 63–71. https://doi.org/10.1038/s41893-019-0462-4.

Weerarathne, L. V. Y., et al. (2017). Does intercropping play a role in alleviating weeds in cassava as a non-chemical tool of weed management?–A review. *Crop protection*, 95, 81–88.

Werle, I. S., et al. (2022). Allelopathic potential and competitive traits of sweetpotato cultivars. *Front. Agron.* 4. https://doi.org/10.3389/fagro.2022.990879.

Weston, L. A. and Duke, S. O. (2003). Weed and crop allelopathy. *CRC Crit. Rev. Plant Sci.* 22, 367–389. https://doi.org/10.1080/713610861.

Westwood, J. H., et al. (2018). Weed management in 2050: perspectives on the future of weed science. *Weed Sci.* 66, 275–285. https://doi.org/10.1017/wsc.2017.78.

Wu, H., et al. (2007). Autotoxicity of wheat (Triticum aestivum L.) as determined by laboratory bioassays. *Plant Soil* 296, 85–93. https://doi.org/10.1007/s11104-007-9292-7.

Yue, Z., et al. (2022). "Use of plant secondary metabolites to reduce crop biotic and abiotic stresses: A review," in *Secondary metabolites - Trends and reviews* [Working Title]. IntechOpen. https://doi.org/10.5772/intechopen.104553.

Zheng, P. (2020). Bio-Herbicides: global development status and product inventory. http://news.agropages.com/News/NewsDetail---34164.htm.

Zimdahl, R. L. (2018). *Fundamentals of weed science*. Academic Press.

Zimdahl, R.L. (2022). *Agriculture's ethical horizon*. Elsevier.

Index

www.ingramcontent.com/pod-product-compliance
Ingram Content Group UK Ltd.
Pitfield, Milton Keynes, MK11 3LW, UK
UKHW021825100926
8556IPUK00003B/166

9 781801 4672